Praise for *A Golden Thread,* the previous edition of *Let It Shine*

"Western man has been using the sun's rays for useful purposes since the days of ancient Greece, as this comprehensive, carefully researched, clearly written history of solar architecture and technology makes abundantly clear. The illustrations and diagrams that illuminate the text on almost every page are especially fine examples of modern graphic presentations." — *New York Times*

"It is a humbling book. Handsomely illustrated and lucidly written, *A Golden Thread* is a rich mine of information." — *Los Angeles Times*

"The history of mankind's efforts to use the sun's energy is a fascinating story, one told in a lively yet scholarly manner here.... The triumphs and defeats of solar pioneers help us appreciate what a solar future may yet hold."
— *Christian Science Monitor*

"This book is sorely needed in the solar publication field, for although there are any number of how-to books on solar technology, few if any examine the history of this much-neglected energy source in any depth." — *Denver Post*

"I just happened to be carrying *A Golden Thread* in my suitcase at Princeton. On my way home, I devoured it. Richly illustrated and thoroughly documented, it is a feast for the most critical historian's mind and eye." — *Technology Review*

"This book will provide hours of reading pleasure and, at the same time, humble anyone who thinks that solar architecture is a new thing under the sun."
— *American Institute of Architects (AIA) Journal*

"This informative and sobering book traces the evolution of solar applications. Excellent graphics and a highly readable style make this book the best survey available." — *Library Journal*

Praise for *A Forest Journey: The Story of Wood and Civilization* by John Perlin

"Outstanding... This book takes one of those bold imaginative sweeps through world history that leaves you full of excitement, as suddenly events seem to fall

into a pattern for the first time. Perlin not only presents us with a bold hypothesis profusely documented and illustrated, he does it with a storyteller's pace and ability to surprise."
— *BBC World Service*

"'Delight' is not a word one expects to use in connection with deforestation, but John Perlin has certainly written a delightful book.... It deserves to be a classic and should make a welcome present for anyone who enjoys a good read."
— *Forest and Conservation History*

"Well documented and illustrated, it is history at its best." — *American Forests*

"Perlin deftly combines a balance of social and ecological values as well as lessons for the immediate future."
— *Booklist*

"Like some Greek epic poem spanning 4,000 years of civilization ... an impressive array of research and a novel topic."
— *Los Angeles Times*

"The new edition of Perlin's landmark work again brings needed attention to one of the primary concerns of the modern era." — *Forest History Today*

"This work... captures the significant impact of wood on past and present civilizations.... Well written and well illustrated."
— *Choice*

"A journey through time — a sort of Western Civ. 101 with a focus on the crucial role of wood in the rise and fall of states and cultures... Solid survey that adds significant dimension to our picture of the current crisis."
— *Kirkus Reviews*

"Perlin has accumulated what seems every reference to the use and misuse of forests in the period beginning with Gilgamesh and ending with the 1880 U.S. census. In between, he chronicles the deforestation of Asia, the Mediterranean, Europe, the West Indies, and the United States by kings, warlords, and robber barons for purposes ranging from building navies to smelting iron to clearing land for cash crops. The research is exhaustive."
— *Library Journal*

Praise for *From Space to Earth* by John Perlin

"John Perlin's book gives a taste of the tremendous difficulties that early pioneers had to overcome to turn Charles Fritts' 1885 invention of a selenium-based solar module to today's booming photovoltaic business. Perlin gives a vivid and fascinating account of the advances of photovoltaics on Earth. Presenting the development of photovoltaic cells in such a personalized manner makes it a much more lively and interesting read than a mere technical account would have done."
— *Nature*

"The step-by-step progress of photovoltaics has elicited little fanfare. It is my hope that *From Space to Earth* will end the silence. The book is gripping to read and its themes long overdue in book-length form."
— *Photon* magazine

"This 'just in time' story of the development of photovoltaics merits the most serious attention and cannot fail to stimulate the reader's interest in both the episodes recounted and their interdisciplinary applications and prospects. The author has provided us with a 'good read,' and the illustrations enhance one's enjoyment. It is a fascinating story, told so that even an individual without technical training can comprehend the breakthroughs which led to today's widespread and ever increasing adoption of solar power."
— *Interdisciplinary Sciences Review*

"John Perlin's delightful tour through the development of photovoltaics (PV) answers not only the question of what is new under the sun, but most importantly, how we got there. Perlin charts the evolution of the photovoltaic industry from its beginnings to the present. It's the best and most readable book on photovoltaic research, policy, and market growth."
— *Whole Earth*

"Twenty years ago John Perlin published *A Golden Thread*, a comprehensive and authoritative history of solar energy that remains today one of the best books on the subject. Perlin's present book is an equally impressive story of the twentieth-century solar photovoltaics industry. Even diehard opponents of solar energy should find it compelling."
— *Isis*, the official journal of the History of Science Society

LET IT SHINE

Also by John Perlin

A Forest Journey: The Story of Wood and Civilization

From Space to Earth: The Story of Solar Electricity

LET IT SHINE

THE 6,000-YEAR STORY OF SOLAR ENERGY

FULLY REVISED AND EXPANDED

JOHN PERLIN

Foreword by Amory B. Lovins

New World Library
Novato, California

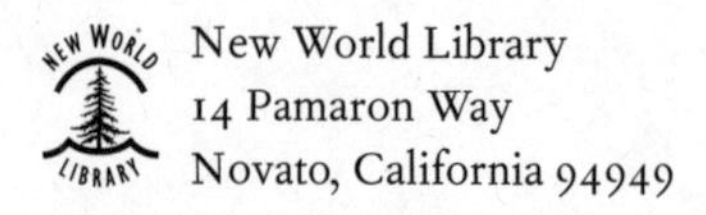
New World Library
14 Pamaron Way
Novato, California 94949

An earlier version of this book was published under the title *A Golden Thread: 2,500 Years of Solar Architecture and Technology* by Cheshire Books / Van Nostrand Reinhold and Company in 1980.

Some of the material in *Let It Shine* appeared in a different form in the *Pacific Standard* magazine.

The permission acknowledgments on page 497 are an extension of the copyright page.

Text design by Tona Pearce Myers

Library of Congress Cataloging-in-Publication Data
Perlin, John.
[Golden thread]
Let it shine : the 6,000-year story of solar energy. — Fully revised and expanded / John Perlin ; foreword by Amory Lovins.
pages cm
Revision of: A golden thread / by Ken Butti and John Perlin. — Palo Alto : Cheshire Books ; New York : Van Nostrand Reinhold, ©1980.
Includes bibliographical references and index.
ISBN 978-1-60868-132-7 (hardback) — ISBN 978-1-60868-133-4 (ebook)
1. Solar energy—History. 2. Architecture and solar radiation—History. I. Title.
TJ810.B88 2013
621.4709—dc23 2013010030

First printing, September 2013
ISBN 978-1-60868-132-7
Printed in Canada on 100% postconsumer-waste recycled paper

New World Library is proud to be a Gold Certified Environmentally Responsible Publisher. Publisher certification awarded by Green Press Initiative. www.greenpressinitiative.org

10 9 8 7 6 5 4 3 2 1

Contents

Part IV. Solar House Heating

Part V. Photovoltaics

Part VI. The Post–Oil Embargo Era

Foreword

The authoritative global network REN21 (Renewable Energy Policy Network for the 21st Century) reports that in 2012, one-fifth of the world's electricity and one-sixth of the world's total delivered energy was renewable. Half the world's new electricity-generating capacity added each year since 2008 has been renewable, and so is one-fourth of global and one-third of European generating capacity. Excluding big hydroelectric dams, modern renewable power (chiefly wind and solar) adds more than 80 billion watts of capacity each year and receives a quarter trillion dollars of annual private investment, and in 2011 invested its trillionth dollar since 2004 — all despite subsidies generally smaller than what its nonrenewable competitors get. Biofuels, too, supply 3 percent of world mobility fuel; over 200 million households, mostly in China, heat their water with solar collectors; and renewable space heating is becoming increasingly common. At least 118 countries, the majority of them developing ones, had renewable-energy targets by early 2012, and an increasing number of countries, cities, and organizations now aim to get all their energy from renewables by 2050; indeed, some communities and firms already do.

In 2013 renewable energy, once derided as a fringe activity, is quietly taking over the world market, driven by overwhelming civic demand and business logic. This is a long game, measured in decades and demanding relentless patience. But its direction is clear, its momentum strong, and its economic, environmental, security, and development case compelling.

Did this broad and deep challenge to fossil and nuclear fuels somehow spring full-blown in recent years from advocates' imaginations, inventors' insights, entrepreneurs' zeal, and investors' quest for profits? Is the emerging renewable age a sudden and unprecedented innovation? *Let It Shine* shows that exactly the opposite is the case.

Today's renewable-energy revolution is the fruit of seeds planted decades, centuries, even millennia ago, then bravely and persistently cultivated through countless vicissitudes. Today's giant photovoltaics industry, growing more than 70 percent a year, simply applies modern materials science and manufacturing methods to an effect discovered by nineteen-year-old Alexandre-Edmond Becquerel in 1839 and refined in Europe and America in 1873–1883. Passive solar buildings were highly developed in ancient China, Greece, and Rome and by the Anasazi. They were described by Socrates millennia before they were rediscovered and applied in central Europe.

I happen to know this astonishing story because in 1979–1980, I had the honor of writing the foreword to Ken Butti and John Perlin's *A Golden Thread: 2,500 Years of Solar Architecture and Technology*. What a delight that a third of a century later Perlin has evolved that pioneering compilation into this new and much expanded edition. He has unearthed much important new material, including documentation of the deliberate suppression of renewable energy by several U.S. administrations — akin to today's intensive disinformation campaigns that gain ferocity as renewables' competition threatens powerful old industries. His new research adds welcome color to an amazing saga.

Yet many of Perlin's fascinating new details embroider a perennial story: how diverse societies around the world have repeatedly invented and refined solar energy, only to have it scuttled, even forgotten, as discoveries of apparently cheap new fuels — coal, oil, gas, nuclear — distracted customers, diverted providers, and befuddled policy makers. Then as those magical substitutions turned out to be transitory, unpleasant, insecure, and increasingly costly, renewables had to be re-created by a new generation of inventors. This rediscovery is flourishing again today. The question is, Now that America has more solar or wind-power jobs than it has coal or steel jobs, now that the capital markets are dramatically shifting investments from coal and nuclear to wind and solar, will renewables finally have the political clout they need to fend off increasingly desperate attacks by endangered rivals? I think so, but only time will tell.

The widely misunderstood U.S. shale-gas boom touched off by fracking technology is a case in point. Many pundits promptly declared renewables dead — supposedly killed by "cheap" natural gas. Yet renewables continued to thrive, inconvenienced but often strengthened by this new price discipline. The gas, it turned out, wasn't really all that cheap: its price will remain volatile, especially if wellhead gas becomes cheap and abundant.[1] (That would drive gas exports, petrochemical pivots to gas, and downstream bottlenecking in infrastructure such as pipelines, all making retail prices more volatile.) Properly counting price volatility makes gas a decade from now about twice as costly as it was at its recent bare commodity price — consistent with the futures market. This expected price rise is reasonable because markets equilibrate (if you want $6–$8 gas, assume $2–$3 gas and use it accordingly), and fracking has at least eight kinds of risk or uncertainty that will take about a decade to resolve. If those wild cards were all satisfactorily resolved, we'd have more, and cheaper, gas with useful optionality, but even if they weren't, that would be okay too, because we didn't need that much gas anyway, and by then we'll be even more acutely conscious of its climate risks.

Ignoring gas's price volatility and uncertainty would be like buying a portfolio of all junk bonds and no treasury bonds because you looked only at the yield or price and not at the risk. You wouldn't do that for long. Yet most commentators and many investors still make this rookie mistake. They may recall that at least three times in the past fifteen years, mistimed gas price forecasts vaporized a total of over $100 billion, but they think this time it'll be different. It won't.

Yet gas's price volatility, like oil's before it and coal's increasingly, actually makes renewables even more valuable. The big, new story about secure, abundant, and stably priced resources for our long-term energy future is less about fracked gas than it is about efficient use and renewable supply. Both can protect you from volatile fuel prices because they have no fuel and hence no price volatility. Savvy investors and buyers are starting to exploit this valuable "price hedge," as Google did when it bought long-term wind-power contracts to lock in the operating cost of its data centers, or as Public Service Company of Colorado did when it cannily bought constant-price wind power instead of slightly cheaper (today) but worryingly volatile gas-fired power.

Another increasingly important economic benefit of most modern renewables is that their small, granular, modular units are very quick to build, so you can build as you need and pay as you go. These attributes cut financial risk, often making value several-fold higher than it is for big, slow, lumpy, inflexible investments. Indeed, such hidden benefits of distributed or decentralized renewables can collectively increase their economic value by even tenfold — enough to flip any investment decision. Rocky Mountain Institute's 2002 *Economist* book of the year, *Small Is Profitable: The Hidden Economic Benefits of Making Electrical Resources the Right Size*, documented this for more than two hundred distributed benefits, which the market is just starting to recognize.[2]

Installing traditional energy resources is rather like building a cathedral, tying up vast, specialized organizations for a decade. The new normal is the opposite: scalable mass production of most renewables is similar to the making of microchips, cell phones, and computers. Each year, 24-7, a single photovoltaic factory churns out billions of watts' worth of cells, comparable in total to a giant power station. Need more? The well-established scaling techniques of the semiconductor industry can stamp out those solar-cell factories like cookies. So at a lower cost than building one huge power plant over a decade, you can build *every year* a factory that *every year thereafter* can make enough solar cells to replace that power plant. With such scalability come stunning price drops, distressing solar-cell makers but delighting installers, who can win ever bigger markets and thus support even greater production at even lower prices.

The holy grail of "grid parity" has already been achieved in many parts of the United States and in many other countries. In much of the United States, entrepreneurs will install solar power on your roof with no down payment and beat your utility bill. New utility-scale solar plants beat new gas-fired plants in California's 2012 auction. German installers, though they source the same equipment, have so streamlined their installation processes that their solar systems cost only half as much as American ones. American installers will learn how to do that too, so by the time the 30 percent U.S. solar tax credit expires in 2016, it won't be needed, even to offset nonrenewables' permanent and generally bigger subsidies.

Where can these steep renewable-energy learning curves — the more you build, the less it costs — take us next? In 2011 Rocky Mountain Institute published *Reinventing Fire*, a rigorous business book.[3] It showed what modern renewables and efficient energy use can do when combined at historically reasonable rates via systematic barrier busting: the United States could run a 2.6-fold bigger economy in 2050 with no oil, no coal, no nuclear energy, and one-third less natural gas; this would be $5 trillion cheaper, with 82–86 percent lower carbon emissions, no new inventions, and no acts of Congress — with the transition led by business for profit.

New ways to integrate variable renewable sources (photovoltaics and wind) reliably into electric grids are no longer theoretical. They've given California 20 percent renewable electricity in 2012, Germany 23 percent, Denmark over 40 percent, and Portugal 70 percent in the windy and rainy first quarter of 2013. Texas had twenty-three days in February 2013 with over 20 percent wind generation (some days approached 30 percent), and the multistate utility Xcel reached 57 percent, nearing Spain's 61 percent record. San Diego Gas & Electric expects burgeoning solar output to reduce its fossil-fueled generation on many afternoons to roughly zero within a few years. On a national scale, as Rocky Mountain Institute[4] and the National Renewable Energy Laboratory[5] have shown, operational and organizational shifts now under way can extend renewables to at least 80–90 percent of total electricity supply, with full reliability and little bulk storage — indeed, with perhaps less storage and backup than utilities have already installed to manage the intermittence of their giant coal and nuclear power stations.

The keys to such high-renewables futures are realizing what modern renewables can do, deploying them sensibly, and teaming them with very efficient use of energy. Proven integrative-design techniques can often make very large energy savings cheaper than can small or no savings, turning diminishing returns into expanding returns. This approach can now triple or quadruple U.S. buildings' energy efficiency with a 33 percent internal rate of return (IRR), saving $1.4 trillion more than they cost, and can double U.S. industry's efficiency with a 21 percent IRR. Similar redesign of vehicles creates a breakthrough competitive strategy that cuts automakers' capital needs by 80 percent, dramatically derisks their business, makes uncompromised electric vehicles affordable, decouples autos from oil and climate, and turns autos into distributed storage that helps the grid accept varying photovoltaic and wind power. Getting all transportation off oil has a 17 percent IRR. Doing all these things together *and* making, by 2050, U.S. electricity 80 percent renewable, 50 percent distributed, and highly resilient has a 14 percent IRR. In short, this transition is not costly; it's highly profitable. And these comparisons value all external or hidden costs and benefits, including carbon emissions, at zero. Counting basic externalities would boost that $5 trillion net present value by several-fold. No wonder the smart money has shifted to renewables, while coal and nuclear orders dwindle because they have too much cost and risk to attract investors.

My own 1984 house illustrates this integration of efficiency with renewables. Its superinsulation, air tightening, and ventilation heat recovery have reached nearly the same standard as that of the thirty thousand passively heated buildings now built in Europe under the Passivhaus movement it helped to inspire — buildings that, like ours, have no heating system but a roughly normal construction cost. Passive solar heating thus provides about 99 percent of our space heat high in the Rocky Mountains, where we used to see temperatures dip below -40°F (or -40°C). The last 1 percent, originally from wood stoves, is now active solar. Combustion has been eliminated. Our passive solar design is akin to that of millennia ago but uses superwindows that insulate as well as fourteen (or in a few cases twenty-two) sheets of glass but look like two and cost less than three. In a further twist, the construction cost saved by eliminating the heating system enabled us to buy a novel solar hot-water system (with the quasi-seasonal storage proved in Lyngby, Denmark, in the 1970s), plus superefficient lights and appliances that stretch photovoltaic output to probably exceed our needs, even including the indoor office and the tropical jungle that the house wraps around. Inspired partly by the Balcomb house mentioned in chapter 25, our jungle supports more than a hundred species of higher plants. This month we harvested 65 pounds of midwinter bananas while it blizzarded outside.

Reinventing Fire's grand synthesis of U.S. energy solutions is now on its way to China, whose energy revolution is starting to spread to Japan and India.[6] China's extraordinary scaling of renewables (they now produce more wind power than nuclear) is rivaled in speed only by Germany's and Portugal's.[7] And in Germany, as in Denmark and Japan, these energy revolutions are led not by environmentalists but by fiscally conservative governments using market forces to strengthen jobs, profits, and security. U.S. renewable deployment is likewise spearheaded by the world's largest buyer of both oil and renewable energy — the Pentagon — both to save money and to carry out its national-security mission.

The fascinating global history chronicled in *Let It Shine* doesn't pretend to be complete. If it were complete, it'd scarcely be readable or even liftable. But it shows beyond a doubt that human ingenuity and persistence can solve the energy problem, just as it has solved many others. We need only see what is possible, figure out what will work, and adopt money-saving solutions, for whatever reasons each of us prefers. Governments can lead, catch up, or get pushed aside. Our most effective institutions — private enterprise, coevolving with civil society, sped by military innovation — can end-run our least effective institutions. With $5 trillion on the table, nobody need wait for Congress.

The resurgence of renewable energy, now being led at last by the developing world, is powerful and by now probably unstoppable. For the first time, the world appears to be breaking out of thousands of years of repeatedly inventing, commercializing, and forgetting renewable energy. The big trend is even more important. Millions of years ago, fire made us human. Then fossil fuels made us modern, built our civilization, and created our wealth. Now, each year, the world burns 4 cubic miles of the rotted remains of primeval swamp goo. But now we have even cheaper and more benign ways to sustain prosperity and security. Switching at last to renewables will be one of the greatest transitions in the history of our species: inventing a new fire that finally makes us safe, secure, healthy, and durable.

Even if we need never again reinvent this renewable-energy wheel, we must understand where it came from. *Let It Shine* shows how today's renewable revolution builds on the tenacious efforts of countless generations of innovators whose vision we may finally be privileged enough to bring to full flower.

— Amory B. Lovins
Cofounder and chief scientist, Rocky Mountain Institute
Old Snowmass, Colorado
April 28, 2013

Introduction

Charles Pope explains that he wrote his book *Solar Heat*, published in 1903, because "SOME CALL TO THE PEOPLE is needed... to arouse interest... [in] 'catching the sunbeams' and extracting gold from them." To accomplish his goal, he tells his readers, he has endeavored "to trace the history of attempts and successes in the utilization of solar heat[;]... discuss ways and means; and attempt to arouse his readers to give to the matter their energy and invention, their brain and capital; that we may very soon see solar enginery take its place by the side of steam enginery and electrical enginery and gas enginery in the public estimation, in technical schools, in mechanical journals, and in myriads of practical, labor-saving constructions."[1] More than one hundred years later I have attempted the same by writing *Let It Shine*.

Many believe that solar energy is a late-twentieth-century phenomenon, yet six *thousand* years ago the Stone Age Chinese built their homes so that every one of them made maximum use of the sun's energy in winter. So begins the story of the genesis of solar energy told in *Let It Shine: The 6,000-Year Story of Solar Energy*, the world's first and only comprehensive history of humanity's use of the sun. Because so few have attempted a comprehensive history of solar energy, page after page of this book brings to light information never before unearthed.

Twenty-five hundred years ago, for example, the sun heated every house in most Greek cities. Years later Roman architects published self-help books about using solar energy to show people how to save on fuel as firewood became scarce and as fleets scoured the known world for much-needed supplies of wood. During the renaissance Galileo and his contemporaries planned the construction of solar-focusing mirrors to serve as the ultimate weapon to burn enemy fleets and towns. Leonardo entertained more peaceful applications. He aimed at making his fortune by building mirrors a mile in diameter to heat water for the woolen industry. Much later, during the Industrial Revolution, engineers devised sun-powered steam engines to save Europe from paralysis should it run out of fossil fuels. In 1767 a Swiss polymath modeled global warming by trapping solar heat in a glass-covered box in the same way that carbon dioxide traps solar heat above the earth. Using the same type of glass-covered box to harvest solar energy, enterprising businessmen established a thriving solar water-heater industry in California beginning in the 1890s! And as electricity began to power cities, the first photovoltaic array was installed on a New York City rooftop in 1884.

A hundred and thirty years later, *Let It Shine* brings to light documents suppressed by the Nixon, Carter, and Reagan administrations that, had the public and Congress known about them at the time, would have permitted solar energy to assume a much larger role in the American energy mix.

Let It Shine presents the step-by-step development of solar architecture and technology. By providing the background for and illuminating the process of discovery, this book permits a deeper understanding of how solar-energy applications have evolved and performed. The book is more than a technological treatise, though. It presents the context in which these developments occurred and the people who made the solar revolution possible, revealing a whole new group of unknown technological pioneers, as well as identifying people famous for accomplishments other than in their work as solar-energy advocates and technologists. No one today thinks of Socrates as a solar-energy promoter, for example. Yet the author Xenophon, in his work *Memorabilia*, records Socrates presenting a basic plan for a solar house. Vitruvius, a Roman still famous today as the architect of architects, transmitted the wisdom of the Greeks on building in relation to the sun. The first aspiring solar-energy entrepreneur was Leonardo da Vinci. And Einstein's famous treatise on light quanta, which won him the Nobel Prize, reveals the great scientist as the father of modern photovoltaics.

Then there are the forgotten people, such as Gustav Vorherr, who in the 1820s opened the first school of solar architecture, in Munich, and his mentor,

Dr. Bernhard Christoph Faust, the first person to write a complete book on using solar energy. Thousands of architects — trained in the work of Faust and educated by Vorherr — fanned out to build solar-oriented homes throughout Europe in the nineteenth century. Sympathetic despots of Bavaria and Prussia actually required their subjects to follow the teachings of these men when building, which resulted in one portion of an urban area becoming a "sun city." And who has heard of William Grylls Adams and Richard Evans Day, who discovered in 1876 the photovoltaic effect in solid materials, or Charles Fritts, who put up the first rooftop solar array in the 1880s?

This is but a sampling of what's to be found in *Let It Shine*.

Let It Shine's precursor was *A Golden Thread: 2,500 Years of Solar Architecture and Technology*, which I coauthored with Ken Butti in 1980. Ken went on to follow other pursuits, and I continued to write, producing, for example, *A Golden Thread*'s photovoltaic sequel, *From Space to Earth: The Story of Solar Electricity*. And my book *A Forest Journey: The Story of Wood and Civilization* examines in great detail the fuel shortages, principally caused by forest devastation, that led many civilizations to turn to the sun to heat water and home interiors. *Let It Shine* is distilled from the insights I gained while researching and writing these three books and from new research over the past three decades.

Let It Shine presents the core of *A Golden Thread*, along with twelve new chapters and many additions to the older ones, making it the definitive history of the uses of solar energy throughout time. By giving an in-depth account of past successes and failures in applying the sun's energy to the art of living, the book offers insights into what the future of solar energy may come to be.

The first and most difficult step in crossing over from today's fossil-fueled world to a solar future is realizing that it can be done. *Let It Shine* makes this recognition possible by showing how our predecessors used the sun to better their lives. In doing so, *Let It Shine* demonstrates that the sun can become a major source of power that moves humanity toward living in a world that operates on carbon-free energy.

PART I

EARLY USE OF THE SUN

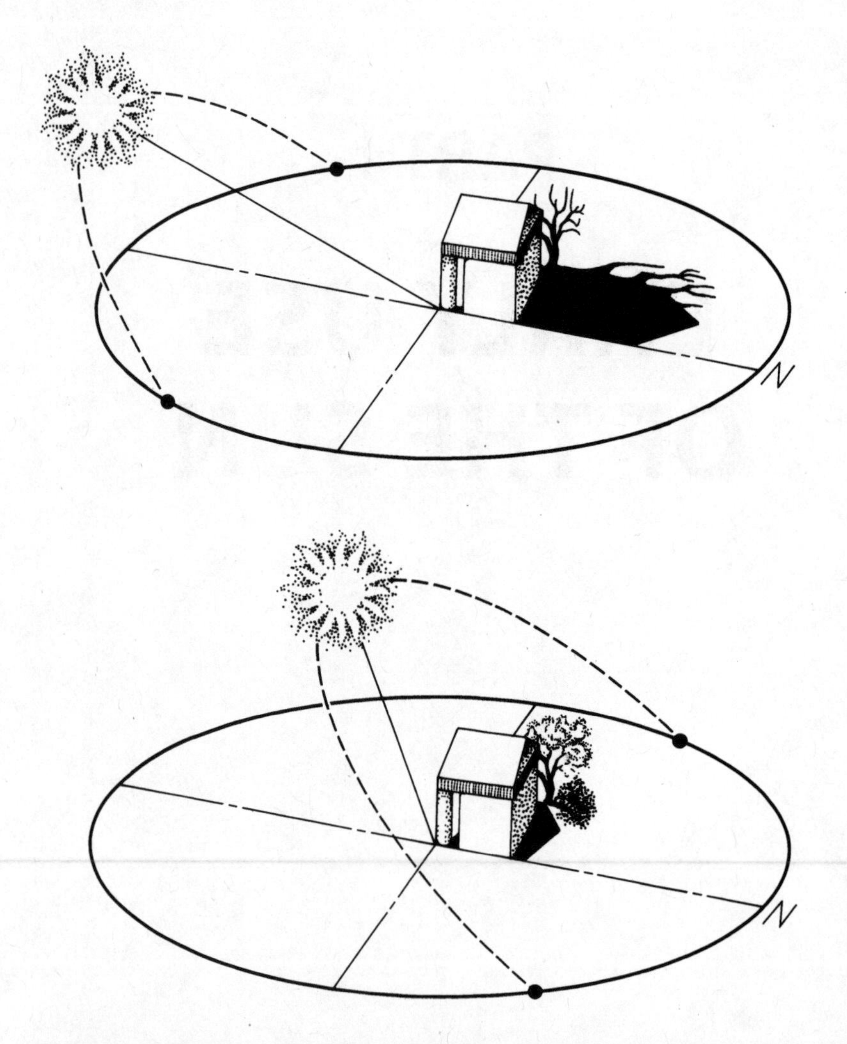

Figure 1.1. Path of the sun at noon on December 21 (top) and June 21 (bottom) in the Northern Hemisphere. In winter, the sun is always low in the sky and remains in the south throughout the day, while in summer it is much higher in the sky and spends the majority of the day in the east and west. These facts have important implications in building a home that is naturally comfortable throughout the year.

[6000 BCE–]

1

Chinese Solar Architecture

When discussing millennia-old techniques for building homes and palaces in a way that permitted the Chinese to take advantage of the sun's position relative to the earth throughout the year, the seventeenth-century Chinese philosopher Yu Li compared a correctly built structure to a person who dresses appropriately according to the changing seasons. "As it is true that clothing should be cool in summer and warm in winter, the same holds true for a house," Yu Li wrote in his book *Xian qing ou ji*.[1] Most important, for warmth in winter, the philosopher explained, "a house must have the correct orientation. It has to face south to catch the heat and light" of the winter's sun.[2] This simple awareness of how to site a home is the first principle of solar-oriented architecture, and no other civilization can boast of having had predominantly solar-oriented houses for as long as China has.

Principles of Solar Architecture

About four thousand years ago, the ancient Chinese began to track the changing position of the sun in relation to the earth by watching the sun throughout the year through openings in a kind of solar observatory.[3] Sometime later, their successors developed a more accurate way to keep track of the sun. They invented the gnomon, a stick or square piece of wood or stone planted perpendicularly (at a right angle) in the ground. Over time, as the sun moved across the sky

Figure 1.2. A Chinese family during the Zhou dynasty checking the shadow cast by their gnomon.

relative to the earth, they could record the shadow cast as the sun's rays struck the gnomon. In this fashion, people could mark the solstices and equinoxes. The great philosopher of science Thomas Kuhn credits the gnomon for allowing "systematic observations of the motion of the sun...[, which] harness[ed] the sun as a time reckoner and calendar keeper."[4]

The long shadow cast by the gnomon in winter would have quantified for the Chinese their observation that, at this time of the year, the sun remained relatively low in the sky throughout the day. That the shadow's direction did not fluctuate very much all day was a result of the sun staying in the south from the time it rose until it set, confirming an observation by an ancient Chinese astronomer: "On the day of the winter solstice, in the exact direction of the east and the west, one does not see the sun."[5]

In summer the gnomon shows the opposite is true: the sun spends the majority of the day in the northeast and northwest, and from 10 AM to 2 PM it is high in the sky toward the south. Knowing the location of the sun throughout the year allowed the Chinese to perfect the art of designing homes and whole cities so that all people could warm their houses with the sun's heat in winter and, during summer, keep the sun out of their houses so they could stay cool and comfortable.

The first account of the use of the gnomon for building comes from the Zhou dynasty, which was established sometime before the twelfth century BCE. Zhou government officials considered proper orientation too important to be left to chance, and so they instructed builders to establish the cardinal points of the compass for exact siting. The book *Zhouli*, which contained the rituals and rules established by the dynasty, explained how this would be accomplished. Builders first had to determine when the equinoxes and solstices occurred, which could be pinpointed by studying the shadows cast by the gnomon. The longest and shortest shadows of the year would mark the winter solstice and summer solstice, respectively. When the shadow cast was half as long as the two solstice shadows, the observer would know that one of the two equinoxes had arrived. At either equinox, the shadow cast by the rising sun would point west, and the shadow cast by the setting sun would point east. Taking note of where the noon shadow fell, the observer would learn where true north and south lay.[6] In this way, sometime in the seventh century BCE, Duke Wan "began to build the Palace at Ts'oo, orienting it...by means of the sun."[7]

How It Worked

In Yu's day, most houses in northern China, whether opulent or simple, conformed to the traditional courtyard style that had prevailed for thousands of years. This type of architecture had either walls or subsidiary buildings surrounding a courtyard; the rectangular main house was recessed into the back of the court, and all its openings faced south. Poet Ban Gu, who lived in the second century CE, saw Chinese solar architecture at work when the palace's south-facing "Door of Established Brightness" was opened in wintertime. "The sun's radiance would flare brilliantly into the palace, heating the rooms inside."[8]

Those living in humble abodes, too, took advantage of the sun's winter location to stay warm. Peasants and workers regarded the southwest nook of the house as "the cozy corner," the most desirable place to nestle, where the warm rays of the afternoon sun poured in, even though it might be freezing outside.[9]

A south-facing building could also stay cool in summer. The Chinese studied the gnomon's shorter summer shadows that resulted from the sun climbing

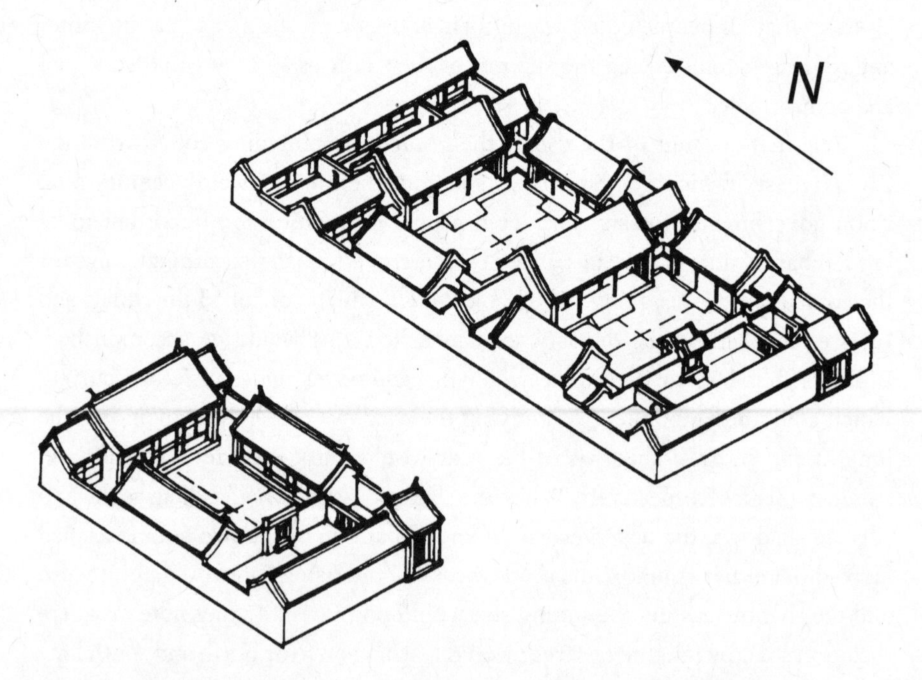

Figure 1.3. A typical Chinese house. The main building faced south, protected from the high summer sun by eaves that extended from the roof's edge on all sides of the building.

much higher in the sky than in winter. They recognized that eaves, if projected over south-facing windows and doors, would keep the high, hot summer sun from entering the buildings throughout the day yet still let in the low winter sun. As Ban Gu observed at the palace complex of the Western Capital in summer: "Their upturned eaves provide a covering mantle [for they] intercept the sun's rays."[10]

The Chinese, Ban Gu noted, immensely appreciated solar architecture for helping to maintain mild temperatures indoors throughout the four seasons, so necessary, the ancients believed, for a long life.[11] The sun's warming rays also reduced people's dependence on charcoal heaters, in this way saving a lot of money. For as temperatures dropped, the price of charcoal would always shoot up.[12]

Deforestation and Solar Architecture

Solar architecture and the relentless war against forests went hand in hand as China's long history unfolded. Unlike Western notions of paradise, where trees and human beings and animals happily coexisted before the Fall, the Chinese saw only chaos in primordial times, when "vegetation was luxuriant and birds and beasts swarmed."[13] Paradise on earth began, as far as the Chinese were concerned, when the legendary Shun brought order to nature and civilization to China. Under his orders, the forests were "set fire to and consumed, opening up the Middle Kingdom for cultivation."[14] In the four thousand years since then, the Chinese have continued Shun's work, chopping their way to new lands, leaving most of China to resemble the slopes of Niu Mountain, once covered with lovely woods but by 300 BCE so bare that people believed nothing had ever grown there.[15] By the fourth century BCE, if not earlier, it was seemingly axiomatic that wooded areas near large population centers would be deforested.[16] Without easy access to forests, charcoal no doubt became harder to get, changing solar architecture from a matter of choice to one of necessity.

Archaeological Evidence for Solar Architecture

Archaeological discoveries, too, reveal that building with the sun in mind began very early in China. More than six thousand years ago, entrances to the homes at Banpo, in the north, were, according to one Asian scholar, "deliberately oriented toward the mid-afternoon sun when at its warmest a month or so after the solstice," which was the coldest time of the year.[17] Overhanging thatched roofs kept the unwanted sunshine off the structures during the hotter months. Submerging the main living spaces below ground level at Banpo moderated the temperature inside throughout the year.

Recent excavations at Erlitou, also in northern China, show solar architecture in full bloom. Settled two thousand years after Banpo, Erlitou is characterized by a house type and a form of urban planning that would continue basically unchanged for the next four thousand years, well into the twentieth century. A reconstruction of the Erlitou palace shows it set on the north side of a south-facing courtyard, so that it could face south.[18] Qinhua Guo, the lead archaeologist at Erlitou, states, "The new discovery reveals that many city construction rules [principles] in the later dynasties can be dated back to the Erlitou site. This includes the crisscrossing streets [running perpendicular to each other] and constructions [buildings] facing south."[19]

Figure 1.4. Model of the Erlitou Palace.

Though laid out thousands of years after Erlitou, Beijing too was given the same type of grid pattern. "Its streets are all so straight, so long, so broad and well proportioned," remarked a European visitor to Beijing in the 1600s.[20] The perpendicular layout with streets running east-west assured a southern orientation for anyone wishing one. "You shall rarely see a palace or house of any great person which does not face that point of the compass," the visitor remarked.[21]

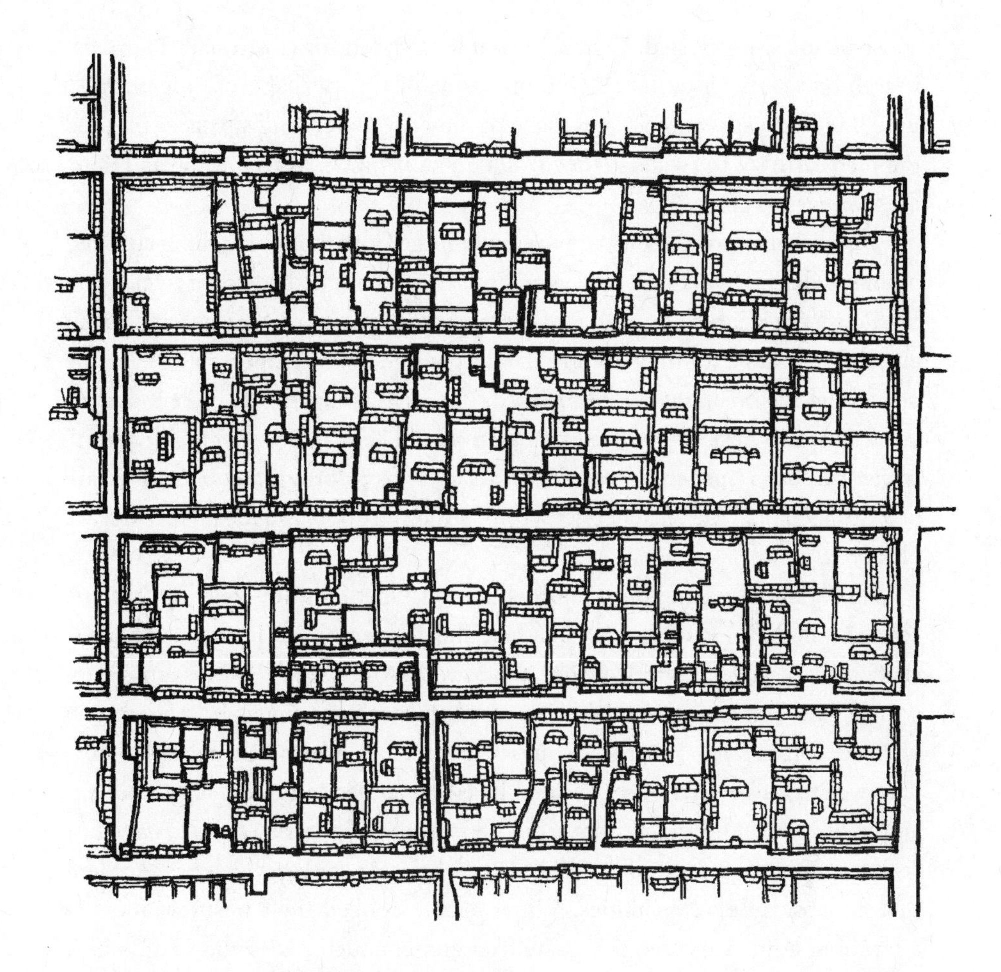

Figure 1.5. Layout of Beijing in a grid pattern.

Early Energy Conservation

Winter in the Middle Kingdom could be very harsh. Dew on grass, leaves, and rooftops began to freeze in late autumn, and heavy snows fell by early December. When putting up a structure such as the palace at Ts'oo, builders had to think of the season when determining the best material for its walls, just like people considering what type of coat would suit them with winter on its way. As the royal residence went up in 1300 BCE, "crowds of [laborers] brought the [wet] earth in baskets" to "dress" it in proper clothing, an onlooker at the construction site reported. "They threw it with shouts into the frame. They beat it with responsive blows.... Five thousand cubits of [walls] arose together."[22] Five hundred years passed, and those putting up a later king's palace still used rammed earth for the walls so they would remain "impervious to wind and rain" for the years to come.[23]

As crickets retreated indoors, seeking refuge from the approaching cold and hiding under people's beds, the peasantry knew the time had come to close up the house for winter. No different from getting winter clothes out and fixing any rips or tears a few days before the ugly weather strikes, peasants sealed the cracks and the openings not needed. An ode written more than three thousand years ago gives an account of the preparations. "The windows that face [north]," which had been opened during the hotter months to bring in cool drafts, "are now stopped up," the ancient poet wrote. "And the doors [on the north side] are plastered."[24]

The rich also employed more luxurious methods for keeping out the cold. At night or during bad weather, the Emperor Wu had goose-feather curtains covering his south-facing windows.[25] And while the peasantry slept under plain blankets, the wealthy covered themselves using fabric filled with wadded cotton and, before retiring, had curtains drawn around the bed.[26] The wealthy also used *kangs* (heated beds) for added comfort — another example of interest by the ancient Chinese in energy conservation. A *kang* consisted of a platform built of material such as brick or adobe, which has excellent thermal-absorption capabilities. When people cooked their meals in the late afternoon, a flue captured the waste heat and conducted it to the *kang*, which then radiated sufficient heat throughout the night for the comfort of those who slept in it.[27]

Linguistic and Ritualistic Evidence for the Importance of Orientation of Buildings

As the importance of solar architecture grew, the southern aspect took on great stature in Chinese life. Ancient wisdom associated the south with fire and summer — in other words, with warmth — while the north came to be synonymous with winter and somberness, with things cold and dark.[28] The southwest corner became known as the seat of honor, where "the elder and respected members of the family reside."[29] Ritual also required the emperor to face south whenever in the company of an audience. Sages explained the custom by pointing out that "the diagram of the south conveys the idea of brightness." By facing south the emperor shunned darkness and embraced enlightenment while governing.[30]

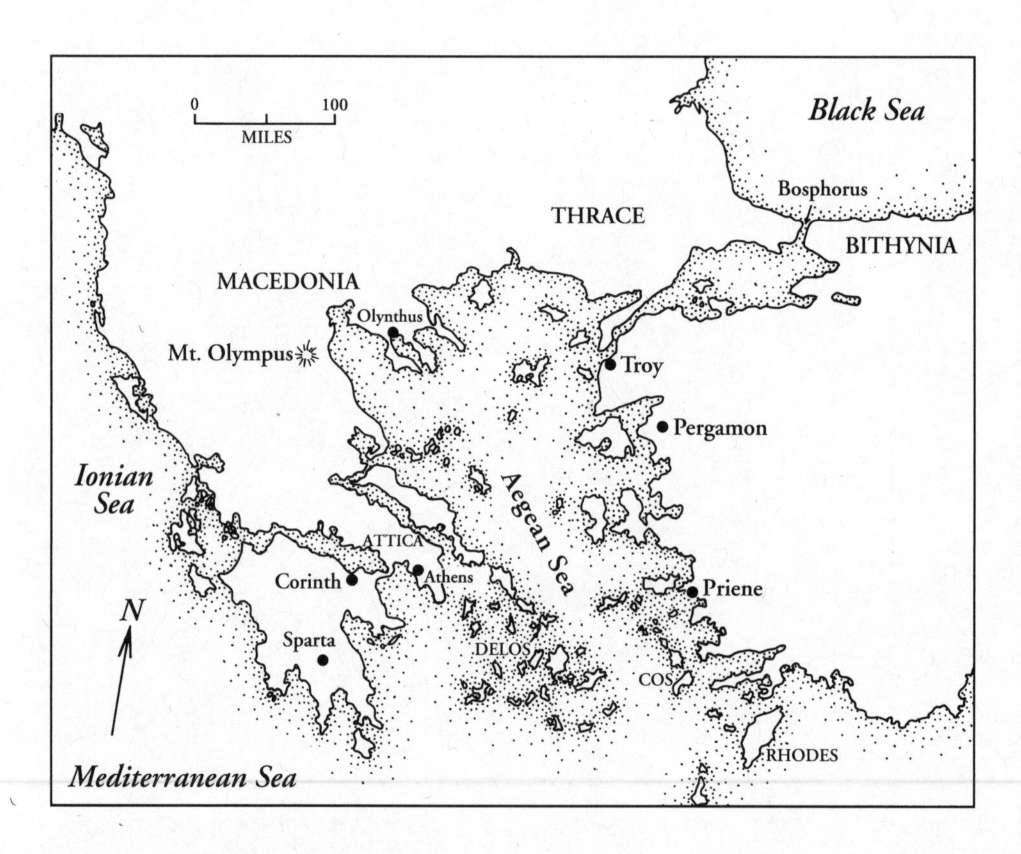

Figure 2.1. Ancient Greece and Asia Minor.

2

Solar Architecture in Ancient Greece

Never in history has there been a stronger or more eloquent advocate for the use of solar building principles than Socrates. Xenophon, a disciple of Socrates, presented in a Socratic dialogue the philosopher's belief that the art of "building houses as they ought to be" was firmly based on the principle "that the same house must be both beautiful and useful." In what has become known as the Socratic method, Socrates began his discourse by asking a question: "When someone wishes to build the proper house, must he make it as pleasant to live in and as useful as it can be?" After his student answered in the affirmative, the master then asked, "Is it not pleasant to have the house cool in summer and warm in winter?" And when the student assented to this as well, Socrates then closed the discussion by affirming, "Now in houses with a southern orientation, the sun's rays penetrate into the porticoes [covered porches,] but in summer the path of the sun is right over our heads and above the roof, so we have shade.... To put it succinctly, the house in which the owner can find a pleasant retreat in all seasons...is at once the most useful and the most beautiful."[1]

The second great sage of Greece, Aristotle, provided additional details not mentioned by Socrates. He, too, started with a question: "What type of housing are we to build for slaves and freemen, for women and men, for foreigners and citizens?" And then he answered, "For well-being and health, the homestead should be airy in summer and sunny in winter. A homestead possessing these qualities would be longer than it is deep; and its main front would face south."[2]

Socrates's father worked in the construction industry as a stonemason, and many believe Socrates followed in his father's footsteps. If he did, he would have known about the remodeling of two houses in downtown Athens — the Agora — where workmen changed the arrangement of rooms so the most important ones would face onto a southern courtyard.[3] The fact that Socrates spent the majority of his life in the neighborhood assures that he knew about the solar project in Athens. And less than 2 miles outside of the city, another example of solar design was built in Socrates's time. This was a large rectangular house that sat on the foot of the northerly slope of Mount Aegaleo, one of several mountains near Athens. According to the archaeologists who excavated it, the farmhouse they discovered, "which faces south and has its entrance and court on this side and its main rooms on the north," corresponds exactly to the ideas expounded by the great philosopher as recorded by Xenophon.[4]

Olynthus: A Planned Solar City

Socrates surely also knew about the solar district planned and built as part of the city of Olynthus, northeast of Athens, since its creation was a consequence of a revolt against his beloved city-state.[5] People from neighboring towns participating in the break with Athens in 432 BCE moved to Olynthus for protection against Athenian retribution. The increase in population forced the Olynthians to establish a new district, which its excavators called North Hill. The latitude was approximately that of New York City and Chicago, and the temperature often dropped below freezing in winter. Approximately twenty-five hundred people settled there.

North Hill was a planned community from the beginning. Starting from scratch, the settlers could more easily implement the principal ideas of solar architecture. The town planners situated the new district of Olynthus atop a sweeping plateau and built the streets perpendicular to each other, just as the Chinese had, with the main streets running east-west. In this way, all the houses on a street could be built with a southern exposure, assuring solar heating and cooling for all residents — in keeping with the democratic ethos of the period. Aristotle later commented that such rational planning was the "modern fashion," which allowed the convenient arrangement of homes so that they could take maximum advantage of the sun.[6]

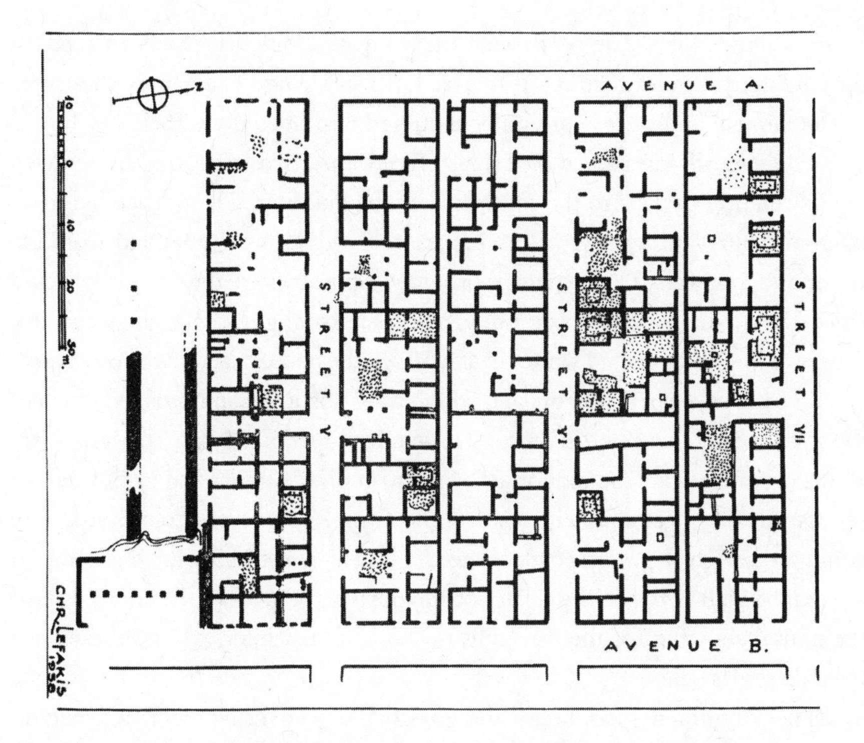

Figure 2.2. Typical street plan in the North Hill section of Olynthus. This ancient community aligned its streets and avenues so that all buildings could face south.

Olynthian builders usually constructed houses in a blocklong row simultaneously. The typical dwelling had six or more rooms on the ground floor and probably as many on the upper floor. These houses were usually a standard square shape and shared a common foundation, roof, and walls with the other

Figure 2.3. Model of Olynthian apartments, south face. Rows were built far enough apart to guarantee that each household would have complete access to the winter sun.

houses on the block. The north wall was made of adobe bricks, which kept out the cold north winds of winter. If this wall had any window openings, they were few in number and were kept tightly shuttered during cold weather.

The main living rooms of a house faced a portico supported by wooden pillars running parallel to the south side of the building. The portico led to an open-air courtyard averaging 320 square feet, which was separated from the street by a low wall. The courtyard provided a place where the occupants could enjoy the outdoors with maximum privacy; and sunlight, the home's primary source of illumination and winter heat, entered the house through the courtyard.

The house's earthen floors and adobe walls absorbed and retained much of the solar energy that came in through its window openings facing the courtyard. In the evening, when the indoor air began to cool, the floors and walls released the stored solar heat and helped warm the house. To prevent cold drafts from coming through the open portico into the house, some builders constructed a low adobe wall between the pillars of the portico, parallel to the south wall of the house, allowing for the warming rays of sun in winter, while shutting out the cold drafts below.

The Olynthian solar house design worked well in summer and winter. When the summer sun was almost directly overhead — from about ten in the morning until two in the afternoon — the portico's eave shaded the openings of the main rooms of the house from the sun's harsh rays. In addition, the closed walls and contiguous dwellings barred the entrance of the morning and afternoon sun into the east and west sides of the homes.

Priene and Delos

About 350 BCE, during Aristotle's lifetime, Mausolus, ruler of the Greek city of Priene, planned to rebuild the city on the steep slopes of Mount Mycale, in what is now western Turkey (Asia Minor). He intended, with the help of Alexander the Great, to make Priene a model city for the Greek world. Luckily for future generations, Priene's ruins are considered the finest surviving example of an ancient Greek city.

Priene is a good example of how the ancient Greek architects and planners coped with adverse topography to create a solar city. To accommodate Mycale's steep slopes, the city's builders devised a checkerboard street plan similar to that used at Olynthus. They terraced the main avenues along the contours of the rocky spur on an east-west axis; the secondary streets ran up the mountain from north to south. Owing to the sharp incline, many of the secondary streets became stairways. Despite Priene's difficult location, all homes, no matter how large or small, were designed according to what Priene's excavator, Theodor Wiegand, called the "solar building principle." The main rooms always opened onto a south-facing covered porch. Even homes belonging to the poorer citizens enjoyed the warmth of the sun in winter and were spared its heat in summer.[7]

Delos, an important trading center in the Aegean, presented an even greater challenge to solar architects. The irregular rocky terrain of this island precluded the division of streets into an orderly pattern as at Olynthus or Priene. It also prevented the use of uniform house plans. Often the contorted topography determined the design of a Delian home. Nevertheless, the main rooms faced south whenever possible. In some parts of Delos, a new adaptation of solar

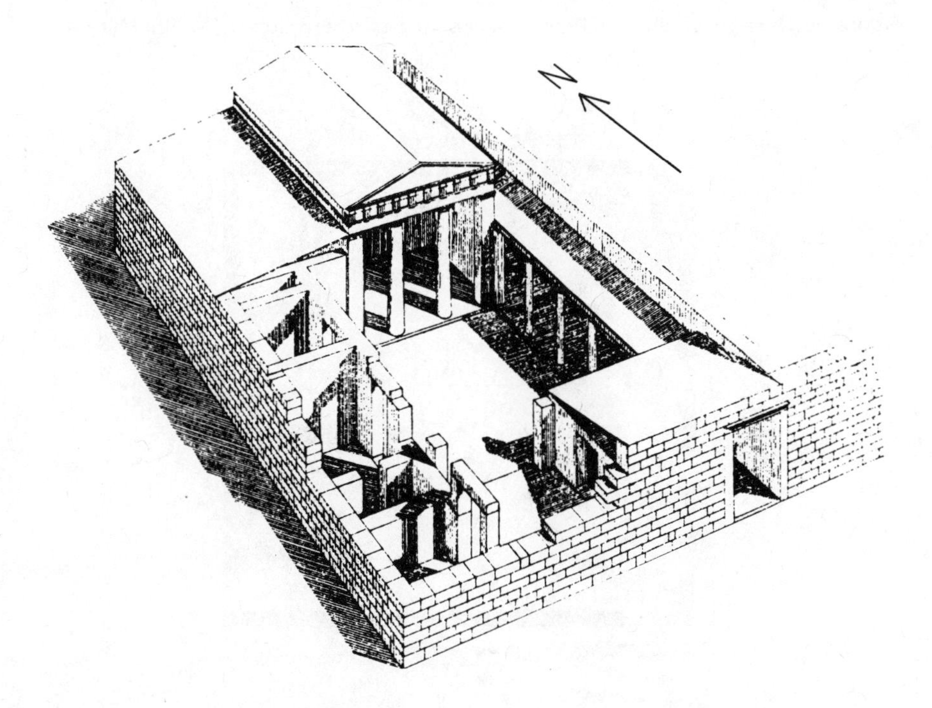

Figure 2.4. Reconstruction of an ancient Greek home, by Theodor Wiegand. The rooms behind the portico faced south.

Figure 2.5. The relocated city of Priene, built on the southern slope of Mount Mycale.

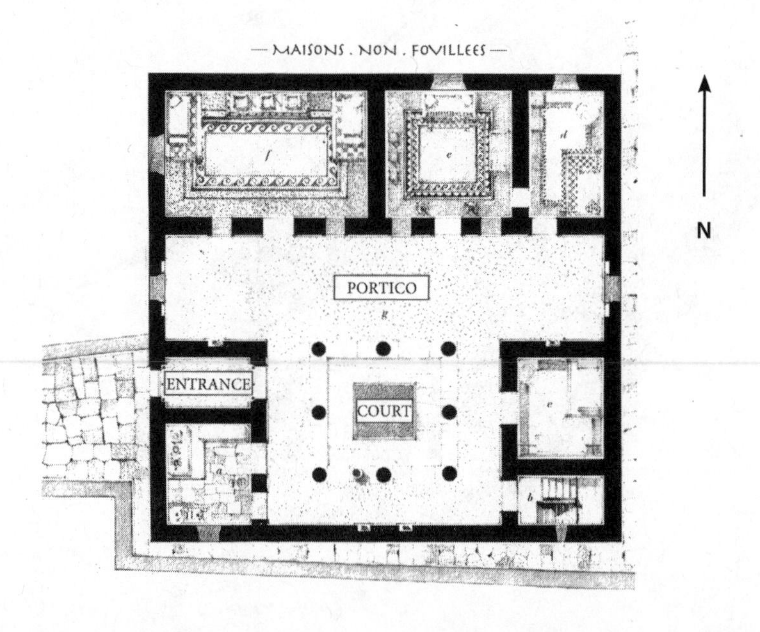

Figure 2.6. Floor plan of a house in Delos. The main rooms, like those in homes built in the North Hill section of Olynthus, faced south and opened onto a portico and courtyard.

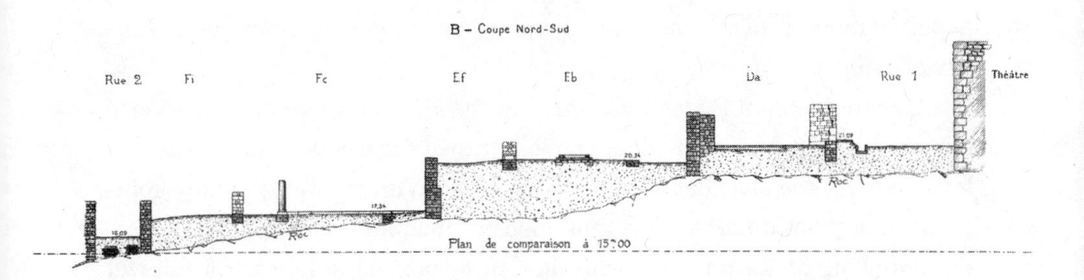

Figure 2.7. Elevation of a terraced Delian house.

design took form. Many residents terraced their homes along the sloping terrain so that the important rooms were on the upper level, with a commanding view of the south.[8]

Diminished Access to Fuel

It is probably no coincidence that construction incorporating solar-building techniques began to appear after miners at Laurion, near Athens, struck a rich vein of silver. The silver mine would lead to the flowering of Athenian wealth and power throughout most of the fifth century BCE. Separating the silver metal from the ore required huge amounts of charcoal, however. At the same time, large numbers of people flocked to Athens, and since everyone cooked meals and heated homes with hibachi-like braziers, the growing population, combined with the surging silver production, took a toll on local forests, the only source of fuel for heating, cooking, and smelting.[9] Surveying the effect of deforestation, Plato remarked about the land, no longer protected by the canopy of the woodland against the erosive forces of rain, sun, and wind, "What now remains compared with what then existed is like the skeleton of a sick man, all fat and soft earth having wasted away, and only the bare framework of the land being left.... There are some mountains which have nothing but food for bees, but they had trees not very long ago, and the rafters from those felled there to roof the largest buildings are still sound."[10]

The people of Delos imported all their charcoal, since no trees grew on the island, creating a seller's market for heating fuel. Realizing the unfair advantage that sellers of charcoal enjoyed, legislators passed laws regulating the trade to protect consumers from price gouging.[11] These measures coincided with the popularization of solar architecture there, no doubt also a response to the scarcity of wood.

Later residents of Priene were fortunate that their predecessors had planned the city for solar heating. Farmers rapidly turned the wooded area around the local river — the Maender — into wheat fields. With the forest canopy gone, rains throughout the watershed sent floods of mud into the river. The consequent infilling of Maender Bay with silt transformed the sea formerly adjacent to Priene into dry land, causing the city to lose its coastal location. Although accelerated erosion dashed any hopes of the city becoming a great port, the solar design of its homes at least made the residents less vulnerable to rising charcoal prices.[12]

How Well Did Solar Design Work?

According to Isomachus, a character in a Socratic dialogue by Xenophon, Greek solar architecture was highly effective. Isomachus brought his bride to his solar-oriented home and "showed her... living rooms for the family that are cool in summer and warm in winter." He told Socrates, "The whole house fronts south so that it is... sunny in winter and shady in summer."[13]

Empirical evidence confirms Isomachus's praise. The Architect Edwin D. Thatcher studied the solar-heating capability of rooms facing south to determine the feasibility of indoor nude sunbathing during the winter. To simulate actual conditions, Thatcher relied on weather data for a climate similar to that of ancient Greece and western Turkey. He found that a naked person sitting in the sunny part of such a room would be relatively comfortable on 67 percent of the days during the colder months of November through March.[14] The room used for this study was not as well protected as an average Greek living room, however — and of course the residents of the latter would have been clothed most of the time. It seems safe to say that for most of the winter the sun would have adequately heated the main rooms of a Greek solar-oriented home during the daytime. When solar heat was insufficient, charcoal braziers could be lit.

The great playwright Aeschylus suggested that a south-facing orientation was a normal characteristic of Greek homes. It was a sign of a "modern" or "civilized" dwelling, he declared, as opposed to houses built by primitives and barbarians, who, "though they had eyes to see, they saw to no avail; they had ears, but understood not. But like shapes in dreams, throughout their time, without purpose they wrought all things in confusion. They lacked knowledge of houses...turned to face the sun, dwelling beneath the ground like swarming ants in sunless caves."[15]

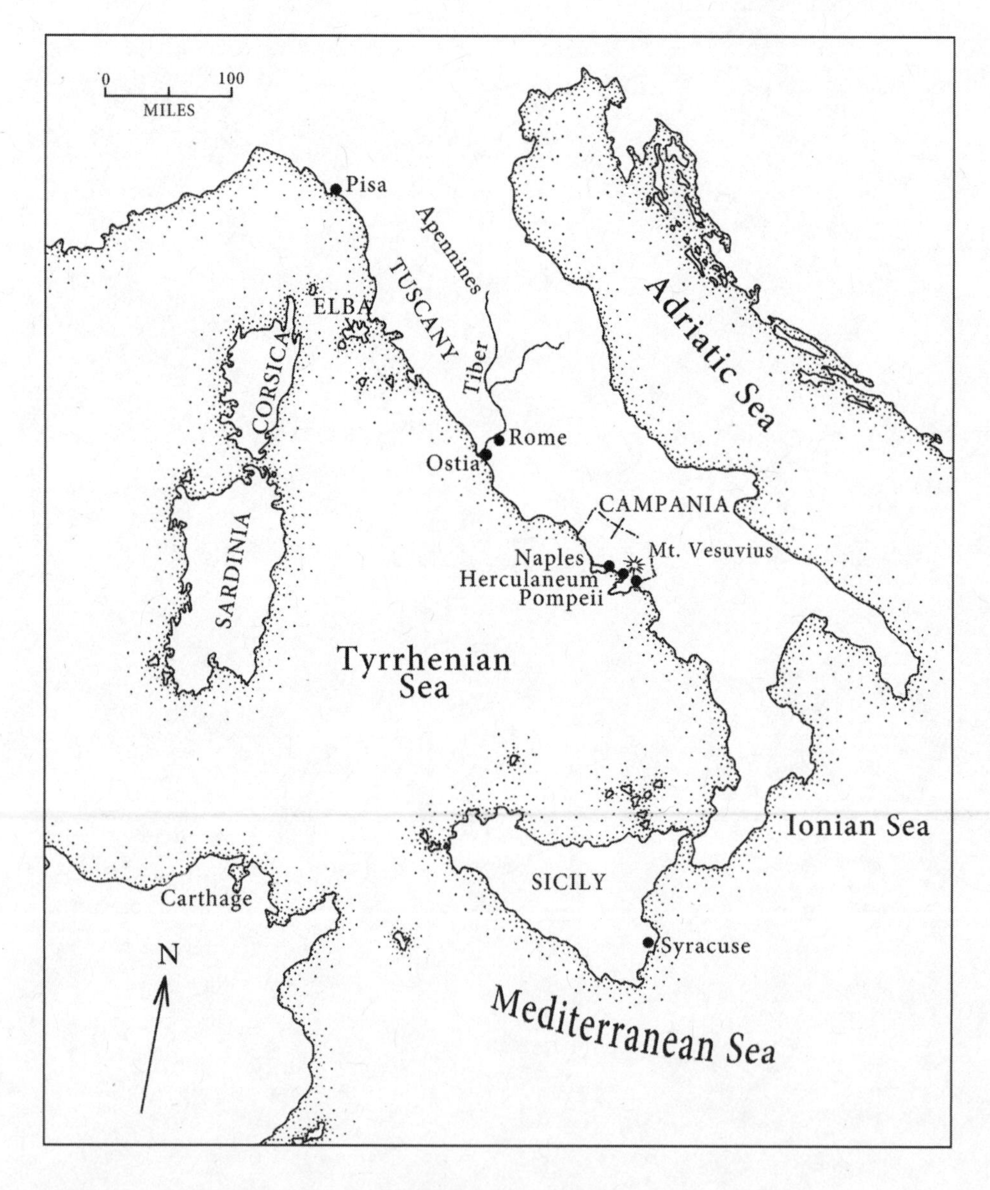

Figure 3.1. Italy during the first century BCE.

3

Ancient Roman Solar Architecture

Some solar houses built in Greek times remained inhabited when the Romans conquered the region in the second century BCE.[1] Vitruvius, the preeminent Roman architect of the first century BCE, apparently visited such structures while serving as a military engineer in Greece. Not surprisingly, when Vitruvius wrote about architecture in his famous, still-extant work *The Ten Books of Architecture*, he applied what he had seen in Greece, advising architects and builders that in more temperate parts of the empire, such as the Italian peninsula, "buildings should be thoroughly shut in rather than exposed toward the north, and the main portion should face the warmer [south] side."[2] But Vitruvius went beyond his Greek predecessors to specify where certain rooms should be placed for optimum comfort. He recommended that Romans living in temperate climates make sure their winter dining rooms looked to the winter sunset, "because when the setting sun faces us with all its splendor, it gives off heat and renders this area warmer in the evening."[3] But summer dining rooms "should have a northern aspect. For while the other aspects, at the solstice, are rendered oppressive by the heat, the northern aspect, because it is turned away from the sun's course, is always cool."[4]

Pliny's Two Villas

Wealthy Romans took Vitruvius's advice. Varro, a contemporary of Vitruvius, who was considered one of the most learned men in Rome at the time, observed that "men of our day aim to have their summer dining rooms face the northeast and their winter dining rooms face the falling sun."[5] A century later, Pliny the Younger, a wealthy official and prolific writer, described in several letters to friends how his two villas were designed to interact with local climate conditions for optimum comfort throughout the year. Pliny chose this strategy, as he told his friend Gallus, because he wanted his villa "large enough for his convenience" but not "expensive to keep up."[6]

Pliny's summer house stood in the Apennine Mountains of north-central Italy, where summers remained fairly cool most of the time and winters could be quite severe. Keeping the house pleasantly warm by building it in relation to the sun stood out in Pliny's plans. He had the main part of the house exposed to the south in order "to invite the sun, from midday in summer but earlier in winter into a wide and proportionally long portico." For the occasional summer day when the rest of the house really heated up, Pliny built an underground room "resembling a crypt, which in the midst of summer heat retains its pent up chilliness," in which he and his guests could escape the scorching weather.[7]

Pliny also had another villa on the seacoast at Laurentum, close enough to Rome that he could occasionally spend time there after a hard day's work in the city. The architect designed the villa with particular care for the portions of it used primarily in winter — for example, joining a large room and adjacent dining room at such an angle that they collected and concentrated the warming rays of the winter sun. Pliny referred to these rooms as his winter retreat. In another part of the house the architect placed a winter dining room that would be "warmed and illuminated not only by the direct rays of the sun, but also by their reflection from the sea." Above and beyond these solar rooms, "crowning the terrace, portico, and garden, stands a detached building which I call my favorite," Pliny writes. "It contains a very warm winter room."[8]

Pliny's motive for building with the sun was apparently to save money. A solar-oriented home would require a smaller furnace and fewer heating ducts than a similar nonsolar, centrally heated villa that burned more fuel for heating. At Pliny's retreat at Laurentum, on the seacoast, the dining room faced southwest, as Vitruvius had advocated, so that the rays of the setting winter sun would heat the room and provide a congenial atmosphere in which to dine. Southeast of the dining room was a large bedroom, and beyond it a smaller one whose windows admitted both the morning and evening sun. The study, where

Pliny spent much of the day reading, was semicircular with a large bay window that let in sunlight from morning to evening.

Glass as a Solar Heat Trap

Pliny called one of his favorite rooms a *heliocaminus* — literally, a "solar furnace."[9] Such a name indicates that this room got much hotter than the other rooms of his villa. Most probably the southwest openings of the heliocaminus were covered with glass or thin sheets of mica or selenite — two types of transparent stone — unlike in Greece, where the openings had no coverings. Such materials act as solar heat traps by admitting sunlight into a room and holding in the heat that accumulates. With such coverings on the windows, the temperatures in Roman solar houses rose well above those in houses earlier built in Greece, making them, in the eyes of the inhabitants, true "solar furnaces."

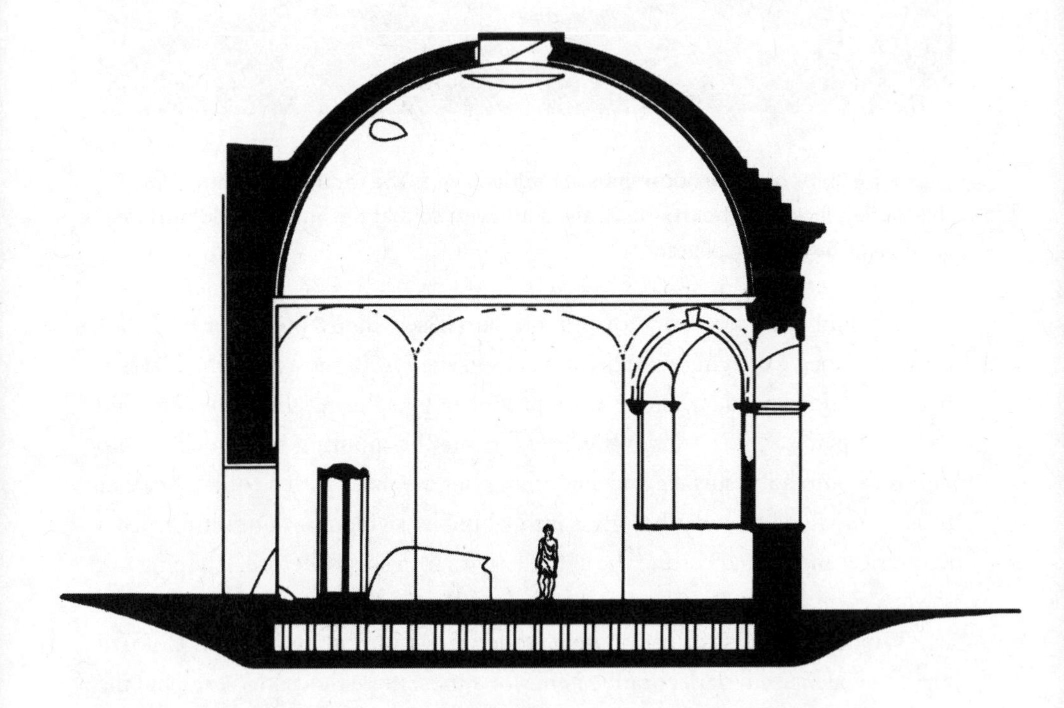

Figure 3.2. Cross section of a Roman heliocaminus. In winter, the large glass or transparent-stone windows admitted plentiful sunlight and kept the solar heat from escaping once inside.

Figure 3.3. Ruins of the heliocaminus at Hadrian's Villa. The windows, as shown in the overlay, faced southeast, south, and southwest so that the house would capture sunlight all day during winter.

It would have been easy for Pliny to purchase windows to cover his heliocaminus, since a thriving window industry existed in Rome at the time. Transparent windows made of stone were produced by splitting the stone into thin sheets or plates. Glass windows were fabricated by pouring molten silica into molds or onto a flat surface and smoothing out the mass with a roller. Another process involved blowing hot glass into a bubble, swinging it until it formed a cylindrical shape, and cutting the cylinder with iron shears before flattening out the pieces on a flat surface covered with sand.[10]

Windows of glass or transparent stone were a radical innovation. Colored glass had been used for decorative items for almost three thousand years, but the Romans were the first — in the first century CE — to use transparent materials to make windows that would let in light but keep out rain, snow, and cold. The philosopher Seneca noted the novelty of this idea in a letter written in 65 CE:

"Certain inventions have come about within our own memory — the use of window panes which admit light through a transparent material, for example."[11]

Excavations at Herculaneum, Pompeii, and elsewhere provide evidence that glass was made into windows for the homes of wealthy Romans — even in the early days of the empire. The great natural historian Pliny the Elder observed in the first century CE that these new windowpanes, unlike opaque window coverings, allowed sunlight to enter a building. By the following century, the use of glass and other transparent materials to cover windows was increasingly common.

The Romans also used the solar-heat-trap principle to cultivate plants. Greenhouses were built to keep plants warm so that they would mature more quickly and produce fruits and vegetables even in winter. Emperor Tiberius, for example, had a penchant for cucumbers year-round. In fact, Pliny the Elder noted that "there was never a day in which he went without." To oblige the ruler's appetite, the kitchen gardeners had cucumber beds mounted on wheels so they could move them into the sun as needed. During the colder days of winter, these gardeners placed "cold frames" glazed with transparent material over these cucumber beds to hold in the solar heat.[12]

Wealthy Romans also used transparent coverings to keep their exotic plants healthy during inclement weather. The favorite plants of the rich often basked in more warmth than did the poorer Roman citizens. Martial, the first-century satirist, could not resist parodying this inequity. He complained that while his patron protected exotic trees from the cold by placing them in a structure with a glazed, south-facing wall that caught the sun's rays, the broken windows of his own room went unrepaired. "Not even the chilly winds will stay in my room," the poet sarcastically complained to his patron in this parody. "Do you wish your old friend to stay until he freezes? I should be better off as the guest of your trees!"[13]

The Roman Baths

Another Roman application of the solar-heat-trap principle was the use of transparent window coverings to help heat the Romans' public baths. Probably no other people in history have cherished baths as much as they. Beginning in the first century CE, the public baths became immensely popular gathering places. People from all walks of life congregated there, usually in the late afternoon when the workday was over, to exercise in the gymnasium; take dips in the cold, warm, and hot baths; sweat in the steam room; or lounge around listening to music, poetry, and gossip.

The Baths of Caracalla, one of the largest Roman baths ever built, held as many as two thousand people at a time. Seneca took a somewhat cynical view of the tumult inside these baths on a typical busy afternoon:

> Picture to yourself the assortment of sounds, which are strange enough to make me hate my very powers of hearing! When your strenuous gentleman, for example, is exercising himself by flourishing leaden weights; when he is working hard, or else pretends to be working hard, I can hear him grunt; and whenever he releases his imprisoned breath I can hear him panting in wheezy and high-pitched tones. Or perhaps I notice some lazy fellow, content with a cheap rub-down, and the crack of the pummeling hand on his shoulder varies in sound as the hand is laid on flat or hollow.... Add to this the arresting of an...occasional pickpocket, the racket of the man who always likes to hear his own voice in the bath, or the enthusiast who plunges into the swimming pool with unconscionable noise and splashing.... Then the cake seller with his different cries, the sausageman, the confectioner, and all the vendors of food hawking their wares, each with his own distinctive intonation.[14]

The Roman baths had not always been so boisterous. When public baths were first introduced in the second century BCE, they were small, modest establishments serving primarily as places for washing one's body. These baths were as dark as caverns. As Seneca remarked, they had had "only tiny chinks — you cannot call them windows — cut out of stone" to let in air and light.[15]

Two hundred years later, baths of great opulence sprang up throughout the empire, including France and Britain. This was the age of Augustus; Rome now ruled most of Europe and large parts of North Africa, the Middle East, and Asia Minor. Roman citizens could afford to enjoy the spoils of conquest, and both public and private baths glittered with large, costly mirrors and mosaics, and Thracian marble lined the pools.

The Romans also demanded that their hot baths — *caldaria* — and sweating rooms be extremely hot. Seneca wryly commented, "Nowadays there is no difference between 'the bath is on fire' and 'the bath is warm,'"[16] adding with a

Figures 3.4 and 3.5. Only a small chink in the ceiling admitted sunlight in these two old-style baths.

little hyperbole that the water in the hot baths was so scalding that condemned slaves could be bathed alive with the same effect as having them burned. To heat water to such temperatures whole tree trunks were burned. In fact, the baths consumed so much firewood that specially designated forests were reserved for their exclusive supply.

Despite such extravagance, the Romans displayed a utilitarian bent toward heating the baths during this time and later periods. Vitruvius, for example, suggested turning to the sun for help: "The site for the baths must be as warm as possible and turned away from the north.... They should look toward the winter sunset because when the setting sun faces us with all its splendor, it radiates heat, rendering this aspect warmer in the late afternoon."[17] From the first century CE onward, the builders of most bathing establishments followed this dictum. In almost every important bathhouse of the time, at least the hot bath faced toward the winter sunset to absorb as much solar heat as possible while the majority of its users were cavorting inside. In addition, the Romans usually glazed the whole south wall of their bathhouses. Seneca wrote that these giant windows trapped so much solar heat that by the late afternoon, bathers would "broil" inside the baths. In some of the more elaborate bathhouses, the sweat room (which was usually semicircular in shape) boasted enormous windows looking to the south and southwest.

Figure 3.6. The Baths of Diocletian during a typical afternoon, as depicted by the nineteenth-century artist Edmond Paulin.

Figure 3.7. Southern window openings measured almost 10 feet high at the main baths of Pompeii.

Among the ruins of Pompeii, as in most Roman cities, are examples of both the old-style Roman baths — structures "buried in darkness," as Seneca described them — and the later baths built to take advantage of the sun. According to August Mau, one of the excavators of Pompeii, there were only "small apertures, high in the walls and ceilings, through which light was admitted in the older baths." In contrast, the *caldarium* of the more "modern" Central Baths had south-facing windows measuring 6 feet 7 inches by 9 feet 10 inches to take maximum advantage of the afternoon sunlight.[18] A physicist at Cornell in the 1990s studying the performance of the glass-covered south-oriented Roman baths concluded that

> the sun alone on sunny days could provide most of the energy to maintain the 100°F temperatures [required for heating the hot baths]. Indeed, even with fires reduced on sunny days, there would probably be some thermal energy [from the sun] stored in the floors and walls that would maintain the temperature as the sun [went] down. On days when the sun [was] obscured by clouds, the hypocaust [the heating

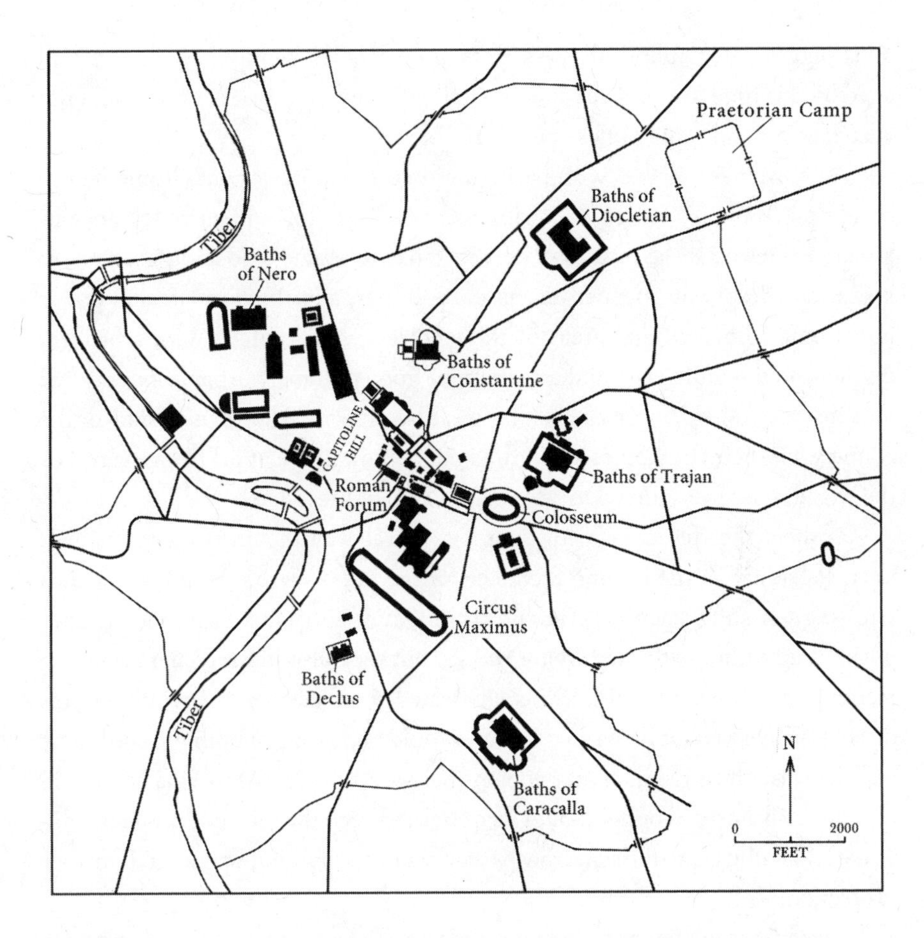

Figure 3.8. A map of imperial Rome shows the locations of the major baths. Most of the newer ones faced south or southwest.

> system in use] with a reduced fire, or turned on only part of the time, could by itself easily maintain the temperature [100°F] even with the outside at 30°F.[19]

Rural Self-Reliance

By the fourth century CE, the fuel situation had become critical. To satisfy the Roman fuel needs, the government commissioned an entire fleet of ships, called the *naviculari lignarii* (literally, "wood ships"), for the sole purpose of making wood runs from France and North Africa to Ostia, ancient Rome's seaport.

This suggests that almost all the forests on the Italian peninsula had been ravaged by this time. Or perhaps Rome, now militarily on the defensive, could not keep its overland supply lines open at all times.[20]

As border skirmishes with barbarians became more serious, Rome had to build up its military forces, requiring increased taxation and currency devaluations. Primarily these measures hurt its citizens. Rome's position grew more precarious despite its stepped-up military efforts, and the flow of vital goods into Rome from outlying areas of the empire was disrupted. With a broken, disenchanted middle class and a scarcity of goods, Rome's urban economy fell into disarray. More and more of the wealthy citizens who owned land in the country left their city homes and settled on their estates. Cut off from the rest of the world, they were forced to adopt a self-reliant lifestyle.

To help the rich cope with their new rural way of life, Faventinus and, later, Palladius — the leading architects of the age — wrote building manuals stressing self-sufficiency. Both architects reiterated Vitruvius's advice on proper positioning of hot baths and living rooms, but they also offered energy-saving techniques of their own. Palladius advocated the placement of winter rooms directly above the hot baths so that they would benefit from both the sun's heat and the waste heat rising from the baths below. And while Vitruvius had noted that olive oil storage rooms should face south to keep the oil from congealing in winter, Palladius added that a transparent window covering would give further protection.

Faventinus and Palladius recommended an ingenious way to make the floor of a sun-heated winter dining room an ideal absorber of solar energy. The technique had been invented earlier by the Greeks and passed on in the writings of Vitruvius. A shallow pit was to be dug under the floor and filled with broken earthenware or other rubble, and atop this a mixture of dark sand, ashes, and lime was spread. This formed a black floor covering that easily absorbed solar heat, especially during the afternoon. The mass of rubble underneath stored large amounts of heat and released it later in the evening when the room temperature cooled. Faventinus assured villa owners that such floors would stay warm during the dining hour and "will please your servants, even those who go barefoot."[21]

Sun-Rights Laws

From the days of Augustus in the first century CE until the fall of Rome, the use of solar energy to heat residences, bathing areas, and greenhouses was apparently widespread throughout the empire. Exactly how much the Romans relied on the sun is impossible to say, but the heliocaminus, or "solar furnace," room was common enough to provoke disputes over sun rights requiring adjudication in the Roman courts. As the population increased, buildings and other objects blocked the solar access of some heliocamini, and their owners sued. Ulpian, a jurist of the second century CE, ruled in favor of the owners, declaring that a heliocaminus's access to sunshine could not be violated. His judgment was incorporated into the great Justinian code of law four centuries later: "If any object is so placed as to take away the sunshine from a heliocaminus, it must be affirmed that this object creates a shadow in a place where sunshine is an absolute necessity. Thus it is in violation of the heliocaminus' right to the sun."[22] That this opinion was written into the Justinian code in the sixth century strongly indicates that the construction of solar-heated buildings continued until this late date.

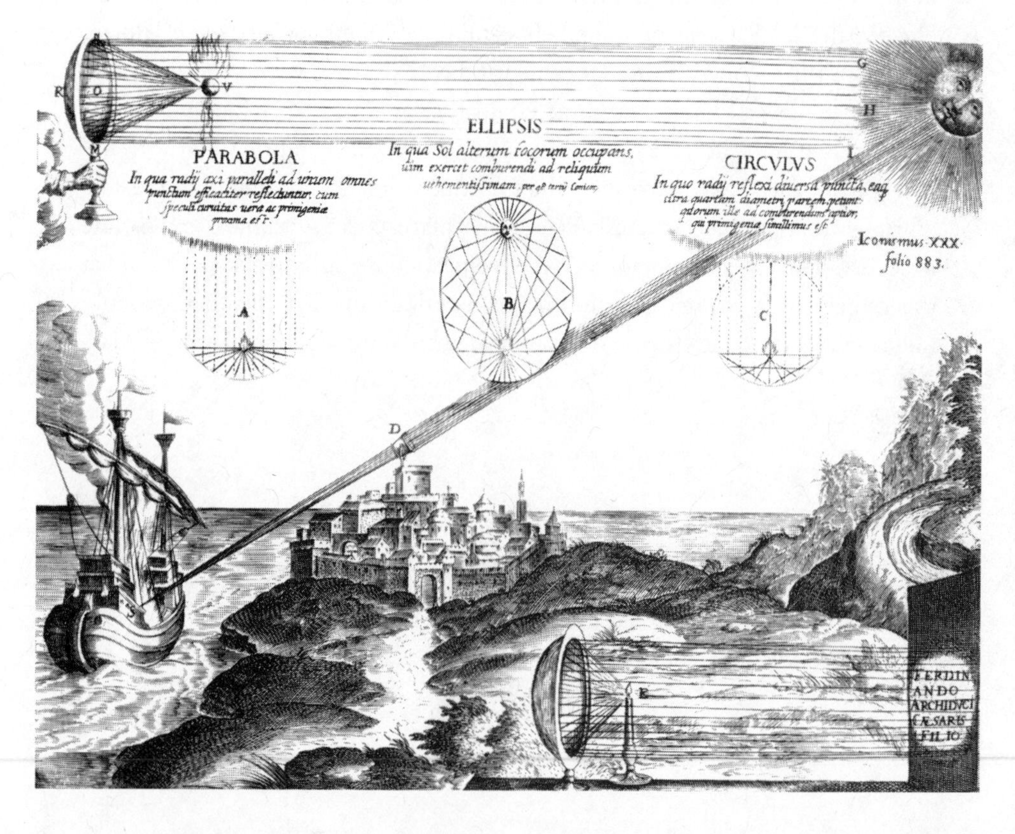

Figure 4.1. Engraving from *Ars Magna Lucis et Umbrae* by Athanasius Kircher, 1646. Because parabolic mirrors and refractive lenses focus the parallel rays of the sun to a point, they strongly concentrate light, which reaches temperatures high enough to ignite combustible materials — in this case an enemy ship.

[400 BCE–1700s]

4

Burning Mirrors

The solar orientation of buildings and use of glass as a solar heat trap were not the only ways the ancients harvested solar energy. The Chinese, Greeks, and Romans also developed curved mirrors made of metal that could concentrate the sun's rays onto an object with enough intensity to make it burst into flames within seconds. These became known as "burning mirrors." Credit for their discovery goes to the ancient Chinese. Confucius, writing of life three thousand years ago, stated that every son who lived at home attached a bronze burning mirror (a *fu-sui*, later called a *yang-sui*) to his belt when he dressed for the day. He would also attach a fire plow, a wooden tool that relied on friction to generate sparks for ignition.[1] On days when the sun shone, the boy would focus the solar rays onto wood and start the family fire; on overcast days he would take out his fire plow and rub its wood stick back and forth in a wooden groove to do the same. The *yang-sui* was as ubiquitous in early China as are matches or lighters today.

Although many ancient Chinese texts written during the Zhou dynasty discuss the *yang-sui*, it wasn't until 1997 that archaeologists confirmed the existence of an extant *yang-sui* from the Chinese Bronze Age. And what a find — the oldest solar device in the history of humanity! It was found in the hand of a skeleton buried in a tomb dated to about three thousand years ago. The local museum took a mold from the original and then cast a copy in bronze. After polishing its curved surface, an archaeologist focused sunlight onto a piece of

Figure 4.2. These *yang-suis* (solar igniters) were removed from Bronze Age excavations.

tinder, just as a possessor of a burning mirror would have done so many years ago, and in a few seconds lit a flame. In the words of the archaeologist: "This verified without a doubt that the purpose of the device is to make fire."[2]

Once archaeologists had proved that this bronze, bowl-shaped object with a knob on the back was a solar fire igniter, they went back to similar objects dug up earlier and identified more than twenty additional ancient *yang-suis*. Multiple molds for *yang-suis* found at a Bronze Age foundry in Shanxi Province, close to the first find, suggest the ongoing mass production of these devices.[3] "The fact that our ancestors could invent and produce a yang-sui around 3,000 years ago is a world-class marvel. It should be considered one of the great inventions of ancient Chinese history," remarked the archaeologist, impressed by the ability of the ancient Chinese to figure out the complex optics required for the optimal performance of the *yang-sui* so early in history.[4]

Westerners discovered burning mirrors much later. Theophrastus, a Greek writing in the fourth century BCE, observed that "substances catch fire from the sun by reflection from smooth surfaces of materials such as copper and silver."[5] A century later, Greek geometers learned that adding curvature to these materials better focused the sun's energy on the substance they wished to burn. In this fashion, they could ignite larger, more heat-resistant objects faster than before.[6] The Greeks' first curved mirrors had a spherical shape, which focused the sun's rays into a line. Then Dositheus, a mathematician and colleague of the legendary Archimedes, constructed the most effective burning mirror ever conceived — the parabolic mirror, which resembled a sliced eggshell. It could make all the rays meet at one point.[7] A century later, in his treatise *On the Burning Mirror*, Diocles gave the first formal geometric proof of the focal properties of parabolic and spherical mirrors.

The Greeks used burning mirrors to light the flame that marked the beginning of their Olympic games. Plutarch, the famous Greek biographer who wrote in the second century CE, stated that when barbarians sacked the Temple of Vesta — the temple tended by the Vestal Virgins at Delphi — and extinguished their sacred flame it had to be relit with the "pure and unpolluted flame from the sun." With "concave vessels of brass" the holy women directed the rays of the sun onto "light and dry matter," which was immediately ignited, and their flame burned anew.[8]

The ancient Chinese too used burning mirrors for religious purposes. According to the *Zhouli*, the book of ceremonies written in approximately 20 CE that describes rituals dating far back into Chinese antiquity, "the Directors of Sun Fire have the duty of receiving, with a concave mirror, brilliant fire from the sun... in order to prepare brilliant torches for sacrifice."[9] Across the sea,

Figure 4.3. A modern-day reenactment of lighting the Olympic flame with a burning mirror.

the Incas likewise lit their holy fires with solar energy. To mark the solstice and New Year, the priests at the Virgins of the Sun temple took out their concave silver mirrors enclosed in gold-and-gem-encrusted frames. They concentrated the sun's rays onto fuel, which would burst into flames.[10]

Discovering that concentrated sunbeams would combust anything burnable, both the Chinese and Greeks came to believe these devices captured "fire from the sun."[11] As result they also came to believe "the sun is fire."[12]

Dreams of Solar Weaponry

As with so many achievements in antiquity, all knowledge of burning mirrors — their powers, methods of construction, and uses — vanished from European culture during the Dark Ages. Fortunately, the Arabs had a great reverence for knowledge during these times. They preserved, translated, and elaborated upon many classical works, including Greek geometry texts that discussed the mathematical properties of parabolic mirrors.

Ibn al-Haytham, an eleventh-century Arab scholar living in Cairo, had many of these ancient works at his disposal. (The geometric proofs of Diocles were probably among them.) Al-Haytham called burning mirrors "one of the noblest things conceived by geometers of ancient times," but he felt that the Greeks did not convincingly explain their proofs. "Since in this matter there is great benefit and general usefulness," he wrote, "[I] have decided to explain and clarify it, so that those who seek truth will know the facts."[13] Al-Haytham's elaborate mathematical proof was translated into Latin and circulated among several European universities in the middle of the thirteenth century. Consequently, his writing served as a bridge between scholars of medieval Europe and the ancient Greeks.

One of those privy to al-Haytham's work was Roger Bacon, a Franciscan monk who taught at Oxford and the University of Paris during the thirteenth century. To Bacon, this essay on the concentrating powers of parabolic mirrors was more than idle academic chatter. It was a clue to the means of building a doomsday weapon that might be wielded by the Antichrist — the Muslims that the Crusaders were then battling in the Holy Land. Bacon advised Pope Clement IV that "this mirror would burn fiercely everything on which it could be focused. We are to believe that the Anti-Christ will use these mirrors to burn up cities, camps and weapons."[14]

Figure 4.4. Roger Bacon.

He warned that an enemy Saracen (al-Haytham) "shows in a book on burning mirrors how this instrument is made."[15] But the specific details of its design were contained in another volume, unavailable to Christendom — suggesting to Bacon a sly deception on the part of the Arabs. Nevertheless, Bacon reassured the pope, there was no need to worry: "The most skillful of Latins is busily

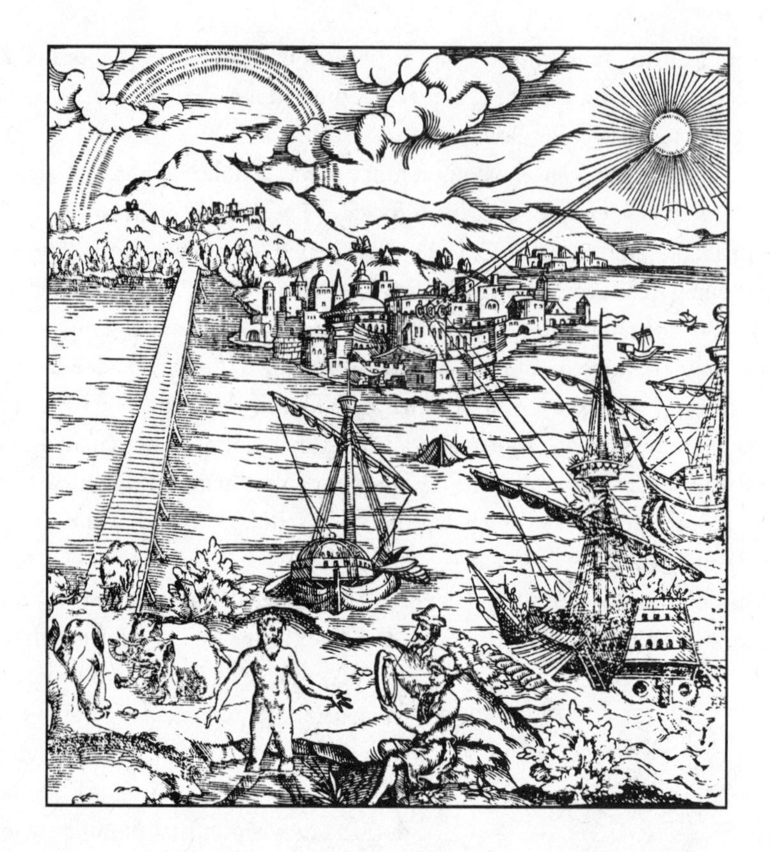

Figure 4.5. Frontispiece of al-Haytham's *Opticae thesaurus*, which showed the many ways optics made use of light. The book aroused Bacon's suspicions. The potential use of burning mirrors for military aims later became a favorite theme during the Renaissance among scientists and people who dabbled in the sciences.

engaged in the construction of this mirror." Whether this "Latin" was a colleague, as suggested by Bacon, or merely an indirect reference to himself is not known.

Several years later Bacon informed the pope that his "colleague" had finally finished building a powerful parabolic mirror after working "many years... at great expense and labor[,]... abandoning his studies and other necessary business." Certainly, if the Christians "living in the Holy Land had twelve such mirrors," Bacon advised the pope, they could "expel the Saracens from their territory, avoiding any casualties on their side" and making it unnecessary for more Crusaders to intervene in the Middle East.[16]

Whether Bacon ever really built this solar weapon or was simply giving free rein to his imagination, we can only speculate. For such a device to be effective beyond the range of spears, arrows, slings, catapults, and other weapons in the hands of enemy troops, it would have to be truly colossal — an unlikely achievement with the technology of the times. The mirror would also have to be moved throughout the day to stay in alignment with the sun. Such a feat would require an army, since a mirror of this size would be enormously heavy. There were other drawbacks as well to Bacon's scheme. The Saracens knew about solar reflectors themselves and so would be most likely to face the sun while attacking, making it impossible for the mirror to reflect the sun's rays in their direction. Furthermore, burning mirrors work only on sunny days. Only the parallel rays of direct sunlight can be concentrated on a small target area, and clouds tend to scatter solar rays in all directions. If the Muslims moved against the Christians on an overcast day — or at night, for that matter — Bacon's solar weapon would have been useless.

In his zeal, Bacon never informed the pope of these limitations. However, this was the first time in medieval European history that someone had advocated the use of empirical science instead of metaphysical speculation. To make such a radical idea palatable, Bacon tried to show how technology could serve the interests of both princes and clerics. But his offer fell on deaf ears. Even though Pope Clement was liberal enough to tolerate the views of a proponent of experimental science, he never read any of Bacon's plans. Once the pope died, a conservative wind swept through the church. Empiricism threatened the entire worldview of most Christians — a view based on the metaphysics of Aristotle and divine revelation. The idea of transforming mild sun rays into a fierce weapon of fire was now condemned as the work of the devil. Instead of receiving the funds to build the ultimate solar weapon that would help defend Christianity, Bacon was branded a heretic and thrown into a dungeon.

The church's suppression of Bacon worked. It successfully quashed all investigations into the use of solar mirrors. Interest in solar weaponry did rekindle several centuries after Bacon's time. Its revival, though, had nothing to do with the Franciscan monk. The Renaissance was now in full bloom, and Greek and Roman classics found their way into libraries. The ideas contained in ancient literature stimulated creative individuals to accomplish great deeds. In works by Lucian, Galen, and Diodorus, people read about Archimedes using burning mirrors to set the invading Roman fleet afire.[17] Though all these accounts had been written many years after the fact, the legend of Archimedes and his solar weapons fired up much interest in and experimentation with focusing mirrors.

During the latter part of the sixteenth century and throughout the seventeenth, almost every European scientist and gentleman experimenter investigated the curious powers of burning mirrors. Giovanni Magini, an Italian astronomer, wrote that he could melt "lead, silver or gold in small quantities such as a coin held firmly with tongs."[18] Magini's mirror, like most of the reflectors built at the time, was spherical, not parabolic. It was also modest in size — less than 2 feet in diameter.

Methods of production limited the size of burning mirrors built in the late 1500s and early 1600s. Most burning mirrors were fabricated from a highly lustrous alloy of copper, tin, and arsenic. Craftsmen melted the alloy into a liquid, poured it into a mold, and then pried the mirror out after it had cooled and taken shape. The last step would have been practically impossible if the mirror were large, because the metal would have been very heavy. A mirror measuring only 2½ feet in diameter weighed at least 110 pounds. Imagine the difficulty of lifting a mirror ten times this size! Moreover, the alloy was brittle, making it nearly impossible to lift a huge mirror out of its mold in one piece. Even if an alloy mirror of tremendous size could have been made, the same problem that Bacon ignored would still have existed: an immense and heavy reflector would require enormous manpower to keep it turned toward the sun.

Still, mirror enthusiasts could dream. If a small device could produce so much heat, imagine the power of a mirror a hundred times its size! Unfettered by strict empiricism, the experimenters of the age presented the theoretically plausible as established fact. Even a scientist of Galileo's caliber was fooled by these claims. He confessed that after "seeing a small mirror melt lead and burn every combustible substance," he believed that "such effects as these render credible to me the marvels accomplished by the mirrors of Archimedes."[19] The legend of the Greek's alleged military feats with burning mirrors found a ready audience at the time.

Galileo's less rigorous contemporaries speculated even more wildly. Giambattista della Porta promised in his book *Natural Magick* to teach his readers to build gigantic burning mirrors that would "cast forth terrible fires, and flames, that are most profitable in warlike expeditions."[20] Armed with solar weapons, people could "burn [their] enemy's ships, gates, bridges," and by igniting strategically placed caches of gunpowder they could "blow up castles [and] towers."[21]

None of these stupendous weapons ever showed up in anyone's arsenal. They were as chimerical as della Porta's recipe for producing silver by combining quicksilver with a toad and stirring it to a boil in a simmering pot.

In less bombastic moments, della Porta also envisioned peaceful uses for burning mirrors. "It is good husbandry," he wrote, to rely on solar energy whenever possible, "for the work is done without wood, or coals, or labor."[22] But while his solar weaponry never saw the light of day, people in his time did useful things with mirror-generated solar heat. Some made perfume with the concentrated rays of the sun. Adam Lonicer, writing in 1561, recorded how this was done: Certain types of flowers were submerged in a clear, water-filled vase, which was placed at the focal point of a spherical concave mirror. The concentrated solar heat caused the aroma of the flowers to diffuse into the water, and, voilà, you had perfume.[23] The extreme heat produced by mirrors also permitted the soldering of metals. Leonardo da Vinci told of the sculptor Andrea del Verrocchio employing this method to put together sections of a copper-ball lantern holder for the Santa Maria del Fiore Cathedral in Florence. He also observed the use of concave mirrors by the public to kindle fuel for their ovens, just as the Chinese had done thousands of years before.[24] In an age without matches, where fire could be ignited only by laborious friction, solar concentrators certainly made sense.

Figure 4.6. Using a focusing mirror and the sun's heat, alchemists produced perfume in a glass vase containing water and aromatic flowers.

Not until the sixteenth century was another fabulous mirror proposed, this time by Leonardo da Vinci. Its purpose would be not military but peaceful — to generate heat and power for industry and recreation. Leonardo proposed building a parabolic mirror 4 miles across that could "supply heat for any boiler in a dyeing factory, and with this a pool can be warmed up, because there will always be boiling water." Why the mirror had to be enormous is not clear from his notes, nor is there any discussion of using many smaller mirrors in tandem to achieve the same end.[25]

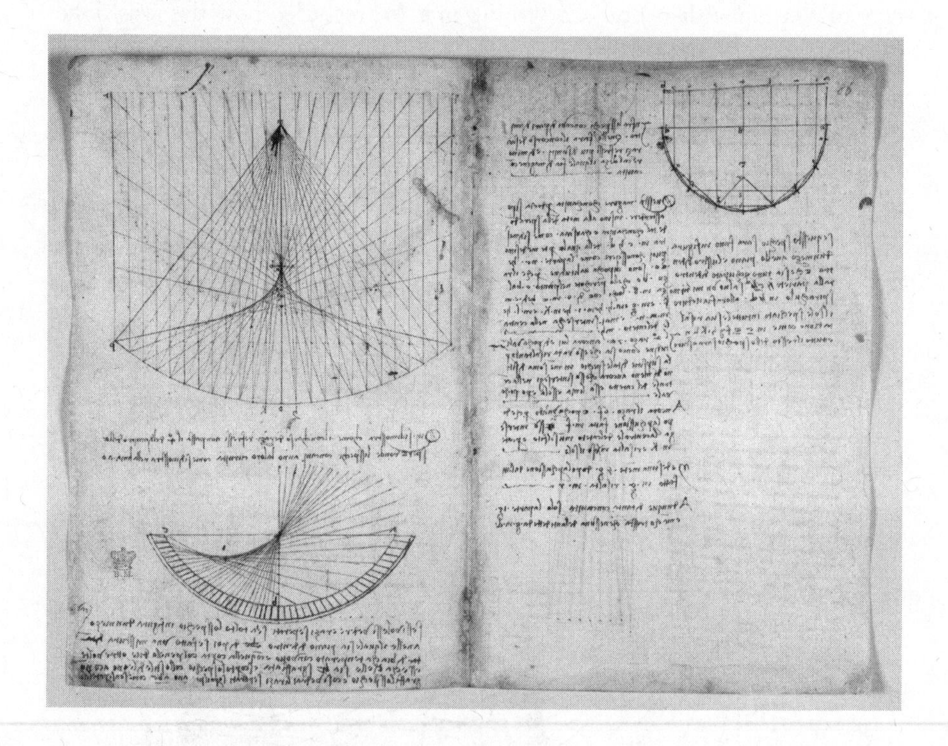

Figure 4.7. A page from Leonardo's workbook shows his design for an enormous concentrating mirror he hoped to build.

Leonardo began building his giant mirror around 1515. Unlike the flamboyant Bacon, the Florentine kept his intentions a closely guarded secret. He valued the project immensely and greatly feared that someone would try to steal his great invention. When a certain individual expressed the desire to work with Leonardo, the great inventor believed that the man's "whole interest was to... get to my work on the mirrors" and to publish the work as if it were his own.[26]

In spite of the importance of this project to Leonardo, his great mirror, like the giant solar weapon Bacon boasted of, never saw daylight.

Leonardo had other reasons for his fascination with concave mirrors. The fire they produce when focused was proof for him that the sun is hot by nature, despite the belief in his day, perpetuated by followers of Aristotle, that the power of the sun's motion creates its heat. For "the sun, which being itself warm, in passing through these cold mirrors, shows great heat," Leonardo observed.[27] His admiration for and emulation of Archimedes also pushed him to embrace burning mirrors. He might have read about what the great geometer had accomplished in the solar field, since he had access to Archimedes's complete set of works, which we lack. If so, Leonardo no doubt wished to have been regarded by his contemporaries as the next Archimedes, or perhaps wished even more that he had built the greatest concentrator of all time.[28]

In Search of the Giant Mirror

The grandiose claims of Leonardo and della Porta led Athanasius Kircher, a Jesuit, to "press with all zeal" during the mid-seventeenth century in his hunt for the elusive giant mirror. While Kircher could not deny that if such "a parabolic mirror were the size of a mountain, it would burn objects at great distances," he wondered who could manufacture such a prodigious machine.[29] To obtain the answer, Kircher traveled throughout Europe, meeting with outstanding craftsmen, "so that they might display anything similar" to the huge burning mirrors he had heard about. But after an extensive search, Kircher came away empty-handed. "Nowhere has such [a powerful mirror] as I sought appeared," he wrote.

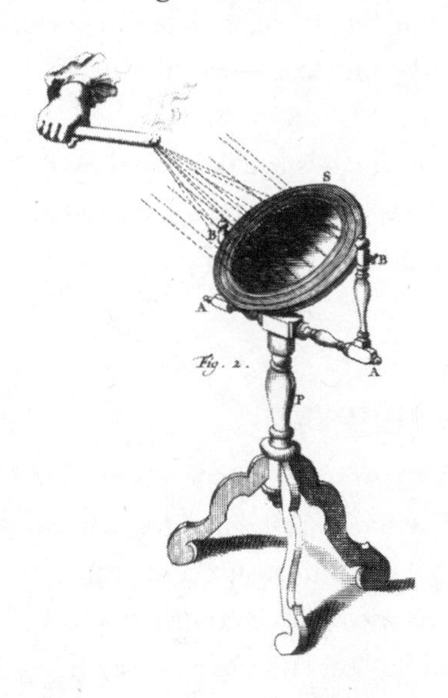

Figure 4.8. A small burning mirror typical of those used by natural scientists during the 1500s and 1600s.

Kircher also tested and had others test the smaller mirrors that he did find. All of them failed to meet the claims of their inventors. For instance, a mirror built by Manfredo Settala of Milan was reputed to quickly reduce to burning embers a piece of wood placed fifteen to sixteen paces (about 40 feet) away. On Kircher's instructions, an independent investigator tested this claim. He wrote back that, yes, it was possible to ignite the piece of wood at that distance, but it required the time it took to say a long Miserere (a very detailed Catholic prayer).

Kircher ended his report with stern and sober words: "Let not mathematicians boast about more things than they can demonstrate, and let them not expose themselves and the noblest art to mocking and jests of men."[30] Kircher's advice was apparently heeded — or perhaps the Jesuit merely echoed the growing empirical sentiment of the seventeenth century. In any event, the claims of mirror builders in later years were more in line with results that could be verified by third-party observers.

The development of powerful burning mirrors proceeded slowly. In the late 1600s, François Villette, an optician from Lyon, built several large spherical mirrors. The largest measured more than 3 feet in diameter, about a third larger than any mirror previously built. Regarding the solar heat generated by one of these mirrors, an observer commented, "One may pass one's hand through it [the focal point], if it be done nimbly; for if it stay there the time of a second,... there is danger of receiving much hurt." In a second article, a second reporter observed that one of Villette's mirrors was able to make tin melt in three seconds, cast iron melt in sixteen, and a diamond lose 87 percent of its weight. "In short," he concluded, "there is hardly any body which is not destroyed by this fire." It burned "most forcibly of any fire we know," even beyond that of a blast furnace, by far the most powerful heat engine then known, and possible could "be of great use..."[31]

Mirror Technology Improves

Although Villette's mirrors were larger than those of his predecessors, they were nowhere near as large as those proposed by Bacon or Leonardo. Reflector size was still limited by the use of heavy, brittle alloys. But late in the seventeenth century, other methods and materials were developed to overcome these problems, and mirrors were produced that were both lighter and easier to handle.

A Dresden mechanic only known to posterity as Gartner constructed mirrors from wood coated on its concave inner surface with wax and pitch and then a layer of shiny gold leaf. But wooden mirrors had their own shortcomings.

"Not only do these mirrors not bear the circumstances that go with burning such as flying sparks, flowing streams of molten metal and slag," one expert pointed out, but the gold leaf easily deteriorated, completely destroying the mirror's reflectivity.[32]

A German noble, the Baron of Tschirnhausen, successfully tried another material more durable than wood and lighter than alloy metal — copper. He hammered a sheet of copper "scarce twice as thick as the back of an ordinary knife" into a mirror the like of which "hath not yet been made[,]...for in magnitude it exceeds even the great one [Villette's mirror] which they shew as a sight in Paris."[33] This was no exaggeration. The baron's device was the largest burning mirror of the seventeenth century — 5½ feet in diameter. Yet this mirror was still manageable because it was made from a fairly thin, relatively lightweight sheet instead of the heavier cast alloy. Moving it to track the sun could be easily handled by one man. This solar reflector was also the most powerful burning mirror of the century. The baron wrote,

> The force of this speculum [mirror] in burning is such [that] even chemists who best know the power of fire will hardly credit their own eyes....A piece of tin or lead three inches thick, as soon as it is put into the focus, melts away in drops....A plate of iron or steel placed in the focus immediately is seen to be red hot on the back side; and soon after a hole is burnt through.

But the baron's brilliant invention had its shortcomings: it was still difficult to fabricate a single sheet of copper large enough to make a mirror of such dimensions. A different tack was taken by Peter Hoesen, an eighteenth-century mechanic and Dresden's royal carpenter. Previously, all methods of mirror building had been inherently restrictive because they used a single piece of material — whether alloy, wood, or copper plate. Hoesen abandoned unitary construction and built his mirrors from sections of hard wood covered with pieces of brass. He carved a parabolic shape from a skeleton made of cross members of very durable wood and then lined the inner concave surface with strips of brass sheet metal measuring 5 feet 5 inches by 2 feet 8 inches. Describing his own process, Hoesen explained,"with skillful use of his hands, he fits it [the brass] to the contours of the depression so that the seams between each sheet are hardly to be perceived."[34]

This method enabled the German craftsman to build mirrors as large as 10 feet in diameter — almost three times the size of Villette's big reflector and almost twice as large as the baron's. Never before had Europeans witnessed such colossal burning devices. By far the most powerful solar reflectors yet developed, Hoesen's mirrors concentrated the rays of the sun in a target area less than 1 inch across. "The hardest stones," said Hoesen, "resist [the mirror's] force for only a few seconds. Things of a vegetable nature burst into flames immediately, turning in a short time to ash.... The bones of animals become calcined."[35] The power of one of Hoesen's mirrors was verified by a research worker who experimented with a reflector 5 feet in diameter. He discovered that copper ore melted in one second, lead melted in the blink of an eye, asbestos changed to a yellowish-green glass after only three seconds, and slate became a black, glassy material in twelve seconds.

Despite its size, a Hoesen mirror was easy to handle. The mirror's mobility was due to its light weight and the clever design of its mounting system. The reflector was supported by two semicircular wooden arms with adjustable screws that allowed the device to be tilted in many different directions. The arms met at the base of a tripod mounted on rollers, making the entire mechanism mobile. "One can put it in all positions using but one hand!" remarked an observer.[36]

Although Hoesen built many burning mirrors of substantial size, Bacon's ominous prophecy was never realized — none were employed as weapons of war. By now gunpowder had given humankind a far more efficient and versatile means of delivering death and destruction over long distances.

But of all burning mirrors ever built, Isaac Newton's outdid them all. He placed six mirrors around a seventh one so that the first six focused sunlight onto the last one, which then shot the bundled sunlight to a point where, according to a contemporary, "it vitrifies brick or tile in a moment, and in about half a minute melts gold."[37]

Burning Glasses

Diocles, the first person to mathematically prove the power of the parabolic mirror, had greater expectations for what he and his contemporaries called burning glasses — solar devices not too different from today's magnifying glasses. Such devices work in a manner converse to that used by burning mirrors. They are shaped to bend light passing through them, and they ignite whatever is placed at their focal point underneath. Because the burning glasses built in antiquity produced greater heat from the sun than burning mirrors did, Diocles hoped to use them to astonish worshippers by instantaneously firing up lamps in temples and setting sacrifices on fire.[38] Several centuries before the time of Diocles, the Greek playwright Aristophanes had placed Socrates in the solar spotlight when he discussed such implements. The playwright's knowledge of them came from local pharmacies in ancient Greece, where burning glasses were sold to the public to start the dinner fire or light lamps.[39] Archaeologists found caches of burning glasses at the ruins of a building in Herculaneum near Rome that

Figure 4.9. Giambattista della Porta using the concave portion of a vase, similar to a ball formed by transparent rock, to focus the sun's rays to ignite material in his laboratory.

had been constructed in the first century CE.[40] Roman doctors treated patients with another type of burning glass: balls of transparent rock crystal that intercepted the sun's rays and were, according to Pliny the Elder, "the most effective method of cauterizing parts that need such treatment."[41]

The ancient Chinese had parallel experiences. Their fire pearls, made of rock crystal, were round and transparent and could be as large as a hen's egg. "When [they are] held against the rays of the sun," according to one ancient writer, "tinder will be ignited."[42] The Vikings, too, produced optical lenses from crystallized rock. Archaeologists found a hoard of finished implements along with unworked rock crystal at the Viking harbor town of Fröjel on the island of Gotland, Sweden, famous as a trading center. Their imaging quality was comparable to that of modern lenses.[43]

During the eighteenth century, wielders of large burning glasses competed with the owners of burning mirrors in captivating audiences with live demonstrations of the power of optics. The Baron of Tschirnhausen, for example, added to his repertoire a double lens "to render the focus [of the burning glass] more vivid" and "its force much augmented." "Every sot of wood, let it be ever so hard," bragged the baron, "or ever so green, nay, tho' soaked in water, will catch fire in a moment."[44] The French used a burning glass so focused that it

Figure 4.10. At its point of focus, a double-lens apparatus ignites a combustible object.

concentrated the sun's energy sufficiently to fire a cannon marking the noon hour every day of the year. The operation of the solar cannon began long before Napoléon set off for his conquest of Europe, and it even predated the guillotine. As late as the end of the nineteenth century, tourists actually set their watches to its firing.[45]

Flat Mirrors

For centuries, natural philosophers, alchemists, and others in western Europe obsessed over using concave reflectors to cast the ultimate death ray. But none came up with the solar weapon they all believed the great Archimedes had devised thousands of years before. Most came to denigrate the belief as mere fable. "The affair of Archimedes setting the fleet on fire has been looked upon as a thing impossible and romantic," judged the Marquis Nicolini, a fellow of the Royal Society of London.[46] Descartes, the last word on the Continent in scientific matters, denied the possibility that such a weapon could be built.

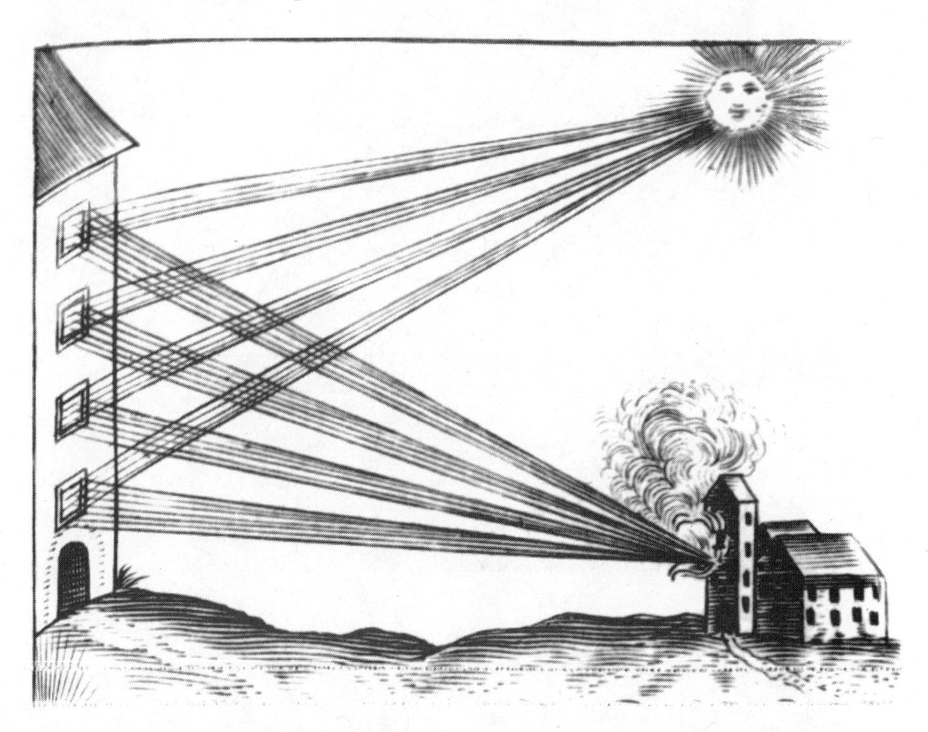

Figure 4.11. More than a century before Buffon, Giambattista della Porta had proposed using a flat, reflecting glass to burn down a nearby castle.

Much to the surprise of the scientific world, however, the lowly director of the Royal Gardens of Paris, Georges-Louis Leclerc, comte de Buffon, proved the great Descartes and the rest of the learned world wrong. Buffon discovered through experimentation that, for his purpose, a concave mirror would be an ineffective instrument of destruction. He observed that "light was able to produce great effects in a focus at a great distance, if one made use of a great number of dishes [flat mirrors], which would reflect so many images of the sun, and fling them into one place." Focusing 168 separate 6-inch-square flat mirrors, Buffon enabled "the faint rays" of an early spring Parisian sun to set ship planks on fire 160 feet from the solar weapon he had built. Watching the spectacle, Nicolini exclaimed, "It is Archimedes revived; and the credit of Antiquity re-established."

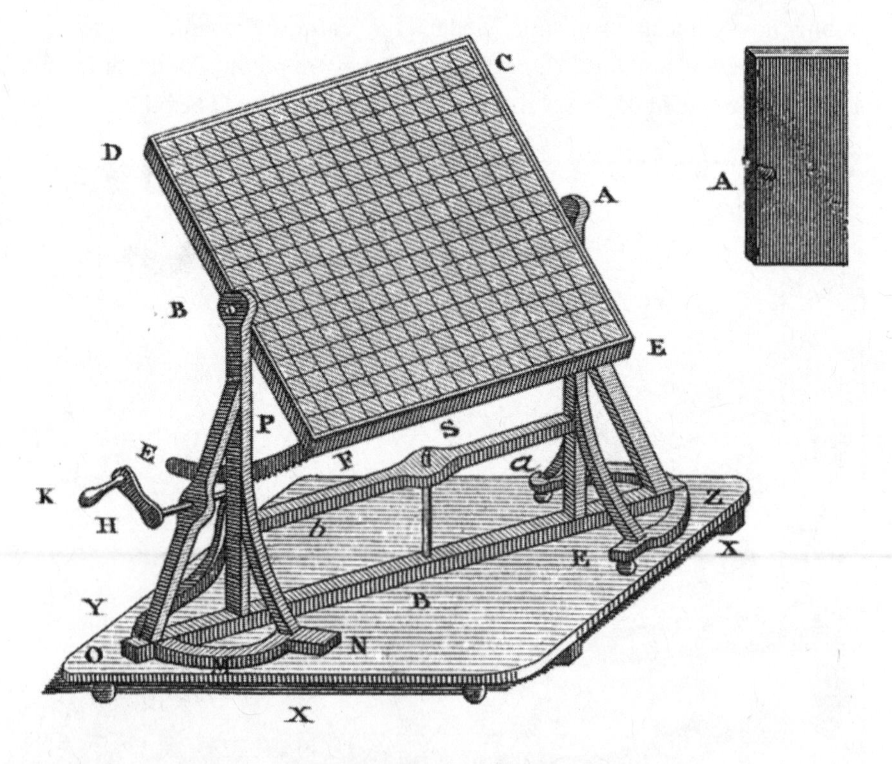

Figure 4.12. One of the array of plane mirrors used by Buffon.

Buffon saw his success, where others greater than he had failed, as a ladder to the heights of scientific fame, and so he played his cards with skill, reenacting multiple times the legend of Archimedes to crowds, one of which included his sovereign, King Louis XV. As his contemporaries saw it, Buffon had not only equaled the great Archimedes but also surpassed him. Paris's leading newspaper wrote, in verse, "Buffon... The miracles of Archimedes / Are merely the games of a studious leisure for you."[47] The famed English historian Edward Gibbon, in his great work *The Decline and Fall of the Roman Empire*, extravagantly praised the French upstart: "Immortal Buffon[,]... what miracles would not his genius have performed for the public service... in the strong sun of Syracuse."[48] In an age when scientists gathered fame through spectacle, Buffon reigned supreme.

Figure 5.1. A conservatory in Victorian England. In the nineteenth century, glass facades became popular additions to many upper-class homes.

5

Heat for Horticulture

Just as the ancient technology of burning mirrors was lost to Europeans during the Middle Ages, the use of glass windows and other transparent materials to trap solar heat was almost unknown in the West for many centuries after the fall of Rome. Window glass at that time was not as common in Europe as it was during the heyday of the Roman Empire. People lived in the midst of constant turmoil, caught between local lawlessness and wars among rival feudal lords and kings. In most regions no central authority existed that could effectively keep the peace and afford protection to citizens, so homes had to be built as defensible units. Instead of installing easily breakable glass windows, people constructed homes with small chinks in the walls to let in light and air. Usually the church was the only building in a medieval village to have glass windows, and these were made of stained glass.[1]

Economic and social conditions discouraged the use of glass or other transparent materials for greenhouses as well. Times were hard. Only a small minority could afford the luxury of raising exotic plants in greenhouses, and plant specimens from abroad were hard to obtain because trade with distant lands had slackened off. Moreover, the threat of periodic warfare did not permit even the rich to devote much attention to such frivolities. Greenhouses also met with strong opposition from the church. Just as Bacon's idea of turning a gigantic solar mirror into a doomsday weapon was suppressed because it was seen as demonic tampering with the divine plan, so the church denounced the growing of plants apart from their natural habitat or out of season. According to legend, a twelfth-century Dominican friar who experimented with forcing fruits and flowers in a greenhouse as the Romans had done was burned at the stake for practicing witchcraft.

Horticulture's Revival

Collecting solar heat for horticulture enjoyed a revival during the sixteenth century. The tide of empirical science had begun to break the bonds of church dogma, and the growing wealth and stability of Europe contributed to an atmosphere that favored scientific exploration. With discovery and trade came an increased flow of money and an appetite for more comfortable living. Ships were now returning from Asia, Africa, and the New World with beautiful flowers, such as African violets, and delicious fruits, such as bananas, pineapples, and coffee. People wanted to grow these exotics at home and enjoy native fruits and vegetables in all seasons — just as the Roman emperor Tiberius had satisfied his year-round craving for cucumbers by raising them in glazed cold frames more than a thousand years before.

The Dutch and Flemish were the first modern northern Europeans to develop horticulture to a level equaling or surpassing that of the Romans. Perhaps their early independence from the authority of the church encouraged their pioneering efforts in the field of scientific gardening. Certainly their great success in world trade helped to provide the new bourgeois merchant class with the means to take up such gentlemanly pursuits as raising exotic plants. J. C. Loudon, a late-eighteenth-century horticulturist, observed that "horticulture ... was in great repute in all the low countries during the seventeenth century."[2]

The French and English followed suit, trying to grow plants from the warmer regions of the world in the inhospitable climate of northern Europe. They also sought to improve the yield of native crops and to grow them out of season. Experts wrote books for the amateur gardener on how to accomplish this.

The unusually short growing seasons during this era were a major problem for gardeners. The period from the year 1550 to about 1850 has been called the "Little Ice Age" because Europe suffered extremely short summers and severe winters. In England, for example, the average temperature from 1680 to 1719

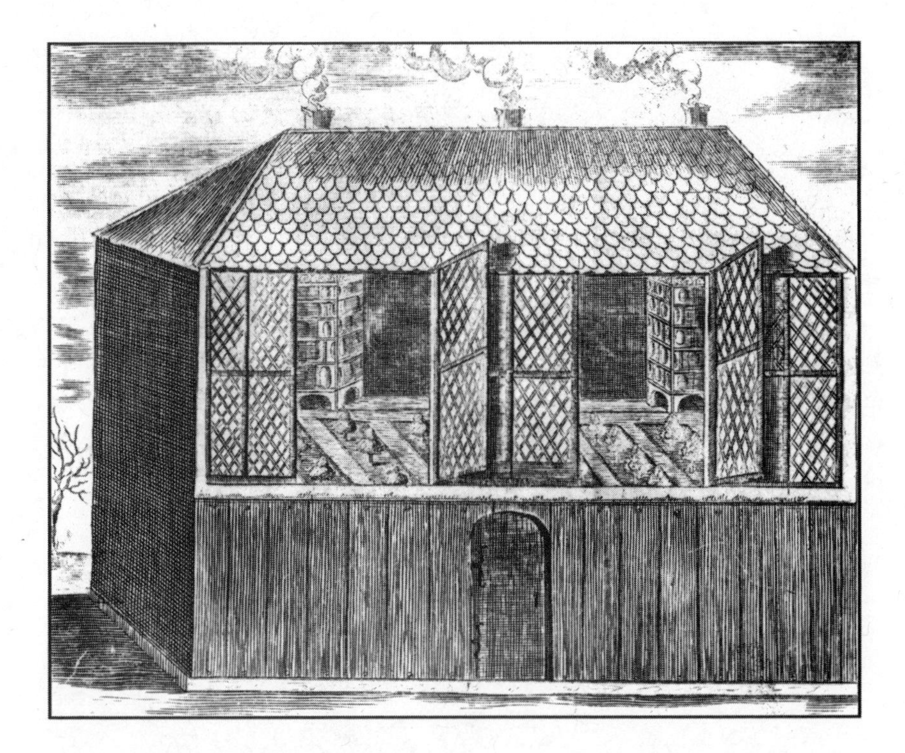

Figure 5.2. Early Dutch greenhouse, circa 1550.

ranged from 58.6°F in summer to 37.9°F in winter. The cold was so extreme that on thirteen occasions during the seventeenth and eighteenth centuries the ice on the Thames could support the weight of people — a rare occurrence during the previous five centuries.[3]

Only if "nature was assisted by art," as one gardening book put it, could a lover of plants "cure this great evil and dangerous enemy," the cold.[4] Solar energy became a favorite tool in this battle. Its power was extolled by Joseph Carpenter in 1717:

> The sun by its heat dissipates the cold and gross humours of the earth; it renders it more refin'd and easier for the vegetation of seed and fruit trees. Tis by the influence of this noble planet that the sap rises up between the wood and bark, producing the first buds, then the leaves and fruits; its beams serve not only to ripen the fruit, but it makes it large, beautiful to the eye, and pleasant to the taste.[5]

The art of harnessing solar heat, of learning how to enhance its beneficent effects, was evolving once again.

Fruit Walls

A technique devised by the French and English to make better use of solar energy in their gardens was to grow plants near a fence or wall heated by the sun. This fence or wall would release its heat to the plants. The branches of fruit trees were literally nailed to sun-heated walls to help ripen the fruit more quickly. One gardening tract pointed out that this practice resulted in substantially greater production and avoided "our entire want [in] some years of the best and latest fruit" of the season.

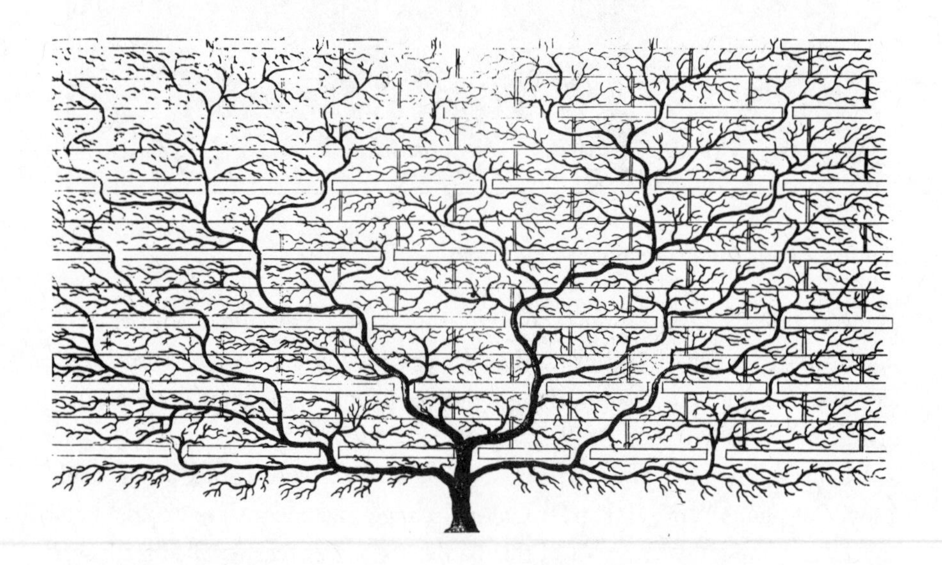

Figure 5.3. Branches of fruit trees were attached to brick "fruit walls" that collected and emitted solar heat to hasten the ripening process of fruit.

As for the heat-absorbing qualities of various materials for use in these "fruit walls," most experts recommended brick. Nicolas Fatio de Duillier, author of the influential 1699 book *Fruit Walls Improved*, explained why: "As to the properest matter of our walls, I think brick to be much better in this countrey

[England] than stone: because they grow much hotter, and keep much longer the heat. By which means they do still warm the plants a good while after the sun is hid under a cloud and in a manner lost to other walls."[6] Early fruit walls were usually built perpendicular to the ground, facing south. But this type of wall had its shortcomings. A vertical wall received the sun's rays on its south face for only a few hours in summer. Because the sun was then high overhead, its rays shone obliquely on the wall, not directly. Fatio de Duillier discussed the difficulty:

> When the days are something long, and the heat of the summer is in its greatest strength, it is late before the sun shines upon them [fruit walls], and the sun leaves them early in the afternoon. When it is about midday the sun is so high that it shines but faintly and very sloping upon them, which makes the heat much less, both because a small quantity of rays falls then upon these walls; and because that very quantity acts with a kind of glancing, and not with full force.

One solution was to face the fruit walls toward the southeast, rather than directly south. As noted by Stephen Switzer, an eighteenth-century gardening authority, "The sun shines early on [the southeast wall]... and never departs from it till about two o'clock in the afternoon."[7] Even so, a vertical southeast wall lacked sunshine during the remainder of the day. Semicircular walls, called "half-rounds," were not much better. Although "every part of the wall, one time of the day or the other, [enjoys] a share of the sun; and the best walls will not fail of being exceedingly hot by the... collection of the sun beams," each section of the wall still had access to the sun for only part of the day, noted Switzer.[8]

Perhaps the most ingenious solution was the "sloping wall" described by Fatio de Duillier. A south-facing perpendicular wall is exposed to only half of the visible sky, but a wall built at an incline of 45 degrees from the northern horizon and facing south is exposed to three-quarters of the sky and can absorb the sun's energy for a longer part of the day. A sloping wall also receives more direct solar rays and so gets hotter. Fatio de Duillier calculated that in England at the summer solstice, the intensity of solar energy striking a sloping wall was three and a half times greater than that hitting a perpendicular wall facing the same direction. Fatio de Duillier confidently predicted that the increased heat

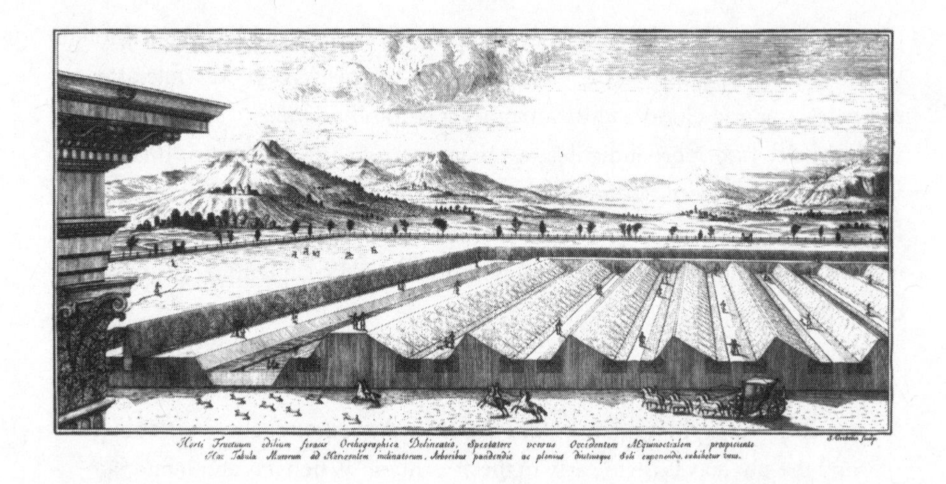

Figure 5.4. Sloping fruit walls in seventeenth-century France, from *Fruit Walls Improved*, by Nicolas Fatio de Duillier. Such walls faced the southern horizon at an angle chosen for optimum year-round solar heat collection.

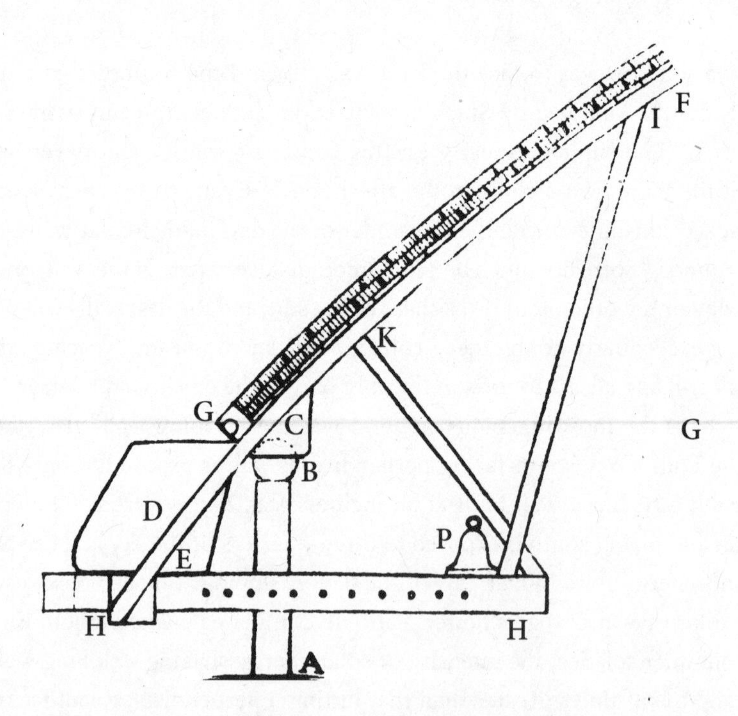

Figure 5.5. Fatio de Duillier's proposed tracking fruit wall. The device would follow the sun's daily path across the sky.

gain of sloping walls would help produce "grapes and figs and other fruits equal here [England] in goodness to those of some much hotter climates."[9] Indeed, the French had cultivated fine crops of grapes for years by growing them on sloping walls facing south. Fatio de Duillier even suggested building a pivoting fruit wall that would follow the sun. Such a wall would "be sure to enjoy almost all the sun's heat."

Greenhouses Come of Age

In 1714 the Duke of Rutland tried to use sloping walls in his English garden during the winter, but he found that they did not provide enough solar heat to keep his plants alive. So he placed glass casements over the walls to keep the collected solar heat from dissipating quickly. The use of glass as a solar heat trap was not entirely new to England. Some forty years earlier, Sir Hugh Plat had suggested that his patrons use glass to protect new seedlings because it would "defend off the cold air and increase the heat of the sun."[10] Cold frames and glass greenhouses soon became immensely popular in England, as well as in Holland — where Europe's first modern greenhouses had been built in the 1500s — and across the Continent.[11] In fact, many have called the eighteenth century the

Figure 5.6. English horticulturists commonly used glass cold frames to extend the growing season. Solar heat trapped by the glass covers allowed these gardeners to grow exotic plants, as well as local plants out of season.

"age of the greenhouse" because having a greenhouse became fashionable for nearly every person of means.

New glass-manufacturing methods allowed the production of large windows for greenhouses and homes. Previously, from the eleventh century to the end of the seventeenth century, windowpanes had usually been made by the crown-glass method. A craftsman blew hot glass into a bubble and then thrust a rod into the top of the bubble directly opposite the blowpipe. He detached the blowpipe, leaving an air hole where the tool had been, and as the glass began to cool he reheated the bubble and twirled the rod until centrifugal force caused the bubble to flatten into a disc. When the disc had cooled and hardened, the glass was cut into small, thin panes.

Locating crown-glass factories near fuel-rich areas helped to lower the price of glass and made windows more readily available to members of the ascendant middle class. Their growing demand for larger and thicker panes was finally met when the French developed the plate-glass process at the end of the seventeenth century. This process was strikingly similar to the Roman method. Glass was melted in a large cauldron, and several workmen carried the molten liquid to a casting table, where they poured it into a rectangular frame. With an iron roller, they flattened the glass to a standard thickness. After the plate had cooled and hardened, it was ground and polished. This plate-glass method produced windowpanes measuring up to 6 feet on an edge, and it quickly eclipsed the old crown-glass process — although crown glass remained common in England until the end of the eighteenth century.[12]

Figure 5.7. Eighteenth-century engraving of the crown-glass method.

Figure 5.8. The plate-glass-manufacturing process was developed in the late 1600s in France. Much larger windowpanes could be fabricated by this method.

Innovative Greenhouse Designs

Scientists sought novel greenhouse designs to enhance their solar heat collection and storage abilities, because they wanted to reduce the amount of fuel needed to keep plants warm at night, on cloudy days, and in the winter. They hoped to save fuel and believed that plants in a solar-heated greenhouse grew better than plants raised in artificial heat.[13]

Like fruit walls, early greenhouses were built with a southern exposure. Scientists soon realized that greenhouse walls should also be sloped to capture more sunlight. Hermann Boerhaave, a seventeenth-century Dutch professor of botany, demonstrated that in Holland's northern latitude the rays of the low-lying winter sun would enter a greenhouse more directly if the glass walls were steeply inclined. He determined that an angle of 75.5 degrees from the northern horizon would be best for a latitude of 52.5 degrees.[14] Most directors of botanic gardens in Europe took Boerhaave's advice and built their greenhouses accordingly. Greenhouses intended primarily to encourage the growth of larger and more flavorsome fruit during the summer required a less acute angle of incline, because the sun passed more directly overhead. Using this strategy, one English gardener reported "the most abundant crops of grapes perfectly ripened with less time and effort and less expenditure on fuel than I have witnessed in any other instance."[15]

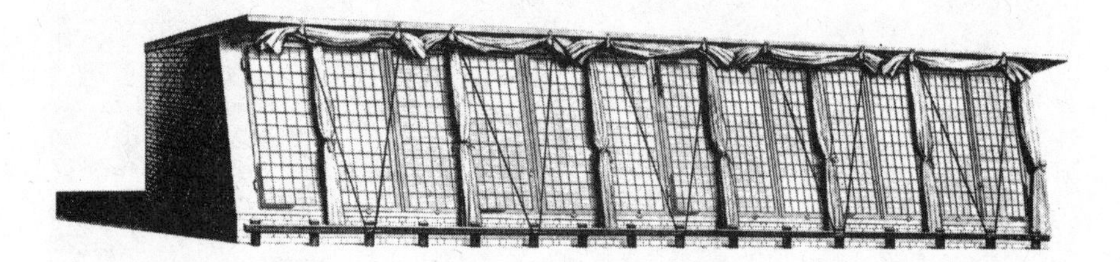

Figure 5.9. Canvas curtains were used for nighttime insulation in this eighteenth-century Dutch greenhouse. The Dutch also used double-paned glass to help control heat losses.

Figure 5.10. Bringing the plants out of the interior during the daytime to catch the rays of sunshine penetrating the greenhouse.

Michel Adanson, an eighteenth-century French scientist, recommended that the floor of the greenhouse, rather than its walls, be sloped. His approach was akin to placing a glass cold frame over a sloping fruit wall. Adanson wrote the first systematic treatise on the theory and construction of greenhouses. He presented rules, tables, and diagrams to be followed for building the most functional greenhouses in every possible location, from the poles to the equator.[16]

Gardeners not only sloped the walls or floors of their greenhouses but also invented insulating techniques to increase heat retention. When the sun was not shining, they placed mats or canvas coverings over the greenhouse to conserve the solar heat that had been collected, as well as the heat generated by the fires that were often lit inside. The Dutch constructed greenhouses with two layers of glass — the dead air space between the layers acted as insulation.

Late in the eighteenth century, Dr. James Anderson took the idea of solar heat storage a step further. Normally when the sun was shining and the greenhouse became too hot, some of the solar-heated air was released by opening windows in the structure. In Anderson's novel design, this hot air was captured and stored for later use. He divided his greenhouse into an upper and lower chamber. During the day, hot air collecting in the lower chamber rose through a pipe to the upper. At night, cold outside air was admitted to the upper chamber, forcing the stored hot air through a duct back into the lower chamber — to heat the plants. How well this storage system worked cannot be ascertained from the records, but it was an early attempt to store solar heat long enough that it could be used when the sun was not shining.[17]

The Development of Conservatories

As personal wealth accumulated during the nineteenth century in England and other European countries, the greenhouse began to assume a more lavish form — the conservatory. In this glassed-in garden, the well-to-do could leisurely amble with their guests through lush, junglelike foliage. *The Gentleman's House*, an architectural guide for the wealthy landowner, pointed out the difference between the greenhouse and the conservatory: "The greenhouse is a structure in which plants are cultivated as distinguished from the conservatory in which they are placed for display."[18] The greenhouse was primarily functional, whereas the conservatory was a place where exotic plants were displayed for atmosphere.

Gardening manuals described the fuel-saving features of a southern orientation for the conservatory.[19]

Architects such as the British designer Humphry Repton brought the sunlit ambience of the conservatory right into the home by attaching this glass garden to the south side of a living room or library. On sunny winter days the doors separating the conservatory and the house were opened to allow moist, sun-warmed air to circulate freely into the otherwise gloomy, chilly rooms. Some contended that the conservatory also gave families a healthier way of spending their free time. As the nineteenth-century book *The English Gardener* exclaimed, "How much better during the long and dreary winter for daughters and even sons to assist their mother in a greenhouse than to be seated at cards or in the blubberings over a stupid novel!"[20]

By the late 1800s the country gentry had become so enamored of attached conservatories that these became an important architectural feature of rural estates. According to John Hix, author of *The Glass House*, the conservatory

Figure 5.11. Greenhouse on the roof of a London apartment building.

Figures 5.12 and 5.13. Early solar remodeling. Architect Humphry Repton argued that a dull interior (above) could be transformed into a vibrant home by adding an attached conservatory (below).

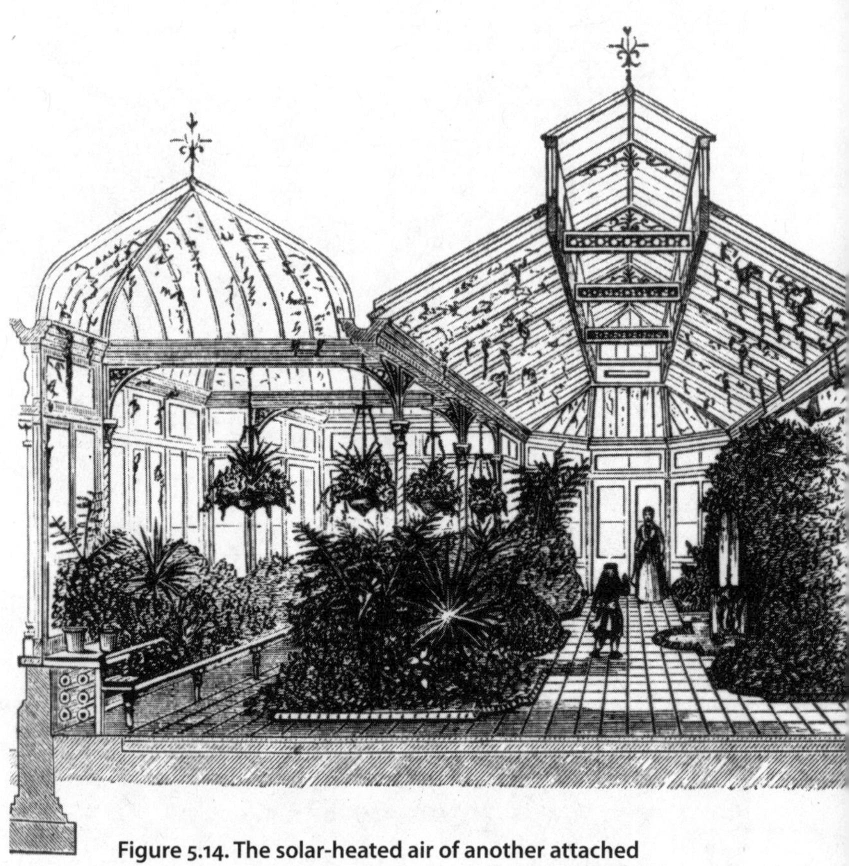

Figure 5.14. The solar-heated air of another attached conservatory added to the enjoyment of playing billiards.

was "no longer...seen as a simple extension to the dwelling, but an integral way of life."[21]

This fashion filtered down to members of the middle class, but on a much smaller scale. Modest conservatories, which took the form of a glass-covered room attached to the south side of the house, were common in urban areas. Rooftops of multistoried buildings also served as sites for greenhouses. As one writer remarked, "A warm greenhouse on the roof [is] a more pleasant thing than a dark parlor."[22] For the crowded city flat, a large window garden on the south wall had to suffice. Jacob Forst, a leading British horticulturist, envisioned that the south side of every urban building could be glassed over for growing

grapes, figs, and cherries. "Such walls would never need paint," Forst argued, and would offer an "admirable arrangement for house ventilation" by trapping sun-heated air and circulating it to the interior of the building.

As conservatories became popular, people grew indifferent to the direction in which they faced. Instead of the sun's rays streaming in from the south, artificial heating systems now provided warmth for the garden houses, and conservatories became fuel consumers rather than fuel savers. One of the chief reasons for the demise of the conservatory in England was the institution of fuel rationing during World War I. The lesson of solar heating had been discovered and then lost.

Figure 6.1. Horace de Saussure, the Swiss French scientist who invented the hot box in 1767.

6

Solar Hot Boxes

The increased use of glass during the eighteenth century made many people aware of its ability to trap solar heat. As Horace de Saussure, one of Europe's foremost naturalists of the period, observed, "It is a known fact, and a fact that has probably been known for a long time, that a room, a carriage, or any other place is hotter when the rays of the sun pass through glass."[1] That no one had ever studied the phenomenon surprised Saussure, whose university studies had focused on solar heat. He found in his research that every other solar scientist had preferred to work with concentrating devices, perhaps for their spectacular results that so attracted the public eye.

While still in his twenties, Saussure set out to determine how effectively glass could collect the energy of the sun. He first built a miniature greenhouse, constructing it from five square boxes of glass that decreased in size from 12 inches wide by 6 inches high to 4 inches by 2 inches. The bases of the boxes were cut out so the five boxes could be stacked one inside the other on a wooden tabletop painted black. Previously he had watched Alpine peasants spread black earth on their meadows to hasten the melting of snow, confirming his belief that black objects absorb more solar heat than objects of any other color. After exposing his miniature greenhouse to the sun for several hours, and rotating the model so that solar rays always struck the glass covers of the boxes perpendicularly, Saussure measured the temperature inside. The outermost box was the coolest, and the temperature increased in each successive box. The bottom of the innermost box registered the highest temperature: 189.5°F. "Fruits ... exposed to this heat were cooked and became juicy," he wrote.[2]

Building a Better Heat Trap

Seeking to even more effectively block the heat from escaping, Saussure made a small rectangular box out of ½-inch pine and lined it with black cork for insulation. Three separate sheets of glass covered the top of the box. When exposed to the sun, the bottom of the box reached a temperature of 228°F, which is 16°F above the boiling point of water and almost 40°F higher than in the first experiment. To achieve even higher temperatures, he wrote, "I caused it to be better insulated and placed at the bottom a tin pan." The inside temperature rose another 100°F higher with these new additions.[3] The device later became known as a hot box because of its boxlike shape and the large amount of solar heat it could retain.

However, Saussure seemed unsure of the physics of how the sun heated the glass-covered boxes. He stated, "Physicists are not unanimous as to the nature of sunlight. Some regard it as the same element as fire, but in the state of its greatest purity. Others envisage it as an entity with a nature completely different from fire, and which, incapable of itself heating, has only the power to give an igneous fluid the movement which produces heat."[4]

Despite Saussure's shaky theoretical underpinnings, his accomplishment is beyond repute. According to current understanding, the short-wave radiation from the sun easily penetrates the layers of glass to the bottom of Saussure's box. But when it strikes the walls and bottom, it is transformed into long-wave radiation — heat waves — which, Saussure observed, "cannot freely traverse the panes of glass which cover the box and accumulates more and more in the space inside," causing the very high temperatures he recorded.[5]

The hot box helped Saussure ascertain why it is cooler in the mountains than in lower-lying regions. His hypothesis was that the same amount of sunlight strikes the mountains as strikes the flatlands, but that because the air in the mountains is more transparent it cannot trap as much solar heat. To test the theory, Saussure carried a hot box to the top of Mount Cramont in the Swiss Alps. The thermometer in the hot box hit 190°F, while the temperature outside was 43°F. The following day he descended to the Plains of Cournier, 4,852 feet below, and repeated the experiment. Although the air temperature was 34°F

Figure 6.2. Sir John Herschel. On an expedition to South Africa in the 1830s, the eminent British astronomer took along a hot box and used it to cook meals.

hotter than on the mountain, the temperature inside the hot box was almost the same as in the previous experiment.

Saussure's hypothesis was confirmed: the sun shines with almost equal force at higher and lower elevations, as proven by the equal temperatures in the hot box on the mountain and on the plains. Joseph Fourier, a prominent physicist of the early nineteenth century, lauded Saussure's hot-box experiments for illuminating the role that the atmosphere plays in influencing the earth's temperature.[6]

Later Hot-Box Experiments

Several nineteenth-century scientists conducted experiments with hot boxes and obtained comparable results. Sir John Herschel, the noted astronomer, made a hot box in the 1830s that he took with him on an expedition to the Cape of Good Hope in South Africa. It was a small mahogany container blackened on the inside and covered with glass, set into a wooden frame protected by another sheet of glass and by sand that was heaped up along its sides. The outcome of Herschel's experiments with this hot box was not only scientifically interesting but also pleasing to the palate, as his notes indicate: "As these temperatures [up to 240°F] far surpass that of boiling water, some amusing experiments were made by exposing eggs, meat, etc. [to the heat inside the box], all of which, after a moderate length of exposure, were found perfectly cooked.... [On] one occasion a very respectable stew of meat was prepared and eaten with no small relish by the entertained bystanders."[7]

The following day, using a much simpler hot box, Herschel cooked another egg. The astronomer noted, "It was done as hard as a salad egg and I ate it and gave some to my wife and six small children that they might have it to say they had eaten an egg boiled hard in the sun in South Africa."[8] Herschel's account led fellow astronomer Jacques Babinet to remark, "It is astonishing that in countries in which the atmosphere is always clear, as in Egypt, Arabia, and Persia, and where fuel is always scarce and dear, people have never thought of utilizing the concentrated rays of the sun under glass" for tasks where heat is needed, such as cooking.[9]

The story of Sir John's solar cookouts did intrigue Samuel Pierpont Langley, the American astrophysicist who later became head of the Smithsonian

Figure 6.3. Samuel Pierpont Langley, future director of the Smithsonian, experimented with a hot box in 1881 during an expedition to Mount Whitney, California.

Institution. Langley had been fascinated by solar heat ever since he was a child, when he wondered why glass kept the interior of a greenhouse warm. In 1881, Langley took a trip to Mount Whitney to study the effects of solar energy. There he experimented with a hot box. He related his experiences in an 1882 issue of *Nature*: "As we slowly ascended...and the surface temperature of the soil fell to the freezing point, the temperature in a copper vessel, over which lay two sheets of plain window glass, rose above the boiling point of water, and it was certain that we could boil water by the solar rays in such a vessel among the snow fields."[10]

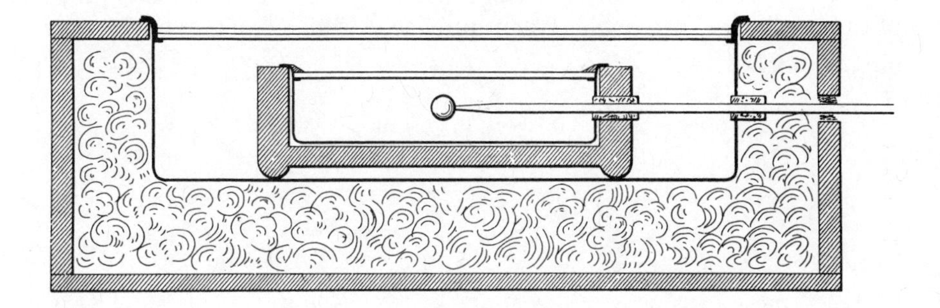

Figure 6.4. Cross section of Langley's hot box, which closely resembled the one used by Saussure. A thermometer penetrating the walls (on the right) measured the air temperature inside the inner box.

Saussure, Herschel, and Langley all demonstrated that temperatures exceeding the boiling point of water could be produced in a glass-covered box exposed to the sun. Saussure assessed the significance of the discovery with great modesty: "As to this application, I did not flatter myself that I could melt metals. I only thought to make this invention serviceable for purposes which only require heat a little above that of boiling water."[11] Still, he realized that the hot box might have important practical applications, stating, "Someday some usefulness might be drawn from this device...[for it] is actually quite small, inexpensive, [and] easy to make."[12] Indeed, his modest hope was more than fulfilled: the hot box became the prototype for the solar collectors of the late nineteenth century and the twentieth and twenty-first centuries — collectors that have been able to supply hot water, make seawater drinkable, and heat air to warm building spaces.

Although created more than two centuries ago, Saussure's hot boxes modeled with amazing precision the dynamics of global warming. As Fourier observed at the time, if the atmosphere solidified and were "exposed to the rays of the sun, [it] would produce an effect" like that of the glass panes of Saussure's hot box. "The heat, arriving as short wave radiation at the surface of the earth, would suddenly lose entirely the faculty which it had of traversing diaphanous [transparent] solids; it would accumulate in the lower levels of the atmosphere, which would thus acquire elevated temperatures."[13] In this statement Fourier suggests what scientists today observe: as greenhouse gases increase in our atmosphere, they duplicate what happened in Saussure's hot box. The clouds of carbon dioxide surrounding Venus provide us with a living example of the ultimate hot box.

PART II

POWER FROM THE SUN

Figure 7.1. Augustin Mouchot's largest sun machine, at the Universal Exposition in Paris, 1878.

[1860–1880]

7

The First Solar Motors

By the early 1800s the sporadic advances of science and technology in previous centuries had begun to snowball, leading to the Industrial Revolution. The use of machines to augment the muscle power of humans and animals meant that goods could be manufactured on an unprecedented scale. But mechanization depended on the production of iron, and to make 1 ton of iron took 7 to 10 tons of coal. Coal was also in demand as a major fuel source to power the newly developed steam engines and supply heat for the factories springing up throughout Europe.

France was at a disadvantage compared to other industrial countries because it had to import almost all of its coal. As a consequence, France lagged far behind rapidly industrializing England. So the French decided to pursue an aggressive program to step up domestic coal production. The plan worked, and output doubled in two decades — providing resources and power for iron smelters, textile plants, flour mills, and the many other new industries that began to appear by the second half of the nineteenth century.

Many of the French now felt secure about the nation's energy situation. But not everyone shared this complacency. In 1860 Augustin Mouchot, a professor of mathematics at the Lycée de Tours, cautioned, "One cannot help coming to the conclusion that it would be prudent and wise not to fall asleep regarding this quasi-security. Eventually industry will no longer find in Europe the resources to satisfy its prodigious expansion.... Coal will undoubtedly be used up. What will industry do then?" Mouchot's answer was: "Reap the rays of the sun!"[1] To show that solar power could be harnessed to run the machines of the industrial age, he embarked upon two decades of pioneering research.

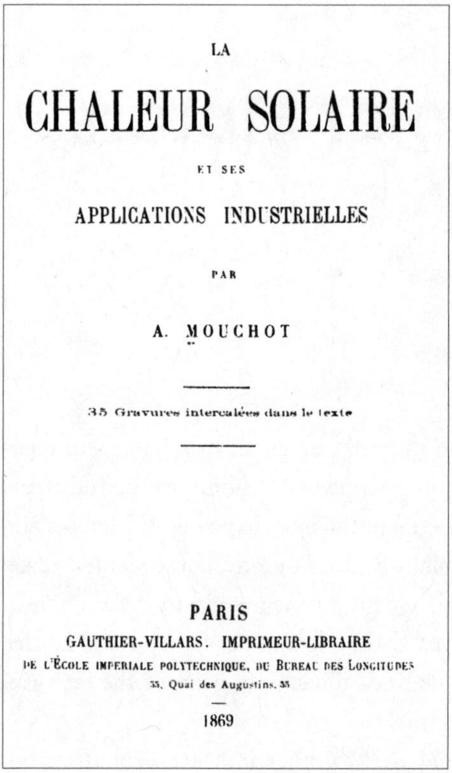

LA

CHALEUR SOLAIRE

ET SES

APPLICATIONS INDUSTRIELLES

PAR

A. MOUCHOT

35 Gravures intercalées dans le texte

PARIS

GAUTHIER-VILLARS, IMPRIMEUR-LIBRAIRE

DE L'ÉCOLE IMPÉRIALE POLYTECHNIQUE, DU BUREAU DES LONGITUDES

55, Quai des Augustins, 55

1869

Figure 7.2. Title page of Mouchot's visionary work *The Heat of the Sun and Its Industrial Applications*.

Early Uses of Sun Power

Mouchot began investigating the potential of solar machinery with a study of its historical roots. His findings surprised his contemporaries, who thought solar power was a new concept. As Mouchot put it: "One must not believe, despite the silence of modern writings, that the idea of using solar heat for mechanical operations is recent. On the contrary, one must recognize that this idea is very ancient and in its slow development across the centuries it has given birth to various curious devices."[2]

The first of what Mouchot called "curious devices," powered by the sun, was built by Hero of Alexandria in the first century of the Christian era. Hero invented a solar siphon that could transfer water from one container to another when it was placed in the sun. Solar energy heated air inside a closed sphere; the heated air expanded and exerted pressure on water inside the sphere, forcing it out.[3]

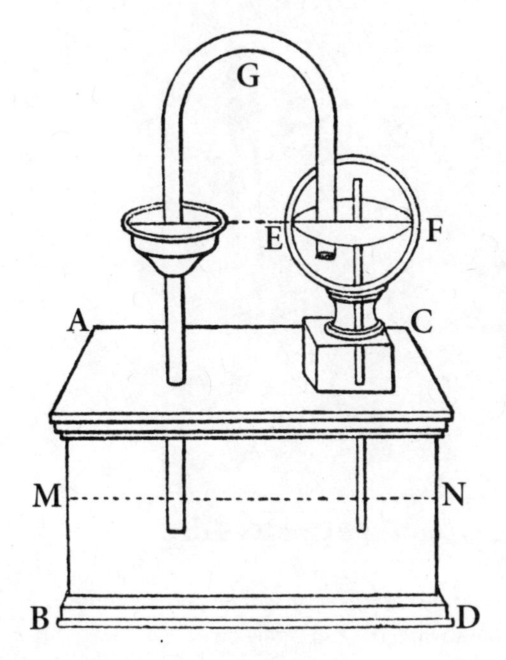

Figure 7.3. Solar siphon built by Hero of Alexandria, first century BCE. Solar-heated air in the globe (E–F) expanded, forcing water to travel through a tube (G).

During the sixteenth and seventeenth centuries, many natural philosophers proposed solar machines based on the same principle. Athanasius Kircher, the Jesuit priest who searched the Continent looking for giant solar reflectors, claimed he had developed a solar clock — although it is not clear whether his complex invention really worked.[4] In his book *New and Rare Inventions of Water Works*, published in 1659, Isaac de Caus told how to make "an admirable engine, the which being placed at the foot of a statue, shall send forth sound when the sun shineth upon it, so as it shall seem that the statue makes the said sound."[5]

This device imitated the legendary "voice of Memnon," a wailing sound that issued from a Theban statue of the Ethiopian king Memnon when the morning sun struck it. The ancient historian Tacitus and the Greek geographer Pausanias described this wonder in their writings. De Caus recreated this voice by connecting a solar siphon to a whistle. The siphon consisted of two adjacent metal boxes made of copper or lead. One box was partly filled with water and

Figure 7.4. Solar whistle developed by Isaac de Caus in 1659. Working on the same principle as Hero's siphon, the whistle blew when the sun shone on it.

Figure 7.5. Caus also designed a solar pump using a series of lenses to focus sunlight onto two tanks of water.

the other was empty. As the sun's rays heated the water-filled box, the air inside expanded and forced the water out of the container through a curved tube and into the second box. The water gradually displaced the air in the second box, and as the air flowed upward into two organ pipes atop the apparatus, it produced sound.[6]

Mouchot's First Attempts

Although such solar inventions aroused Mouchot's curiosity, he complained that "no one has adapted them in a practical way."[7] A true child of the industrial age, he was not content to see solar energy simply used for amusing contraptions. The practical development of solar power to serve industry became his principal pursuit.

Mouchot was thirty-five when he began his research at Tours in 1860. His objective was to find a way to collect the sun's energy efficiently enough to drive industrial steam engines economically. A hot box resembling Saussure's seemed promising because it could generate temperatures high enough to produce steam. But Mouchot's initial experiments proved disappointing. He felt that a hot box large enough to run an industrial machine would take up a great deal of area and be much too expensive.

Mouchot's second solar collector too was based on the hot-box concept, but its design provided greater exposure to the sun's rays. A bell-shaped copper cauldron coated on the outside with lampblack was covered by concentric glass

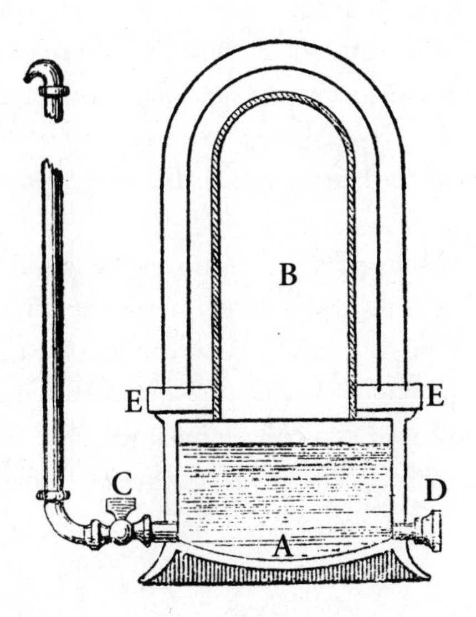

Figure 7.6. Cross section of Mouchot's first solar pump, patented in 1861. Sunlight heated air inside a copper cauldron (B); the air expanded and forced water out of the tank below it (A) and through an escape valve (C).

bell jars "to retain, as in a trap, the heat of the sun." Because glass covered the entire heat-collecting surface, replacing the glass top and wood walls of the old-style hot box, the solar collection area was greater. At all times of the day, the sun struck some part of the bell jar perpendicularly. By contrast, the hot box had to be moved constantly to keep its glass top oriented toward the sun. Mouchot found that the apparatus could collect "practically all the rays falling upon the exterior bell, which is to say a rather large sum of heat relative to the volume of the apparatus."[8] Nevertheless, an impractically large device would still be required to produce enough heat for industrial purposes.

The solution Mouchot tried next ingeniously combined two solar developments that had, up to this point, evolved independently: the glass heat trap and the burning mirror. A solar reflector could concentrate more sunlight on the collector than the collector could receive on its own. The glass heat trap could then be kept to a manageable size and still produce sufficient heat to drive engines. Consequently, Mouchot considered a mirror "indispensable to making the solar device practical."[9] Linking the two approaches led to several successful inventions: a solar oven, a solar still, and a solar pump.

The solar oven had a tall, blackened cylinder of copper surrounded by a cylinder of glass, with a 1-inch air space in between. Food went inside the copper cylinder, which was then covered by a wooden lid. The solar mirror was shaped like a vertical trough; it faced south and reflected a band of sunlight onto the cylinder. The mirror was made of polished silver sheets attached to a wooden frame. Mouchot cooked "excellent" dinners in this apparatus, just as Herschel had done several decades earlier. Mouchot claimed that "this new oven allowed me, for example, to make a fine pot roast in the sun. This pot roast was made out of a kilogram of beef and an assortment of vegetables. At the end of four hours the whole dinner was perfectly cooked, despite the passage of a few clouds over the sun, and the stew was all the better since the heat had been very steady."[10]

With a few modifications Mouchot converted the solar oven into a still that could turn wine into brandy. Whereas stills of the period normally relied on coal or wood, Mouchot boiled the alcohol by means of solar energy. The glass-covered copper cauldron served as the boiler in which wine was heated to a vapor. The vapor then cooled and collected in a conventional receiver. In his first experiment, Mouchot filled the cauldron with 2 quarts of wine. A few hours

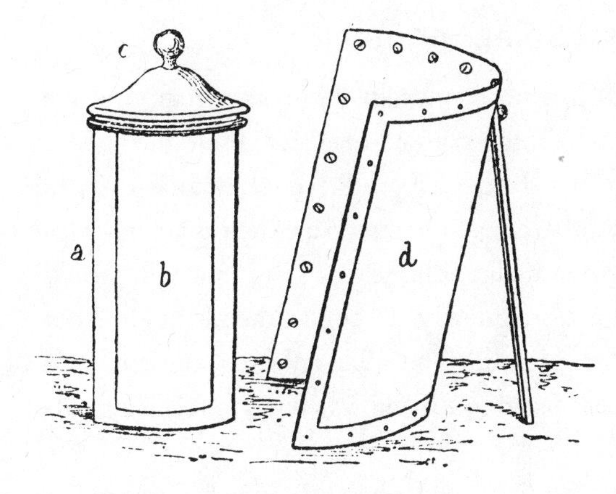

Figure 7.7. Mouchot's first solar cooker. A blackened copper cylinder (b) covered by a glass sleeve absorbed concentrated sunlight reflected onto it by a mirror (d).

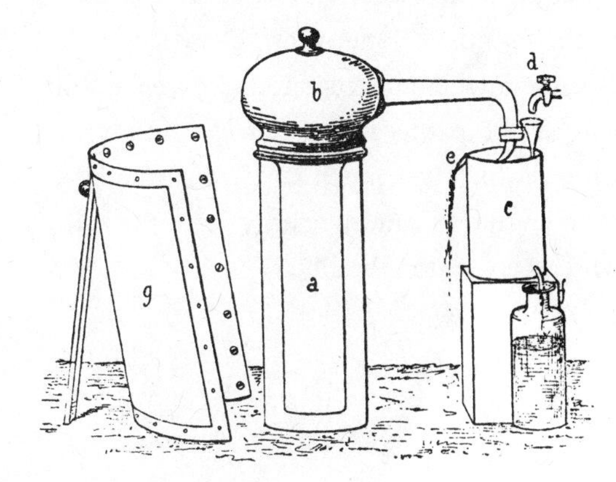

Figure 7.8. Solar still built by Mouchot. Alcohol evaporated in the copper cylinder (a), and its vapor was delivered to the condenser (c).

later he had the pleasure of drinking the first brandy ever distilled by the heat of the sun, remarking that it had a "most agreeable flavor."[11]

Mouchot's solar pump was similar to the basic design of the solar oven and still. A tall, hollow copper cauldron was soldered on top of a short tank filled with water, and the cauldron was enclosed by two glass covers. A cylindrical reflector concentrated the rays of the sun on the cauldron, rapidly heating the air inside. The expanding air exerted pressure on the water in the container below. Within twenty minutes enough pressure had built up to shoot a jet of water through a nozzle attached to the container, producing a spray 10 feet long that lasted over half an hour. During succeeding attempts, it pumped a continuous stream of water 20 feet. Mouchot patented his first primitive sun machine on March 4, 1861.

The First Solar Engines

Despite these successes, Mouchot had not yet attained his main goal: to drive a steam engine with sun power. The large volume of water inside the copper cauldron took a long time to boil, and the device produced steam too slowly to drive an industrial motor. Mouchot therefore substituted a 1-inch-diameter copper tube for the cauldron, so that the smaller volume of water in the tube would heat much faster, generating steam more quickly. To collect the steam Mouchot soldered a metal tank to the top of the tube. The solar reflector consisted of a parabolic trough-shaped mirror that faced south and was tilted to receive maximum solar exposure.

Excitedly, Mouchot reported what happened when he connected this boiler to a specially designed engine: "In the month of June, 1866, I saw it function marvelously after an hour of exposure to the sun. Its success exceeded our expectations, because the same solar receptor [that is, the reflector and boiler] was sufficient to run a second machine, which was much larger than the first." Mouchot had invented the first steam engine to run on energy from the sun.[12] He presented it to Napoléon III, who received it favorably.

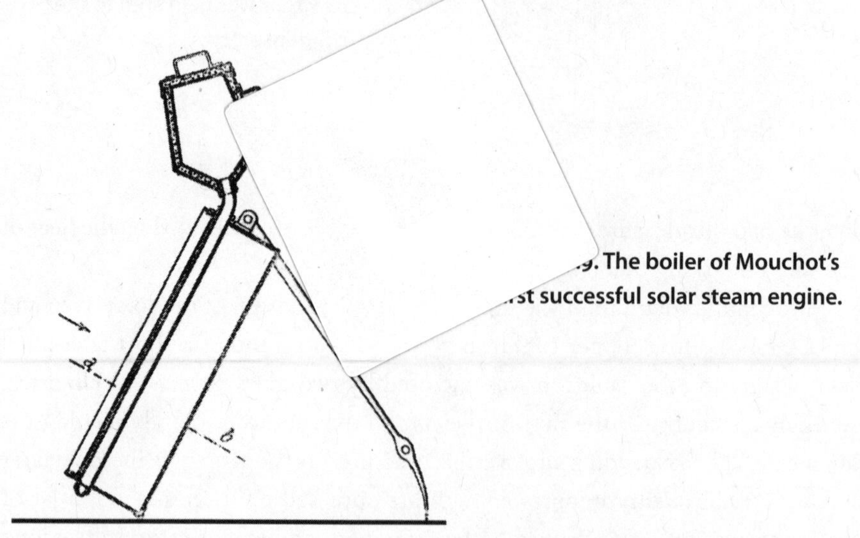

… The boiler of Mouchot's … successful solar steam engine.

Over the next three years Mouchot continued to refine his solar motor. To increase the boiler's steam-generating capacity so that it could run a large, industrial-sized machine, he replaced the copper tube with two bell-shaped copper cauldrons, one inside the other. The double cauldron was sheathed in glass.

The space between the copper shells held a somewhat greater volume of water than had the previous model, but the layer of water was thin enough to heat rapidly.

The French government gave Mouchot the financial backing to construct an industrial-scale solar engine along these lines. For this project he built a 7-foot-long cylindrical boiler based on the double-cauldron design. It had a tall, trough-shaped reflector that faced south. Mouchot added a clock mechanism that automatically moved this device from east to west to follow the daily course of the sun, which had previously been done manually.

On the whole, the boiler's performance satisfied Mouchot: it vaporized water into steam at a pressure of 45 pounds per square inch. But he recognized that the mirror needed improvement. First, because the mirror's tilt could not be adjusted for seasonal shifts in the sun's path, it did not reflect an optimum amount of sunlight throughout the year. Second, the mirror concentrated sunlight onto only one side of the boiler — the opposite side remained cooler, lowering the overall efficiency of the machine. Third, the reflector was constructed of silver plates and wood. Mouchot was worried that a mirror large enough to run industrial equipment would be too heavy for the clock mechanism to move.

An Improved Sun-Powered Motor

International conflict halted further experimentation. Napoléon III declared war on Prussia in July of 1870; soon afterward the enemy marched into Paris and the French suffered an ignominious defeat. One of the victims of the conflict was Mouchot's solar reflector — disappearing, as he put it, "in the midst of our disasters."[13] After the government collapsed and Mouchot's funds were cut off, he had to seek other sources of support. But Mouchot had a hard time obtaining backing, which he attributed partly to the "bias and specious objections that engineers pronounced on a question too foreign to their own studies for them to judge."[14]

Eventually he took the advice of a colleague and approached the regional government of Indre-et-Loire, the wine-producing district in which he lived. The solar device he showed them "so pleased the general counsel that they gave me 1,500 francs on the spot," remarked a happy Mouchot, "making available to me the means for completing the construction of a large solar receptor capable of distilling alcohol and producing a sufficient amount of power for mechanical applications."[15]

With customary alacrity, Mouchot finished his new solar machine by 1874. He put it on public display in Tours, the district capital, where the journalist Leon Simonin saw it and reported: "The traveller who visits the library of Tours sees in the courtyard in front a strange-looking apparatus. Imagine an immense truncated cone, a mammoth lamp-shade, with its concavity directed skyward." This "strange-looking apparatus" was Mouchot's redesigned solar reflector.

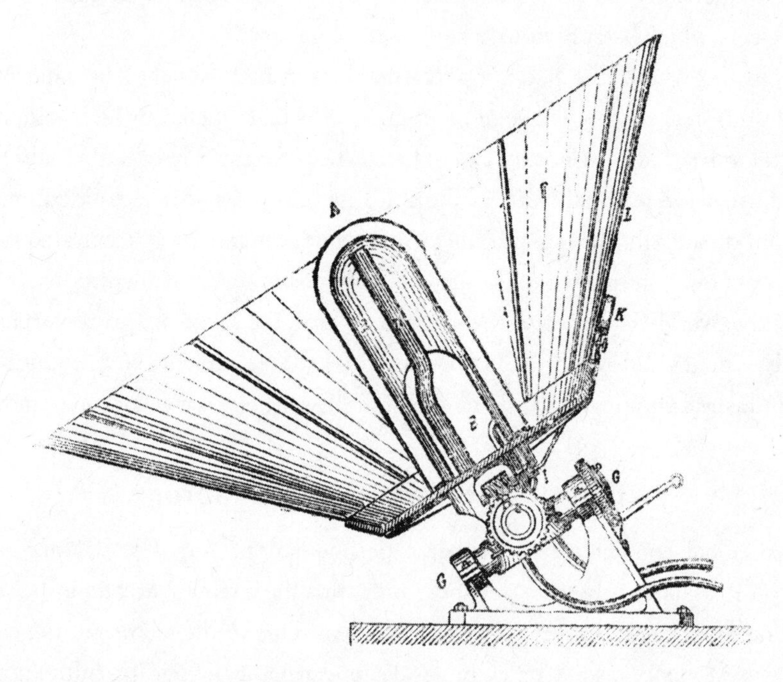

Figure 7.10. The Tours solar machine of 1874. The boiler (A) in the center of the reflector pivoted to follow the sun's daily and seasonal motion.

Made of copper sheets coated with burnished silver, the inverted cone-shaped mirror measured 8½ feet in diameter at its mouth and had a total reflecting surface of 56 square feet. The mirror's huge size required it to be constructed in sections, a technique developed by Peter Hoesen in Dresden a century before. The reflector's novel shape had the advantage of correcting one of the major defects of previous models: its surface could reflect sunlight at right angles to all sides of the boiler, which was located along the axis of the cone. Simonin described the boiler's configuration:

> On the small base of the truncated cone rests a copper cylinder; blackened on the outside, its vertical axis is identical with that of the cone. This cylinder, surrounded as it were by a great collar, terminates in a hemispherical cap, so that it looks like an enormous thimble.... This curious apparatus is nothing else but a solar receiver...or in other words, a boiler in which water is made to boil by the rays of the sun.[16]

The machine could generate enough steam to drive a ½-horsepower engine at 80 strokes per minute. When Mouchot put it on display, the reaction was one of amazement — a motor that ran without fuel, on nothing more than sunbeams! On one occasion the crowd became somewhat apprehensive when the steam pressure rose to over 75 pounds per square inch. One nervous bystander exclaimed, "It would have been dangerous to have proceeded further, as the whole apparatus might have been blown to pieces." People were also impressed that the boiler could operate a commercial distiller capable of vaporizing 5 gallons of wine per minute.[17]

The success of the Tours machine bolstered Mouchot's belief in sun power. But he also became aware of the limits of its practical application in a country like France. First of all, he realized, "sun machines would take up too much space in our cities and therefore could not be profitably used."[18] The Tours motor occupied a 20-foot-square area and produced only ½ horsepower. Two hundred such machines would be needed to drive a typical 100-horsepower industrial motor. If two hundred solar machines were arranged in four lines, with enough space between each device to prevent them from casting shadows on one another, a total of 100,000 square feet would be needed.

Sun Power for the Colonies

An additional impediment to the commercial use of solar engines in France was the problem of intermittent sunshine — especially during winter. But Mouchot believed that France's sun-baked colonies in North Africa and Asia, many of which had recently been conquered and were just being opened up to French exploration and settlement, offered unlimited possibilities. For example, "in torrid zones such as Cochin-China [South Vietnam], the matter of hygiene comes to the fore. In Saigon, water has to be boiled to be made potable. What a savings in fuel one could realize using a [solar still] in the ardent heat of those climates!"[19] The combination of nearly constant sunshine and abundant open space convinced Mouchot of the commercial viability of solar power in these colonies, and so he sought financial aid to develop his machines for such areas.

Mouchot's opportunity came a year after the demonstration at Tours, when he had the good fortune to meet the Baron of Wattville. The baron became a strong advocate of solar power, and through his influence the French Ministry of Public Education agreed to pay for a scientific expedition to the colony of Algeria so that Mouchot could determine whether solar cooking, distilling, and pumping would be practical there. The French government wanted to aid the new wave of colonial settlement in Algeria that followed its suppression of the insurrection of 1871. The subsequent confiscation of tribal lands and commercial holdings made it easier for Europeans to acquire property, but a lack of indigenous fuel supplies hindered economic development. Algeria had to import all its fuel from Europe, 85 percent of it from England in the form of relatively expensive coal. Lack of a railroad system in the colony drove the cost of coal up even further, especially in remote districts, where prices were nearly ten times higher than in more accessible regions. The French government hoped that solar power could be a great boon to Algeria's economy.

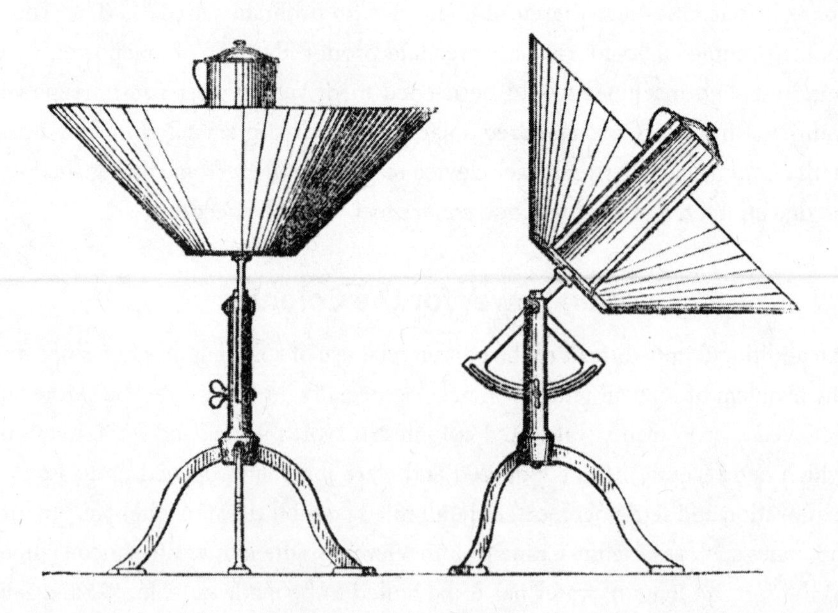

Figure 7.11. Mouchot developed this portable solar oven for French troops in Africa.

Mouchot arrived in Algiers on March 6, 1877, and immediately set to work testing an improved solar oven. "It seems possible without any great cost to provide our soldiers in Africa with a small and simple portable [solar] stove, requiring no fuel for the cooking of food," he wrote. "It would be a big help in the sands of the desert as well as the snows of the Atlas [Mountains]."[20] The oven Mouchot designed had a truncated conical reflector, like the one at Tours, with a glass-enclosed cylindrical metal pot — which served as the boiler — sitting at the focus of the reflector. The entire apparatus weighed only 30 to 40 pounds and could be collapsed and packed into a 20-square-inch box. Before a group of appreciative spectators, Mouchot baked a pound of bread in forty-five minutes, over 2 pounds of potatoes in one hour, a beef stew in three, and a perfect roast — "whose juices fell to the bottom" of the pot — in less than half an hour.

By removing the pot and replacing it with bottles of freshly processed wine, Mouchot turned the oven into a pasteurizer. The sun heated the bottles, killing any traces of bacteria that might later multiply when the wine was being shipped over long distances. Mouchot envisioned that Algeria would be able to use the sun to not only ripen its grapevines but also improve its wines and make them transportable, which would give that country a new source of prosperity. Mouchot also tested a solar still, similar in size and design to the solar machine exhibited at Tours. He wrote that the brandy he made from wine was "the subject of astonishment . . . [, for] it is undeniable that the alcohol comes out of the solar still bold [and] agreeable to the taste, and with an appropriate wine it offers the savor and bouquet of an aged 'eau-de-vie.'" The device could distill freshwater and salt water as well.[21]

Mouchot traveled from the Sahara to the Mediterranean to test the feasibility of using solar water pumps for irrigation. Irrigation was crucial to agricultural development in Algeria, and coal-powered pumps were too expensive. Mouchot experimented with a solar pump similar to the Tours machine and found that it worked more reliably in Algeria's sunny weather than in France's variable climate.[22]

Following a year of testing, Mouchot presented his findings to the authorities in Algiers. They were so impressed that they awarded him five thousand francs to construct "the largest mirror ever built in the world" for a huge sun machine that would represent Algeria in the upcoming Universal Exposition in Paris.[23] Afterward, it was to be shipped back to Africa and used commercially. With the help of his assistant, Abel Pifre, Mouchot completed this new solar machine in September 1878. At its widest point the cone-shaped mirror measured twice the diameter of the device shown at Tours the previous year, and its total reflecting surface was four times greater. The boiler had an unusual design: a group of long vertical tubes were fastened side by side to form a circular column at the focus of the reflector.

Mouchot's solar machine astounded exposition visitors by pumping more than 500 gallons of water per hour, distilling alcohol, and cooking food.[24] But the most remarkable demonstration occurred on September 22, as Mouchot recounted: "Under a slightly veiled but continually shining sun, I was able to raise the pressure in the boiler to 91 pounds... [, and] in spite of the seeming paradox of the statement, [it was] possible to utilize the rays of the sun to make ice."[25] He had connected the solar motor to a heat-powered refrigeration device invented by Ferdinand Carré in the 1850s. Mouchot saw an important future for solar refrigerators in hot climates, where sun-generated ice would help prevent perishables from spoiling.

Solar Electricity

The following year Mouchot returned to Algeria to resume his research. He spent much of his time trying to resolve a difficult question: how could solar heat be stored so that sun machines would be able to work during cloudy weather or at night? A colleague suggested using heat-absorbing materials capable of withstanding the high temperatures produced by a solar reflector. Placing these solar-heated substances in an insulated container would result in solar heat being retained for later use.

But Mouchot discovered what he thought was a better alternative. If solar energy were used to break down water into hydrogen and oxygen, the gases could be stored in separate cylinders. When heat was needed, the chemical reaction resulting from recombining the two would produce very high temperatures. Or the gases could be used separately — the hydrogen as fuel and the oxygen for industrial purposes. As for the method of separating water into its components, Mouchot decided to try "an instrument already in excellent condition[,]... the thermoelectric device."[26] The principle behind its operation was simple: when two different metals, such as copper and iron, are soldered together and heat is applied to the juncture between them, an electrical current results. Mouchot

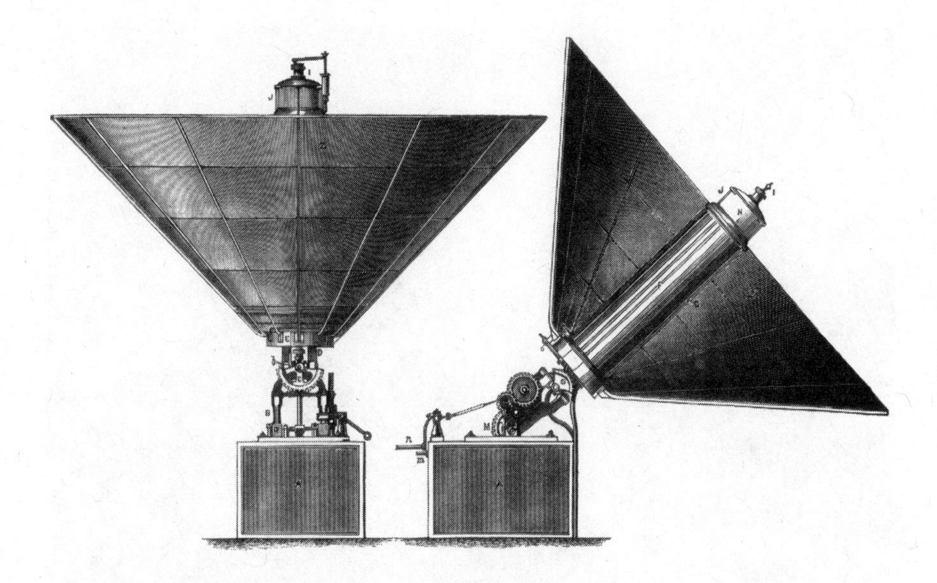

Figure 7.12. Two views of the solar motor Mouchot built for the Universal Exposition in Paris. The giant machine pumped 500 gallons of water per hour and even powered an ice maker.

planned to heat a hundred such metallic couplings with a solar reflector and in this way generate enough electricity to change water into its constituents.

By 1879 Mouchot had "already made a few experiments which bode well for this procedure.... Some very primitive devices have given me significant amounts of electricity," he wrote. Mouchot had great expectations and hoped to decompose water and produce "a reserve of fuel that would be as precious as it is abundant."[27] But for all his efforts, he could not compete with the more efficient methods of electrical generation rapidly being perfected about the same time.

In 1880, Mouchot returned to his mathematical studies. His assistant, Abel Pifre, took over the solar research. Pifre built several sun motors and conducted public demonstrations to gain support for solar power. At the Gardens of the Tuileries in Paris, he exhibited a solar generator that drove a press that printed five hundred copies of the *Journal Soleil* (Solar journal).[28]

But the time wasn't right for solar energy in France. The advent of better coal-mining techniques and an improved railroad system (most of France's coal lay at its borders) increased coal production and reduced fuel prices. In 1881 the government took one final look at the potential for commercial use of

Figure 7.13. Abel Pifre's solar-powered printing press, 1880. While exhibiting it at the Gardens of the Tuileries, he printed five hundred copies of the *Solar Journal*.

solar energy: it sponsored a yearlong test of two solar motors — one designed by Mouchot and the other by Pifre. The report concluded, "In our temperate climate [France] the sun does not shine continuously enough to be able to use these devices practically. In very hot and dry climates, the possibility of their use depends on the difficulty of obtaining fuel and the cost and ease of transporting these solar devices."[29]

Furthermore, the cost of constructing silver-plated mirrors and keeping them highly polished proved economically prohibitive for most uses. During ensuing years, however, the French Foreign Legion made some use of solar ovens in Africa. In remote areas of Algeria, settlers and explorers also used Mouchot's solar stills to obtain potable water, as their only source was brine water heavily charged with magnesium salts.

Although Mouchot did not succeed in bringing France into the "sun age," his pioneering efforts demonstrated a great variety of ways that solar energy could be used to benefit humankind and laid the foundations for future solar development.

Figure 8.1. John Ericsson's final sun motor, which he had hoped would provide cheap energy for the world's fuel-short regions.

[1872–1904]

8

Two American Pioneers

Eight years after Augustin Mouchot began his first experiments, an American engineer named John Ericsson also voiced the hope that someday solar energy would fuel the machines of the industrial age. Born in Sweden in 1803, Ericsson emigrated to America in 1839. By that time he was already well known for such inventions as the screw propeller, which made steam navigation practical. He later achieved widespread recognition for designing the ironclad battleship *Monitor*, which saved the Union blockade, essential for defeating the Confederacy. Six years later, Ericsson set his sights on a more peaceful goal: to bring his dream of sun power to fruition.

Like Mouchot, Ericsson was deeply concerned about the rapid consumption of coal. In a paper written in 1868, he contended that only the development of solar power would avert an eventual global fuel crisis. He pointed to the fuel-saving example of solar evaporation of brine and seawater to produce salt — the most successful application of the sun ever known to the world in his time. Over the centuries the solar production of salt saved the burning of hundreds of millions of tons of wood and coal.[1]

Solar Salt

In 1857 alone, using solar power to produce the 4 million bushels of salt consumed in the United States avoided the burning of 8 million tons of coal.[2] Salt was essential for flavoring foods; and over the millennia, until the advent of refrigeration in the latter part of the nineteenth century, it had been essential for preserving foods. Consequently, for thousands of years, salt had ranked among the commodities indispensable for commerce and human existence. No one would have disputed the claim about salt made by the *Syracuse Courier* in 1861: "There is no article which enters more largely into the preparation and preservation of the food of man" than salt.[3]

To obtain salt, people had to evaporate naturally occurring salty water found either in the sea or in salt lakes and brine springs, requiring a large fire lit under metal kettles holding salt water burning until all the liquid was evaporated. Salt left at the bottom would be scraped out and used. In Elizabethan times, wood was the primary fuel consumed in salt making. Sixteenth-century England did not have a lot of wood available, so Lord Cecil, Queen Elizabeth's chief economic adviser, sent emissaries to look into other ways of producing salt. They reported that refiners of salt in Flanders used the sun as their fuel.

The Flemish salt manufacturers, Cecil's men explained, built long, shallow, watertight troughs that opened to the sea. When operators wanted to fill them, windmills opened the floodgates to let in the incoming tide. Once enough seawater had run in, the gates were shut. Then solar heat went to work on the water, which evaporated after several days in the sun. Only salt remained. Workers shoveled the salt out and more water was let in to repeat the process. The English praised the solar saltworks as "a great help for the sparing of firewood."[4]

After the Revolutionary War, Americans along the coast of Cape Cod and Long Island adopted the salt-making methods of the Flemish. At first, however, instead of using wind power they used buckets to pour seawater into nearby shallow evaporating troughs. They also added a significant component: covers to keep rain out of the troughs. Rollers permitted easy covering of the troughs when bad weather approached, and their easy uncovering as the sun came out.[5]

Inland near where Syracuse, New York, now stands, Joshua Forman, an enterprising British immigrant, noticed on a walk in the early nineteenth century patches of white crust covering the ground. Though they glistened like crystals, they couldn't be ice. The weather was too sultry for that. After asking around, Forman found out that in certain low spots salt springs formed pools of brine. When the sun shone, the water dried up, leaving salt deposits. Always keen on making a buck, Forman thought to himself, why not imitate nature's work on a grand scale and pump up the brine now slowly oozing out? Getting free fuel supplied by the sun to make salt seemed more sensible then using artificial heat as was being done in 1821 in the area when Forman considered putting together a company to realize his dream.

To achieve his plan, he oversaw the construction of much larger troughs than had been previously constructed from planks. His had a depth of 12 to 14 inches and a width of 18 feet, and their lengths extended toward the sea 200 to 2,000 feet, depending on the area available. The covers were made in sections, each one measuring 16 by 18 feet.[6]

Figure 8.2. Raking the salt from troughs after the sun's heat had evaporated the brine.

Figure 8.3. A view of the solar saltworks at Syracuse with the troughs exposed and the covers rolled back.

By the 1850s about half the salt made in America came from the Syracuse area.[7] Syracuse salt preserved most of America's meat products at this time. Evaporating salt at peak levels to serve the meatpacking industry required boilers that burned nearly 120,000 cords of wood annually, destroying almost 3,000 acres of woodlands per year.[8] As a consequence, "timber of all kinds," one local report warned, "is fast disappearing from the vicinity."[9]

USE SYRACUSE, N. Y.,
SOLAR SUN-MADE
COARSE SALT,
MANUFACTURED BY
The Onondaga Coarse Salt Association,
IF YOU WANT CLEAN HIDES AND SKINS
FREE FROM SALT STAINS.

It is mellow, honest Salt--the cleanest and purest for Hides and Skins—made from pure brine pumped from wells and evaporated by the sun—that is all, and is especially suitable for keeping the grain of Hides clean and bright. Salt-stained Hides are a nuisance and dear at any price.

Figure 8.4. Advertisement for Syracuse Solar Salt.

Ebenezer Merriam, an authority on salt making, therefore concluded that "the period is approaching" when "some method must soon be adapted by which fuel may be saved."[10] A way out of the dilemma, Merriam suggested, was to expand the solar saltworks in the area. Suddenly, people of influence in Syracuse realized that solar energy could help save the salt industry, which was the lifeblood of their town, from declining. The *Syracuse Daily Standard*, dubbed the "official paper of the city," praised "Old Sol [for] furnish[ing] a cheap substitute for the common fuel used in salt manufacturing." The sun, the paper said, "should have due credit.[11]

With the price of fuel "much enhanced," the manufacture of salt by solar heat dramatically grew, increasing seventeenfold in sixteen years.[12]

Salt producers who wished to continue using "artificial" heat deemed coal their savior. But by 1864, according to official records, those burning coal "made little or no salt after the beginning of June for want of fuel[, while] the crop of solar salt exceeds that of 1863."[13] Paradoxically, the sun in Syracuse — a far northerly spot not noted for its sunny weather — proved more reliable than expectations for cheap coal. Syracuse's boiled salt was eventually priced out of the market by coal's ever-rising cost, unable to compete with salt made in Michigan and Canada with inexpensive fuel from their readily available supply of wood. In 1887, the amount of sun-made salt surpassed the quantity manufactured with coal. Solar manufacture of salt had grown to the point that visitors to Syracuse, looking down from a hill in the city, could see a wide and shallow valley all covered with brown wooden troughs open to the sun.

Solar salt also played a part in India's march to independence. Indians in the state of Gujarat along India's west coast also used solar heat to distill salt. But under English rule the British suppressed solar salt production by taxing it at a high rate to enable English salt to compete on the Indian market. Such a move adversely affected every Indian household, and Gandhi chose to fight an evil perpetrated by the empire that every household had come to hate. India's road to independence began when Gandhi marched to the seashore, picked up a crust of salt evaporated by the sun, and sprinkled some on a piece of food, demonstrating the Indian right to harvest and use what nature had bestowed upon them. No better act could show his fellow citizens, the English, and the world the inequity of British rule and India's ability to defy it.[14]

Solar Salt Inspires Ericsson

If proponents of solar power in Syracuse could accomplish so much solely by exposing brine water to the summer sun that far north in an inclement clime, just consider, Ericsson thought, what could be accomplished with concentrating devices in the sunnier parts of the world. "A great portion of our planet enjoys perpetual sunshine," the great inventor wrote. "The field therefore awaiting the application of the solar engine is almost beyond computation, while the source of its power is boundless. Who can foresee what influence an inexhaustible motive power will exercise on civilization, and the capability of the earth to supply the wants of our race?"[15]

Like others, Ericsson recognized that "a couple of thousand years dropped in the ocean of time will completely exhaust the coal fields of Europe, unless, in the meantime, the heat of the sun be employed."[16] And so he declared that he would dedicate the balance of his life to averting such a disaster by making solar engines an economical alternative.

Ericsson completed the construction of a solar-powered steam engine in 1870. He erroneously claimed it was the world's first, brusquely dismissing Mouchot's solar-driven motor built four years earlier as a "mere toy."[17] However, Ericsson's own invention — which, ironically, he intended to present to the French Academy — bore a striking resemblance to Mouchot's. Both had three components: a concentrating mirror, a boiler, and a steam engine. Ericsson had ruled out the feasibility of a hot-box collector on grounds similar to Mouchot's. A hot box, it appeared, could not produce sufficient steam to drive an engine unless it were enormous. Unlike Mouchot, though, Ericsson decided to abandon the concept of a glass heat trap entirely. Instead, bare metal tubes serving as the boiler were placed at the focus of a series of parabolic, trough-shaped solar reflectors. This collector generated enough steam to run a small conventional engine.

Little more is known about this sun motor, which Ericsson did not patent. He was emphatic about keeping its details secret: "Drawings and descriptions of the mechanism... will not be presented, nor will the form of the generator... be delineated or described," he stated. He had had some unfortunate experiences with "enterprising persons" who filed patents on slightly modified versions of his inventions, preventing him from making similar improvements on his own devices.[18]

A Solar Hot-Air Engine

In 1872, Ericsson tried a very different approach for his next solar motor: it was powered by solar-heated air rather than steam. This unique invention was a logical next step, for he had spent much of his early career developing the hot-air external combustion engine. This type of motor worked on the following principle: heat externally applied to a cylinder caused air inside the cylinder to expand and push down a piston; inrushing cold air pushed the piston up again, and the cycle repeated itself. Normally burning wood or coal provided the heat. Ericsson modified the process by placing the cylinder upright along the axis of a curved, dish-shaped reflector. The mirror concentrated the rays of the sun onto the top of the cylinder, heating the air inside. For this motor, as for all of Ericsson's other solar engines, the reflector had to be moved manually to stay aligned with the sun.[19]

Elated at the hot-air engine's performance, Ericsson wrote a letter to his close friend and associate Harry Delamater: "The world moves — I have this day seen a machine actuated by solar heat applied directly to atmospheric air. In less than two minutes after turning the reflector toward the sun the engine was in operation. . . . As a working model I claim that it has never been equalled; while

Figure 8.5. Swedish American John Ericsson, renowned inventor and solar pioneer.

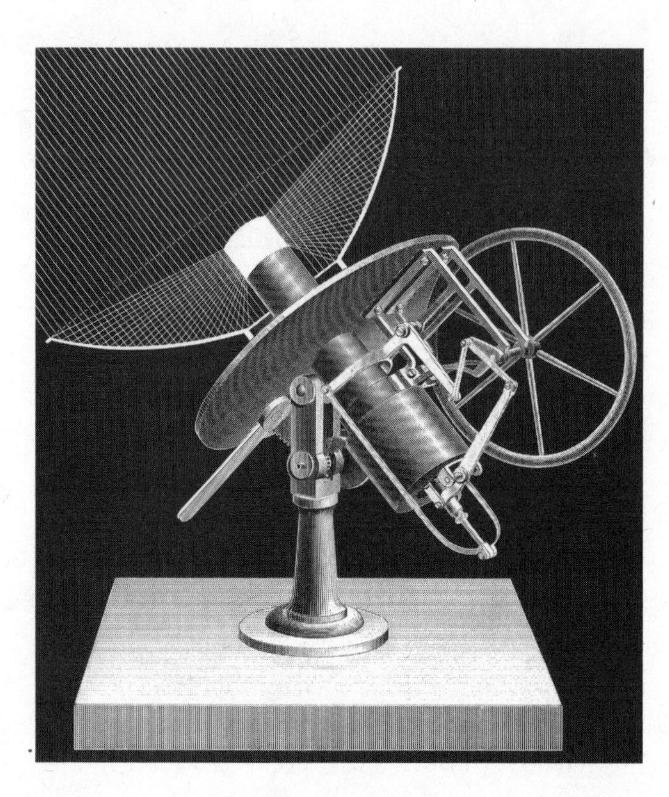

Figure 8.6. Engraving of the world's first engine powered by solar-heated air. The curved mirror concentrated sunlight upon a piston.

on account of its operating by a direct application of the sun's rays it marks an era in the world's mechanical history."[20] But three years and five more experimental engines later, Ericsson's enthusiasm had been tempered. He realized that "although the heat is obtained for nothing, so extensive, costly, and complex is the concentrating apparatus" that engines powered by solar energy were actually more expensive than similar coal-fueled motors. The concentrating mirror, usually made of silver sheets or some other silvered metal, cost too much to manufacture and keep from tarnishing.

Another obstacle to the commercial viability of solar motors confronted him, just as it had hindered Mouchot: the use of solar energy was restricted to daylight hours in areas enjoying almost constant sun. Ericsson felt he could not "recommend the erection of solar engines in places where there is not steady sunshine until proper means shall have been devised for storing up the radiant energy in such a manner that regular power may be obtained from irregular solar radiation."[21] But his decades-long search for a way to store solar heat was of no avail.

An Inexpensive Solar Reflector

Ericsson had better luck in developing a way to cut down the cost of the solar reflector. He replaced the silver-coated mirror with window glass silvered on its underside. Because this silver finish was not exposed to the elements and was further protected by a special coating, the mirror did not tarnish. "Hence, it will only be necessary to remove dust from the mirror" to maintain its reflectivity, Ericsson pointed out, "an operation readily performed by feather dusters."[22] The reflector could be purchased for less than sixty cents per square foot — much cheaper than a polished silver mirror — and the cost of upkeep was minimal. The design of this reflector also marked a departure from Ericsson's previous models. Individual silvered glass mirrors were attached to a metal frame to form a parabolic trough-shaped reflector. A tubular boiler was mounted above the mirror.

Figure 8.7. The lightweight parabolic trough-shaped reflector developed by Ericsson in 1884.

In 1884 Ericsson unveiled his new reflector design, claiming that its cost would not exceed that of a conventional steam boiler. A number of California farmers contacted him about buying the motor, hoping to obtain a cheap power source for their irrigation pumps. But Ericsson's announcement had been premature — some problems remained to be worked out. "Consequently," he explained, "no contracts for building sun motors could then be entered into." After four years of refining the model, Ericsson felt he had made enough progress to offer the solar pump for sale. In the fall of 1888 he confidently stated that "a simple method of concentrating [solar] power...has been perfected." Because "first-class manufacturing establishments [are] ready to manufacture such machines, owners of the sun-burnt lands on the Pacific Coast may now with propriety reconsider their grand scheme of irrigation by means of sun power."[23]

But only seven months later, before any of these plans had materialized, Ericsson died at age eighty-six. According to his obituary in *Science*, Ericsson's interest in solar motors obsessed him even on his deathbed: "He continued to labor at his sun motors until within two weeks of his death. As he saw his end approaching, he expressed regret only because he could not live to give this invention to the world in completed form. It occupied his thoughts to the last hour." Because of Ericsson's penchant for secrecy, the details of what he called his "perfected solar motor" followed him to his grave. A grain of pessimism also lingered at his death, as he came to realize that "until the coal mines are exhausted [the] value [of solar] will not be fully acknowledged."[24]

Interest in Solar Energy Grows

Ericsson's vision of harnessing solar power remained very much alive in the minds of other American engineers and scientists. Many had warned of an impending fuel crisis — warnings largely ignored by the public. But the devastating effects of a series of coal strikes around the turn of the century, culminating in a massive strike in the winter of 1902, threw "a new and lurid light on [these predictions,]...for many a home has been fireless and many a factory has closed its doors," according to *Harper's Weekly*.[25] Charles Pope, author of *Solar Heat*, one of the first books on solar energy, agreed: "The year of 1902 has added an awful chapter to the history of our need of a new source of heat and power," he wrote.[26]

To Pope and others, solar energy appeared to be the most promising alternative to coal for powering machines such as water pumps. In an article reprinted in the *Smithsonian Institution Annual Report* of 1901, Robert Thurston, a renowned engineer, wrote, "Engineers and men of science are studying the art of... harness[ing] the direct rays of the sun, and the solar engine is exciting special interest.... Probably at no time in the past has this matter assumed importance to so many thoughtful and intelligent men or excited so much general interest."[27]

Indeed, by the turn of the century numerous solar inventors and entrepreneurs crowded the field. At least twenty-two patents for solar motors had been filed. But few of these patents ever led to workable solar engines. As Charles Pope wrote, "There comes a strong suspicion that the patent office admitted essayists, rather than inventors to their lists; and that these men were not actually makers of machines which did what they claimed."[28] Mixed in this whirlwind of solar enthusiasm were charlatans and dreamers and some of the most talented inventors of the time.

Eneas's First Solar Pump

One of the more successful leaders of the turn-of-the-century solar movement was Aubrey Eneas. An English inventor and engineer residing in Massachusetts, he began studying sun power in 1892. With the aid of investors, he founded the Solar Motor Company of Boston to design and manufacture commercial solar engines. Like Ericsson, Eneas was especially interested in marketing solar-powered irrigation pumps in the American Southwest, where conventional fuels such as wood and coal were scarce and expensive. This part of the country seemed ideal for solar power, because 75 percent of the days were sunny.

Not much detailed information on solar motors had been published up to this time. "Though I have had technical training and considerable engineering experience," Eneas wrote, "I find so much that is new and confusing, and so little data of a practical nature."[29] Probably all that he had at his disposal were the meager accounts of Ericsson's work printed in the scientific periodical *Nature*.

In 1898 Eneas built his first solar motor, almost an exact replica of Ericsson's 1884 model — with silvered glass mirrors forming a trough-shaped reflector, and a bare metal cylinder at its focus. This engine was many times larger than Ericsson's, and preliminary experiments led him to believe that it was "a step in advance of what has already been accomplished, of practical value to our great arid west, in affording to this district an irrigation engine that requires but little care and no fuel."[30]

But Eneas had fallen prey to the same premature optimism as Ericsson. Although plans were made to sell the motor for fifteen hundred dollars, more extensive tests proved disappointing. After a full summer of operation, Eneas concluded that the reflector had several serious drawbacks. According to engineering knowledge of the day, generating sufficient steam to efficiently run a fair-sized engine required boiler temperatures exceeding 1,000°F. But the parabolic mirror Eneas had been using could not produce such a high temperature, and increasing the reflector's size would make it too cumbersome to support and difficult to maneuver. Furthermore, the reflector's shape was inefficient. As Mouchot had discovered earlier, the reflector concentrated sunlight only onto the side of the boiler it faced; the opposite side of the boiler merely received diffuse solar rays, and more heat actually escaped from this side than was collected there.

Back to the Conical Reflector

Eneas decided to abandon his first design and build a reflector shaped like an inverted, truncated cone with a boiler standing upright along its axis — much the same as Mouchot had done thirty years earlier. Perhaps he had read of Mouchot's work in an abridged account published by Ericsson in 1876.

In subsequent experiments, Eneas made some important observations about the maximum efficiency of the conical reflector. Because the mouth of the reflector has the widest diameter, the mirror surface is greater at this end and more sunlight is concentrated onto the upper end of the boiler. Conversely, less sunlight is concentrated onto the lower end of the boiler near the bottom of the reflector. In fact, the lowest part of the boiler inside Mouchot's reflector received less heat from the sun's rays than it lost to the outside air. For maximum steam production, Eneas calculated, the smallest diameter of the mirror should be at least eight times the diameter of the boiler, and the larger diameter should be at least 32 feet wide — nearly two and a half times wider than Mouchot's reflector.

Accordingly, Eneas built a truncated-cone reflector measuring 33 feet in diameter at its mouth and 15 feet in diameter at its bottom. The ratio of reflecting surface to boiler area was twenty-five to one, so that nearly twice as much sunlight was concentrated on the boiler, compared to Eneas's previous model. The reflector consisted of more than eighteen hundred small, silvered-glass mirrors. The boiler resembled one of Mouchot's later designs: two concentric metal shells held a total of 100 gallons of water. Although Eneas first covered the boiler with a glass jacket, the final version of the motor had a bare metal boiler.

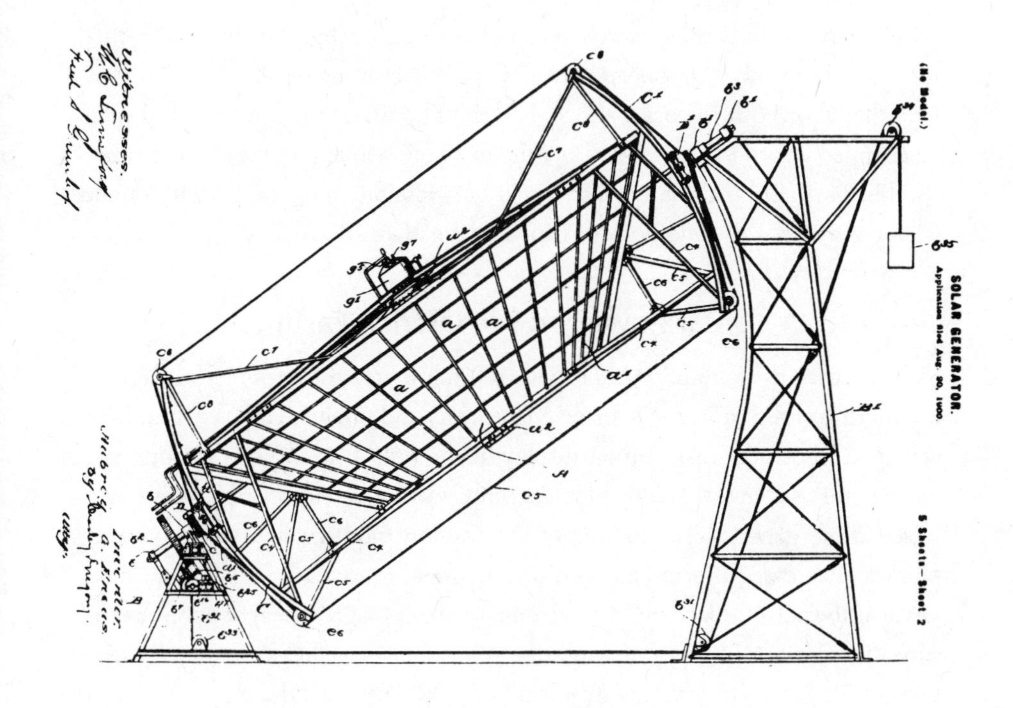

Figure 8.8. Patent drawing of Aubrey Eneas's solar motor.

Eneas made a lightweight but strong metal tower to support the 8,000-pound machine so it could be angled to receive optimum sunlight. The high point of the reflector slid up or down a track in the vertical tower, directing the mouth toward the high summer or low winter sun. A clock mechanism automatically rotated the mirror from east to west so that it would follow the daily motion of the sun.

By 1899 Eneas had finished building this solar engine. But the New England sunshine was insufficient to produce satisfactory test results. He decided to ship the motor to Denver, Colorado, where the rays of the sun are very hot due to the high altitude. The Denver tests convinced Eneas and his backers that they were on the right track, although they discovered one major flaw. The machine was unevenly balanced — it required great force to raise the massive device and adjust its angle to the sun's seasonal motion. Eneas solved this problem by mounting the mirror in a cradlelike frame that distributed the mirror's weight evenly. At opposite ends of the reflector, he installed arched runners that allowed it to rock back and forth. The tilt of the mirror could then be changed effortlessly. In this way "the machine will at all times be in proper equilibrium," Eneas declared, and it will be "possible to operate it with minor power and under all conditions of wind and weather."[31]

Solar Power on the Ostrich Farm

Satisfied with this alteration, Eneas moved the motor to the sunny climate of Southern California in 1901 for extensive tests and public demonstrations. He arranged to put the machine on display at the only ostrich farm in America, located in Pasadena. It belonged to Edwin Cawston, a fellow Englishman who raised these gangly birds to supply the fashion industry with their popular plumes. The ostrich farm had become a national tourist attraction, and Eneas saw it as the perfect place for his invention to receive a great deal of public exposure. On his part, Cawston thought the solar motor would garner extra attention for his farm. He advertised on handbills: "NO EXTRA CHARGE TO SEE THE SOLAR MOTOR — The only machine of its kind in the world in daily operation. 15-horsepower engine worked by the heat of the sun."

As Cawston and Eneas hoped, thousands saw the sun machine at the ranch. Anna Laura Meyers, who visited the ostrich farm many times during her childhood, recalled, "It was one of southern California's wonders. A strange thing, just as strange as an ostrich. People would just look at it and wonder about it."[32]

As news of the sun motor spread, more than a dozen popular and scientific publications sent reporters to cover the story of Eneas and his machine. "Sun power is now at hand!" was the message that many enthusiastically proclaimed. Even the more conservative articles indicated that sun power was at last "within the domain of what may be called practical science." F. B. Millard, one of the journalists who came out to the Cawston ranch, witnessed the solar motor operate a pump capable of irrigating 300 acres of citrus trees by drawing 1,400 gallons of water per minute from a reservoir 16 feet deep. Millard was convinced that "solar motors will before long be seen all over the desert as thick as windmills in Holland, and that they will make the desert blossom as the rose — a phrase that literally represents the possibilities of the machine." He also saw the sun motor as a source of inexpensive power for domestic purposes, which meant

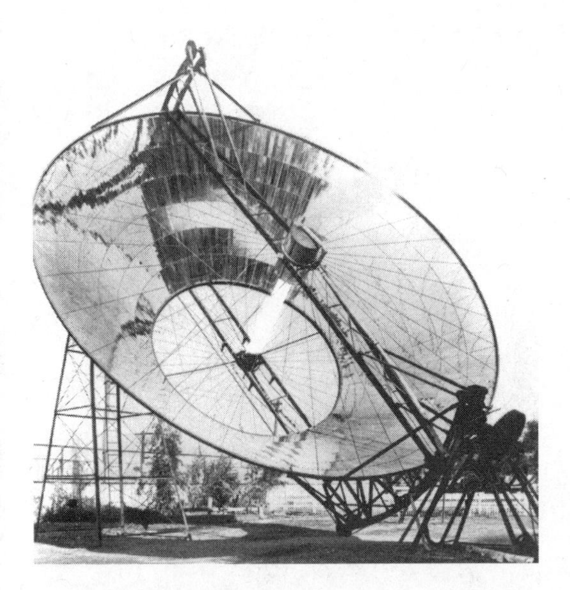

Figure 8.9. The Pasadena sun motor in operation. Notice the glistening white boiler and the steam escaping from a relief valve.

Figure 8.10. A workman nimbly climbing the ladder inside the reflector avoids being caught in its focus, which creates visible "hot spots" on the boiler behind him.

VISIT THE

OSTRICH FARM

100 GIGANTIC BIRDS

One of the strangest sights in the United States.—N. Y. Journal.
One of the features of Southern California.—L. A. Times.

PASADENA ELECTRIC CARS PASS THE ENTRANCE

No Extra Charge to see

THE SOLAR MOTOR

The only machine of its kind in the world in daily operation. 15-horsepower engine worked by the heat of the sun.

OPEN TO VISITORS EVERY DAY

Figure 8.11. Handbill advertising Cawston's ostrich farm and Eneas's solar motor.

Figure 8.12. Eneas's solar motor in the background at work on the ostrich farm.

"cheap homes in the arid regions[,]...homes for millions of men where there are now only hundreds."[33]

On a typical day the machine began operating one and a half hours after sunrise and continued until an hour after sunset. To start the motor the attendant turned the reflector toward the rising sun. As the rays struck the mirror, the boiler heated up. Millard described the spectacle:

> At first the morning dew is seen slowly to ascend in a wreath of vapour from the gigantic mouth. Then the bright glasses glitter in the morning sun, and the heatlines begin to quiver inside the circle, the greatest commotion being about the long, black boiler, which, as the intensity of the focused rays increases, begins to glisten so that in any photograph taken of the machine, the boiler is shown almost pure white. Within an hour of the turning of the crank and getting the focus there is a jet of steam from the escape valve. The engineer moves the throttle, there is a succession of

> hisses from the boiler, a "clank-clankety-clank!" and the sun is drawing water in a way which he little dreamed a few months ago.

The reflector automatically followed the sun's course throughout the day. Steam pressure and the water supply to the boiler were also self-regulated, and the motor even oiled itself. So little attention did the engine require that, according to Millard, the operator had enough time off to "hoe his garden, or read his novel, or eat oranges, or go to sleep."[34]

Another reporter, impressed by the almost incredible heat reflected by the mirror, wrote, "Should a man climb upon the [reflecting] disk and cross it, he would literally be burned to a crisp in a few seconds. And a pole of wood thrust into the magic circle flames up like a match."[35]

Marketing the Solar Motor

Buoyed by the glowing publicity and two years of successful operation of the solar pump at the ostrich farm, Eneas felt ready to offer his invention for sale. In 1903 he incorporated the Solar Motor Company in California and opened offices in Los Angeles's prestigious Bradbury Building. The complete machine — reflector, boiler, steam engine, and pump — sold for $2,160.

Eneas made his first sale to Dr. A. J. Chandler, who owned a large tract of land just outside of Mesa, Arizona, 35 miles southeast of Phoenix. By the summer of 1903 the company had set up the machine on Chandler's land. Unfortunately, after a short period of operation the frame supporting the mirror weakened, and under the force of a windstorm the reflector toppled and was badly damaged. As Pete Estrada, who was twelve at the time, remembered, "The metal frame that held it up was still intact, but the top of it, the mirror, was down all over the ground, broken, every bit of it."[36]

But the accident did not discourage interest in sun power. If California needed an economical method of irrigation, Arizona required one desperately. The following year the Santa Fe Railroad, seeking to encourage the development of the American Southwest, persuaded the Solar Motor Company to put a sun machine on public display. By March the firm had installed one in Tempe, a small town 8 miles southeast of Phoenix. The idea of such a large and unusual contraption being set up near the local high school did not rest well with

Tempe's provincial residents, many of whom had not yet seen a car. But the demonstration proceeded as planned, opening on March 21, 1904.

In July, John May purchased the motor for his farm, located in the hot and desolate Sulphur Springs Valley, not far from Wilcox, Arizona. With the help of Bert Parker, a local mechanic, construction was completed by September. On the opening day Parker wore his Sunday best to launch the machine's maiden run. Before a crowd of people from all over the county, believers as well as skeptics, Parker stepped up to the platform. He turned the huge mirror toward the rising sun, and soon the steam pressure in the boiler shot up to 110 pounds. As he gradually opened the steam valve until the boiler was going full blast, the motor — wrote one journalist — looked "like a bucking horse that had to be tamed."[37] It took a few more minutes to adjust the machine properly and get it running smoothly. A powerful 4-inch spray of water jetted out of the pump's nozzle, and the water broke through a makeshift dam built to contain the flow. Even the doubters were converted, and the spectators left the farm confident that a new era had begun in the Southwest. One account of the occasion likened it to the day at Kitty Hawk three years earlier when the Wright brothers proved that humanity could loosen its earthly bonds.

On sunny days the machine steadily pumped 1,500 gallons of water per minute, enabling Mr. May to grow "corn that would do Iowa credit; watermelons that were so big and luscious that they could have taken a prize at any man's fair," as one reporter observed. "It was a sight to see in arid Arizona." But the sun motor's performance was not the only point a potential customer had to consider. The destruction of the pump at the May ranch by a hailstorm some months later added to growing skepticism about the machine's ability to withstand the harsh weather of the Southwest.[38] Unscrupulous rivals in the solar business also tainted Eneas's reputation. *Harper's Weekly* referred to such people as "wild-eyed wizards with companies to promote." But the major obstacle to marketing Eneas's engines was their cost — $2,500 plus $500 for installation. This came to $196 per horsepower — two to five times the cost of a conventional steam plant. Even though no fuel expenses were incurred after the initial investment, the high purchase price deterred buyers. In the following years, the commercial prospects for Eneas's solar machines dimmed. "My...experience with large reflectors has convinced me," he wrote, "that even where the greatest efficiency is obtained, the cost of construction, even on an extensive scale, is too great to permit of their use in a commercial way, except in a few instances."[39]

While Eneas felt he had reached a dead end with reflectors, others began to consider other, more economical ways to produce solar power for industry and agriculture. Low-temperature solar engines, requiring neither expensive mirrors nor the high heat they generated, appeared to be one solution.

Figure 8.13. Eneas's sun motor at John May's farm.

Figure 9.1. Charles Tellier's solar pump, attached to his shop near Paris. The engine ran on sun-heated ammonia instead of water.

[1885–1915]

9

Low-Temperature Solar Engines

The solar engines developed by Mouchot, Ericsson, and Eneas all relied on the use of reflectors to concentrate the sun's rays. In their eager attempts to build a commercially viable engine, these early pioneers assumed that some concentration of solar energy would be needed to achieve the high steam temperatures they thought were essential. Because conventional wisdom taught that the higher the steam temperatures, the more efficiently an engine would run, they sought to produce steam at temperatures in excess of 1,000°F.

Unfortunately, this emphasis on concentrating collectors and high temperatures led to a number of critical drawbacks. As Saussure had demonstrated a century earlier, high temperatures inside a collector inevitably cause large heat losses. So, even though high temperatures meant greater engine efficiency, solar collection efficiency dropped substantially, bringing down the overall efficiency of converting solar energy into mechanical power.

There were other drawbacks too. The reflectors used by Mouchot, Ericsson, and Eneas were complex and expensive pieces of equipment. Once installed they were vulnerable to high winds and inclement weather. To make matters worse, they always had to face the sun, which required either a full-time attendant or a delicate mechanism to move the reflectors automatically. And when there was no direct sunlight on hazy or cloudy days, these concentrating collectors could not function at all.

Solar reflectors would be unnecessary in an engine designed to operate at lower temperatures. Simple, inexpensive hot boxes, which Mouchot and Ericsson had rejected, or even bare metal plates could be used instead to eliminate most of the problems. Because they did not reach such high temperatures, these collectors would not lose as much heat. Furthermore, the collecting surface would not have to follow the sun's motion. Such devices could absorb even diffuse sunlight during hazy or cloudy weather. Better solar-collection efficiency and lower construction costs appeared to outweigh the loss (due to lower operating temperatures) in engine efficiency. Late in the nineteenth century, a number of inventors began to realize these advantages and started to develop low-temperature solar engines.

A Low-Temperature Solar Pump

Charles Tellier, a French engineer, was the first person in modern times to develop a low-temperature solar collector to drive machines. Whereas conventional engines used pressurized steam, Tellier's devices used pressurized vapor from certain liquids that had boiling temperatures well below that of water. Ammonia hydrate, for example, will boil at 28°F; sulfur dioxide will boil at 14°. These substances vaporize rapidly when exposed to higher temperatures.

Tellier discovered the unusual properties of low-boiling-point liquids during his research in cold food storage. Called the "father of refrigeration," Tellier radically transformed international trade by enabling a ship, the *Frigorifique*, to carry the world's first mechanically refrigerated cargo — chilled meat exported from Rouen, France, to Buenos Aires. He used a low-boiling-point liquid for the refrigeration system, just as most refrigerators today rely on similar liquids to transfer heat away from a container storing food.

Tellier began experimenting with these refrigerants as a means of powering a solar pump. On the sloping porch roof of his shop in Auteuil, an exclusive suburb of Paris, he set up a row of ten solar collectors. They were metal plates, each 4 feet wide by 11 feet high, made of two sheets of corrugated iron riveted together. The grooves in each set of sheets were aligned to form a series of hollow channels, through which the ammonia hydrate flowed. The bottom metal sheet of the collector was insulated to help prevent heat loss.

As the sun struck the tops of the collectors, the metal conducted solar heat to the liquid inside. As a result, the ammonia vaporized and exerted a pressure of

Figure 9.2. Watering the garden by means of a solar pump. Pressure in the pump was supplied by the expansion of a fluid that heated while passing through hot metal plates, seen next to the horse.

40 pounds per square inch. The vapor circulated through pipes to a water pump consisting of a spherical chamber submerged in a well. The pressurized ammonia gas pushed a diaphragm in the chamber downward, forcing water out of the pump in a jet. Afterward the gas traveled through metal tubes set in a tank of cold water. The vapor condensed to a liquid again, and the ammonia was ready to repeat another cycle.

According to Tellier, the pump drew more than 300 gallons of water an hour. But he realized that France's climate was "not favorable to the operation of such a device."[1] He estimated that the capacity of the pump could more than double (to 792 gallons of water per hour) in the sunnier tropical regions of the world. Like his countryman Mouchot, Tellier looked south to the African colonies as the real proving ground for solar power. In 1890 he published a book titled *La conquête pacifique de l'Afrique Occidental par le soleil* (The peaceful conquest of Africa by the sun). Its publication coincided with the partitioning of Africa by the European powers. Tellier understood that a lack of cheap fuel hindered the underdeveloped continent's progress in industry and agriculture. He foresaw wide-reaching economic and social benefits from using low-temperature sun machines to make these lands more productive.

With this goal in mind, Tellier began testing designs for low-temperature solar engines. To increase the amount of solar heat collected, he put each metal plate inside a shallow wooden box covered with a single layer of glass. The hot box increased the collector's performance enough to power an engine, but more precise data on its operation are unavailable.[2] A contemporary of Tellier, the noted English engineer A. S. E. Ackermann, was one of those frustrated by the difficulty of ascertaining exactly how well the French inventor's solar devices worked. After reading Tellier's book, he commented, "With so much detail it is disappointing that [I] could not find the results of a single experiment with the [power] plant."[3]

And when Tellier later dropped his promising solar research to return to the field of refrigeration, his reasons also remained something of a mystery. While his experiments with low-temperature solar pumps had opened up a whole new approach to harnessing sun power, it remained for others to continue the work he had begun.

Willsie and Boyle

Two American engineers, H. E. Willsie and John Boyle, took up where Tellier had left off. Between 1892 and 1908 they explored the potential of low-temperature solar power plants based on the French inventor's design. In the May 13, 1909 issue of *Engineering News*, Willsie described what he and his colleague had accomplished. They had first begun to look into solar energy, he wrote, because they realized — as had Ericsson and Eneas before them — that the sun-drenched American Southwest desperately needed a source of cheap power for its irrigation pumps and mines.[4]

At the outset both men agreed to avoid publicizing their solar experiments until they had achieved some tangible results. They did not want to jeopardize their careers, for "the building of sun motors has not been an especially good recommendation for engineers." Frauds, failures, and eccentrics had given sun power a bad name. For instance, one priest proclaimed that his solar machine proved the truth of Genesis, and that its use would serve as the cornerstone of a new social order.

Willsie and Boyle began their research with a review of the solar motors built by their predecessors. They discovered that "the state of the art most developed was with reflectors to concentrate the sun's rays upon some sort of boiler." But they also knew that every reflector-powered motor had been a commercial failure. Consequently they chose to work with a Tellier-type motor using a low-boiling-point liquid, so that a sophisticated and expensive reflector would be unnecessary.

The partners spent a decade drawing up blueprints and conducting small experiments. In 1902, they decided the time had come to implement their plan for a full-scale sun-power plant. Willsie explained the impetus for this decision: "Boyle was then in Arizona, surrounded by conditions which daily remind one of the desirability of converting the over-abundant solar heat into much needed ...power. To the Southwesterner 'cheap power' brings visions of green growing things about his home to stop the burn of the desert wind, and of the working of the idle mine up the mountain side."[5]

Before moving to the Southwest the two men built a crude prototype of their solar power plant in Olney, Illinois, to test an important modification of Tellier's original design. In Tellier's process the solar-collector plate had to be strong enough to contain the high-pressure vapor it produced. Willsie and Boyle felt that this would probably make the collector too expensive. Instead they decided to use water to transfer solar heat from the collector to a low-boiling-point liquid in a separate system of pipes. Solar-heated water did not require such a heavy-duty collector, and the high-pressure vapor could be kept in a more limited circuit.

This innovation allowed them to eliminate Tellier's metal plate–hot box combination and substitute a simple hot-box collector — a shallow, rectangular wooden box covered by two panes of glass. The exterior walls and bottom of the box were insulated with hay; the hot-box interior was covered with black tar paper. Three inches of water filled the box. Even in the cold October weather of the American Midwest, the solar-heated water was hot enough to vaporize sulfur dioxide, which then drove a low-temperature engine. This test proved that a sun machine did not need a solar reflector — surprising even Willsie and Boyle's friends, who, Willsie noted, "were skeptical about window glass being able to take the place of mirrors."[6]

In December of that year, they repeated the experiment in Hardyville,

Arizona, where "over 50 percent of the heat reaching the glass was absorbed by the water."[7] After a few additional tests, Willsie and Boyle patented the system in 1903. They also incorporated the Willsie Sun Power Company that year, which enabled them to procure capital by selling stock and obtaining loans.

The First Solar Power Plants

Saint Louis became the site of a full-size solar power plant built by the Willsie Sun Power Company in the spring of 1904. A set of large, shallow, rectangular boxes covered with glass served as solar collectors. The bottoms of the collectors were inclined toward the south and held a thin film of water. As the sun warmed the water it traveled to a boiler, where ammonia was heated to produce a high-pressure vapor that drove a 6-horsepower engine. Through condensation the ammonia returned to its liquid state and flowed back to the boiler. The water circulated back to the collectors in a separate cycle.

The plant ran on sunless days and at night as well, when an auxiliary boiler powered by conventional fuel took over. Newspapers in Saint Louis and New York announced the success of this twenty-four-hours-a-day solar-powered generator. Pleased with the results, Willsie and Boyle next moved their operation to the Southwest. They bought some land in the Mojave Desert just outside of Needles, California — one of the hottest places in the country, where the sun shines 85 percent of the time.

Over the following years they built and rebuilt the Needles plant several times, each time making design improvements. Lack of money brought their work to a temporary halt in 1906. Although the interruption was unwelcome, they did learn some valuable lessons about the effects of the harsh desert climate on their apparatus. After two years of exposure, only 2 percent of the glass collector covers had broken, and a few of the panes had turned purple in the sun. Aside from some minor damage, the power plant had weathered well. In the spring of 1908 additional funds enabled them to continue, and by midyear the fourth and final version of the plant was completed.

Their final Needles plant utilized a dual water-heating process that generated higher water temperatures. Angled to the south, two groups of hot boxes had a solar-collecting area of more than 1,000 square feet. The first group of hot boxes, with a single glass cover, raised the water to 150°F. The second group, with two glass covers separated by an air space, increased the water temperature another 30°F.

The hot water went to a boiler, where coils of pipe containing liquid sulfur dioxide were heated as the water passed over them. Willsie and Boyle chose sulfur dioxide and not ammonia for this motor because a "sulphur dioxide engine had been carefully tried out in Germany," said Willsie, "and because it required less heavy pipe and fittings to withstand the pressure...obtained [in] the heater."[8] But sulfur dioxide had its dangers, too. If the chemical ever mixed with the heated water, a powerful acid would be formed that could quickly corrode the metal parts of the machine. But this did not present any immediate difficulties, and in test runs the vaporized sulfur dioxide performed well, generating a maximum of 15 horsepower.

The Needles plant also boasted one feature that other inventors had long dreamed of — a solar-energy storage system. Mouchot had tried decomposing water into hydrogen and oxygen. Others had suggested that excess solar energy collected during the day could be used to lift weights or water; the force of their descent at night would generate power. Aware that increasing the temperature of 1 pound of water by 1°F stores as much energy as raising a similar amount of water 778 feet, Willsie and Boyle chose to store solar-heated water as the simpler of the two methods.

The solar-heated water produced during daytime operation flowed from the collectors into an insulated tank. The amount of water needed at the time went on to the boiler; the rest was held in reserve. After dark, when a valve to the storage tank was opened, hot water flowed out and passed over the pipes containing sulfur dioxide, and the engine could continue working. Willsie could rightly claim, "This is the first sun power plant...ever operated at night with solar heat collected during the day."[9]

The Needles plant worked better than any solar generator previously built. But one critical question remained: Was it economically superior to conventional generators? Willsie noted that the sun plant would cost $164 per horsepower to build, versus only $40 to $90 for a conventional plant. But operating costs favored the solar plant, especially in the American Southwest, where coal was very expensive. According to Willsie's calculations, running a conventional coal-fed steam engine cost $1.54 per kilowatt-hour, as opposed to only $0.45 per kilowatt-hour to operate the solar plant. As a result of the fuel savings, the solar plant could pay for itself in less than two years.

However, a couple of years after Willsie made this optimistic comparison, the gas-producer engine was introduced to the Southwest. This type of engine burned coal to produce artificial gas, and motors using artificial gas were two to four times more efficient than conventional coal-burning engines.

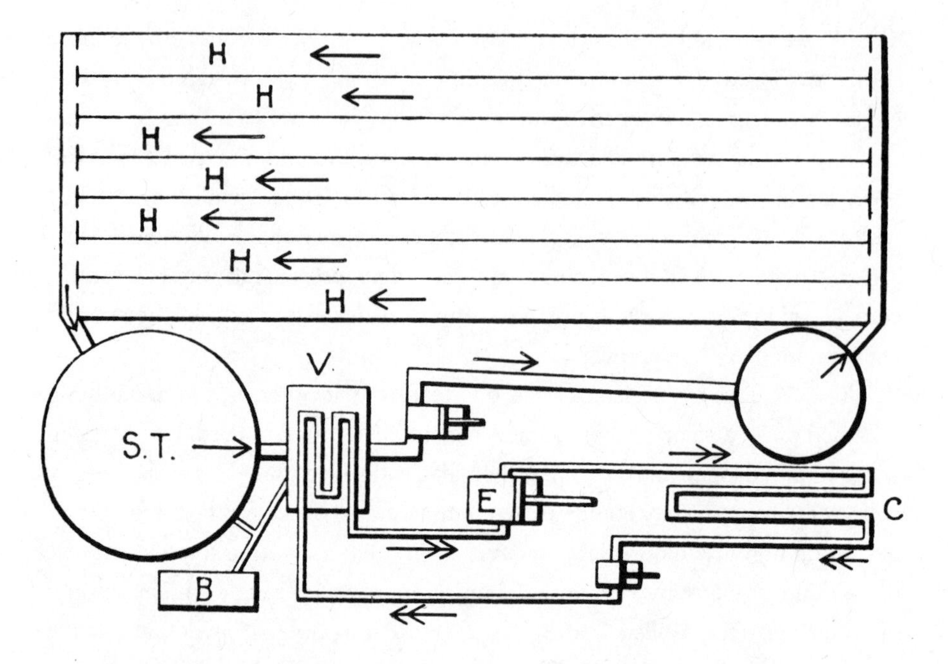

Figure 9.3. Flow diagram of the power plant at Needles, California. Water (single-headed arrows) flowed through the collectors (H) and, once heated, went to a storage tank (S.T.) before being fed into a vaporizer (V). There it gave up its heat to sulfur dioxide, which drove the engine (E).

Figure 9.4. Willsie and Boyle's fourth and final sun-power plant, built in Needles, California, in 1909. The hot-box collectors sit at the right. Next to them are the boiler room and cooling tower.

Figure 9.5. Willsie stands inside the boiler room by the boiler.

Consequently, the solar engine became less economically attractive where coal could be cheaply shipped. Nevertheless, there were many areas where coal had to be transported from the railroad line by horse-drawn coach, raising the cost of operating a gas-producer engine. Willsie mentioned the example of a mine in Mojave County, California, where "the fuel is hauled 20 miles over the desert and then 5 miles over a mountain range,"[10] resulting in a coal bill of about $72,000 annually. In such remote areas the solar power plant would be highly marketable if mining companies and farmers could be convinced that the large investment was worthwhile in the long run. In the Mojave, Willsie noted, it took less than eight years for their solar plant to pay for itself through the fuel saved. He also pointed out that "during the summer months there will be an excess of sun power... that may be profitably used for the manufacture of artificial ice."

Despite the favorable prognosis, there is no record of the two inventors expanding their operations. And it is not known why their experiments ended with the Needles plant. Nevertheless, Willsie and Boyle made a giant stride toward the commercialization of sun power. They demonstrated that a solar reflector was not required to run an engine, and that a hot-box collector could easily drive a low-temperature motor. Their solar-energy storage system, combined with a conventional engine as a backup, enabled a solar power plant to operate around the clock and throughout the year, removing the intermittency issue that hindered earlier solar plants.

Figure 10.1. Shuman's solar plant in Meadi, Egypt, consisting of a parabolic-trough reflector that concentrates the sun's rays on a linear solar heat collector to power a pump.

[1906–1914]

10

The First Practical Solar Engine

While Willsie and Boyle were designing their Needles power plant, a self-taught engineer on America's East Coast was also exploring the use of hot-box collectors to drive low-temperature engines. A resident of the Philadelphia suburb of Tacony, Frank Shuman had an eye toward the technology of the future. When he took his children to that turn-of-the-century wonder the picture show, he told them, "Boys, the day will come when you will see this in your living room!"[1] Not only did Shuman see television on the horizon, but like others before him he also foresaw a time when fossil fuels would be extremely scarce and solar energy would become the industrial world's only hope. "One thing I feel sure of," he wrote, "and that is that the human race must finally utilize direct sun power or revert to barbarism."[2]

Described by a professional journal of the time as a man "of large practical experience who [has] already made noteworthy and valuable inventions,"[3] Shuman entered the field of solar energy in 1906. After studying the sun machines invented by Mouchot, Ericsson, and Eneas, he reached the same conclusion as Willsie and Boyle had concerning the reasons for their commercial failure: "With the high temperatures involved, the losses by conduction and convection are so great that the power produced was of no commercial value. Where... mirrors are used, the primary cost... and the apparatus necessary to continuously present them to the sun, have rendered them impracticable."[4] Shuman therefore rejected the use of solar reflectors in the initial stages of his research and began experimenting with hot boxes. Like Horace de Saussure, Samuel Pierpont Langley, and others before him, he found that hot boxes could reach temperatures high enough to boil water.

Figure 10.2. Frank Shuman, solar visionary of the early twentieth century.

Next, Shuman began testing a low-temperature solar engine resembling the one used by Willsie and Boyle. He first built a 1-foot-square hot box with blackened tubes inside that held ether, a low-boiling-point liquid. The solar-heated ether vapor drove a tiny engine, the kind that was commonly sold in toy stores at the time for a dollar. Shuman tried using a similar collector to run an engine somewhat larger than the first and was able to produce ⅛ horsepower.

Shuman saw he was on the right track, but that it would take a lot of time and work to refine his ideas. As he put it: "Natural forces are not entirely conquered in a few years."[5] It would also take a substantial amount of capital. To attract investors, Shuman realized, he would first have to build a successful demonstration engine — just as it had taken solid evidence of the possibility of flight to attract investors to the fledgling aviation industry: "You will at once admit that any businessman approached several years ago with a view of purchasing stock in a flying machine company would have feared the sanity of the proposer.

After it has been shown conclusively that it can be done, there is now no difficulty in securing all the money which is wanted, and very rapid progress in aviation is from now on insured. We will have to go through this same course."[6]

With money he had made from other profitable business ventures, Shuman built a demonstration sun motor in his backyard. In August 1907 he printed up handbills inviting the public to "attend an exhibition run of the first practical solar engine — any clear afternoon between 12 and 3 P.M. during the next two weeks." The solar machine continued running well past its original two-week billing, "working steadily on sunny days during the summer of 1907 and 1908," reported *Engineering News*, "pumping thousands of barrels of water."[7] Once Shuman had his solar engine driving the pump, he announced to colleagues that his invention "is the start of a new era in mechanics."[8] The engine also produced power on sunny days during the cold Pennsylvania winters, amid snow piled up around the collectors.

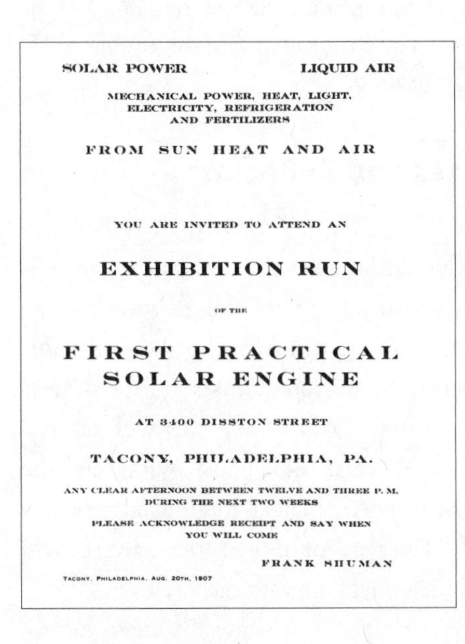

SOLAR POWER LIQUID AIR

MECHANICAL POWER, HEAT, LIGHT, ELECTRICITY, REFRIGERATION AND FERTILIZERS

FROM SUN HEAT AND AIR

YOU ARE INVITED TO ATTEND AN

EXHIBITION RUN

OF THE

FIRST PRACTICAL SOLAR ENGINE

AT 3400 DISSTON STREET

TACONY, PHILADELPHIA, PA.

ANY CLEAR AFTERNOON BETWEEN TWELVE AND THREE P. M. DURING THE NEXT TWO WEEKS

PLEASE ACKNOWLEDGE RECEIPT AND SAY WHEN YOU WILL COME

FRANK SHUMAN

TACONY, PHILADELPHIA, AUG. 20TH, 1907

Figure 10.3. Handbill announcing a demonstration of Shuman's first solar plant.

The power plant was a larger version of the low-temperature motors Shuman had tested earlier. The hot box, totaling 1,080 square feet of solar collection area, lay flat on the ground and contained blackened pipes in which a low-boiling-point liquid circulated. The solar-heated vapor operated an engine with a capacity of 3½ horsepower.

Figure 10.4. Shuman's original solar plant in operation at Tacony, a suburb of Philadelphia. The hot-box solar collector lies flat on the ground, and the engine and condensing coils stand to the right, in the foreground.

Combining Hot Boxes and Reflectors

Although Shuman's sun motor performed no better in Tacony than Willsie and Boyle's plant at Needles, Shuman was much more persuasive than they in selling the idea of solar power to wealthy investors. He painted a glowing picture of solar energy enabling industry and agriculture to expand in the fuel-short but sunny regions of the world — powering the nitrate industry in the deserts of Chile and the borax industry in California's Death Valley, as well as irrigating arid lands in the Australian interior, Eastern India and Ceylon, and the American Southwest. "Throughout most of the Tropical Regions sun power will prove very profitable," he predicted. "Ten percent of the earth's surface will eventually depend on sun power for all mechanical operations."[9]

Shuman's confident, charismatic presentation of the case for solar energy convinced a number of American investors, who had made large profits on his other inventions, to gamble on his solar venture. They formed the Sun Power Company, and for the next several years Shuman was able to obtain enough financial backing to experiment with improvements in his solar-energy system. His long-range goal was to build a large-scale solar power plant, and efficiency and economy were the crucial factors in making this ambitious project practical.

New Jersey

111

THE SUN POWER COMPANY

Shares -4-

This Certifies that Frank Shuman is the registered holder of Four Shares of the Capital Stock of THE SUN POWER COMPANY Transferable only on the Books of the Company by the holder hereof in person or by duly authorized Attorney upon surrender of this Certificate properly endorsed.

Witness the Seal of the Company and the signatures of its President and Treasurer, this 20th day of Jany 1909

Constantine Shuman, Treasurer Frank Shuman, President

Figure 10.5. Stock certificate for the Sun Power Company founded by Frank Shuman, issued to obtain capital for additional solar plants.

He knew that the solar engine he had tested could be improved on both counts. As *Engineering News* observed, "Instead of developing a horsepower with only one or two square feet of heating surface area as in a locomotive boiler, or eight or ten square feet as in a stationary plant, a sun engine built [according to Shuman's first design]...would require one or two hundred square feet."[10] Consequently, a solar plant of industrial size would need so much collection surface in relation to the horsepower produced that the initial capital investment would be prohibitive. Shuman was determined to keep the purchase price of a solar plant low, for exorbitant first costs were to him "the rock on which, thus far, sun power propositions were wrecked."[11]

Shuman developed several ways to increase the efficiency of his solar engine while holding costs down. But he needed an additional infusion of funds to move from the drawing board to actual construction of a large-scale solar plant. He found the necessary capital abroad — a group of British businessmen agreed to form the Sun Power Company (Eastern Hemisphere) in 1910. Shuman announced that "sufficient money is at hand to go into business on a large scale, and there will be great developments in the near future."[12]

The first "great development" was the decision to develop and construct the largest solar plant ever built. The company chose Egypt (then a British protectorate) as the eventual site; land and labor were cheap there, and the desert sun was strong. However, the plant was to be first constructed in the United States to allow Shuman to test it thoroughly before shipment to Africa. He built the plant on two-thirds of an acre near his home in Tacony, Pennsylvania. One of his primary objectives was to increase the amount of heat produced in the collectors. He accomplished this by adding reflectors to the hot boxes and installing a mechanism to adjust the angle of the collectors for optimum solar exposure. He also put metal heat absorbers similar to Tellier's inside the hot boxes and improved their insulation.

In all, there were 572 collectors, with a total area of 10,276 square feet. The reflectors he used were plane mirrors (flat mirrors) made of ordinary silvered glass, and their construction required no expertise in optics. Two mirrors, each 3 feet square, concentrated sunlight onto each horizontal hot box, which had the same dimensions — for a reflector-to-collector-area ratio of two to one. These concentrating collectors sat adjacent to each other in twenty-six long rows, so that each row resembled a trough. In each row there were twenty-two collectors lined up horizontally to form the trough's base; the reflectors on each side formed the trough's walls.

To increase heat collection even further, the rows of collectors were tilted so that the glass covers of the hot boxes were always facing directly into the noonday sun. Notches in the frames on which the collectors were mounted

Figure 10.6. Mirrors on both sides of each hot box reflected additional sunlight onto the collectors.

enabled Shuman to adjust the angle of the hot boxes and mirrors every three weeks so that they stayed in alignment with the sun.[13]

Besides increasing heat collection, Shuman redesigned the engine. He had concluded that his previous engine, which ran on the pressurized vapor produced by a low-boiling-point fluid, did not generate enough power. But substituting high-temperature steam did not seem like a good idea, either, because that would require increasing the temperature in the collectors to the point that the overall efficiency of the system would decrease because of large heat losses.

Instead, Shuman invented a special motor that ran on low-temperature, low-pressure steam. Just as water boils at a lower temperature in the mountains than at sea level because there is less air pressure at higher altitudes, this engine was able to convert hot water into steam at a temperature below 212°F because the steam was kept isolated from atmospheric pressures. Each row of collectors had a feed pipe at one end through which cold water entered the channels of the metal heat-absorbing plates in the collectors. The sun-heated water flowed out through a pipe at the other end of the row, and the hot-water pipes from all the rows emptied into a main duct leading to the engine. After the water was converted to low-temperature steam, which then drove the motor, a condenser converted the steam back into water, which returned to the collectors.[14]

Figure 10.7. The engine room in Shuman's second solar power plant. Low-pressure steam ran the engine.

This engine generated more power than any solar motor previously built. Connected to a pump, the device could deliver 3,000 gallons of water per minute — raising the water 33 feet. Almost 30 percent of the solar energy striking the collectors was converted into useful heat, producing a maximum of 32 horsepower and an average of 14 horsepower on a typical sunny day. The noted engineer A. S. E. Ackermann, sent by the British partners in the company to oversee the design of the plant, estimated that the engine's performance would improve by 25 percent in the hotter, more consistently sunny climate of Egypt.

Figure 10.8. The second Tacony power plant in action.

Solar Energy in Egypt

Before Shuman shipped the solar plant to Africa, the British investors requested that Professor C. V. Boys, an eminent physicist, be brought in to review the project. After investigating the plant's operation, Boys pointed out that only the tops of the hot boxes collected solar energy, while the bottoms not only had no solar exposure but actually lost heat. He suggested that the reflector be extended to cover each hot box on three sides, so that the bottom would receive solar rays as well. This could be accomplished by replacing each row of twenty-two concentrating collectors with a single parabolic, trough-shaped reflector, inside which a long, glass-covered boiler was suspended. As a result of this design change, the ratio of reflector area to hot-box area would be increased to four and a half to one, more than twice the concentration ratio of the previous plant. Shuman and his fellow engineers readily approved the change. At the time they thought Boys's design was original; later Ackermann discovered that it closely resembled Ericsson's parabolic-trough concentrator.

Figure 10.9. Shuman's second solar collector, modified by C. V. Boys.

Rather than rebuild the entire system to incorporate this change and then transport it to Egypt as originally planned, the company decided to construct an entirely new plant on-site. In 1912 Shuman and his crew arrived in Meadi, a small farming community on the Nile 15 miles south of Cairo. Five solar collectors were built on a north-south axis, each one measuring 204 feet long and 13 feet wide, with a distance of 25 feet between them. Shuman used construction materials that were readily obtainable in the industrialized world and sturdy enough to last many years. The collectors rested on a foundation of reinforced concrete and were supported by steel frames. Additional measures were taken to ensure that the collectors could withstand gale-force winds.

Each trough-shaped reflector was made from sections of silvered glass, which were held in place by small brass springs that absorbed the stress of expansion and contraction due to temperature changes. The boiler running down the middle of the trough was held up by rods capable of absorbing additional thermal stress. Covered by glass, the boiler was 15 inches deep with a flat bottom and trapezoidal sides. Inside, in a $3\frac{1}{2}$-inch zinc pipe extending the length of the trough, water was heated to 200°F. This water was converted to low-pressure steam capable of powering an engine similar to the one used in Tacony.

The reflectors shifted automatically throughout the day so that the sun's rays were always focused on the boiler. When the reflector was facing precisely into the sun, a thermostat located directly behind the boiler remained in the shade. As the sun moved westward and its rays struck the thermostat, a tracking

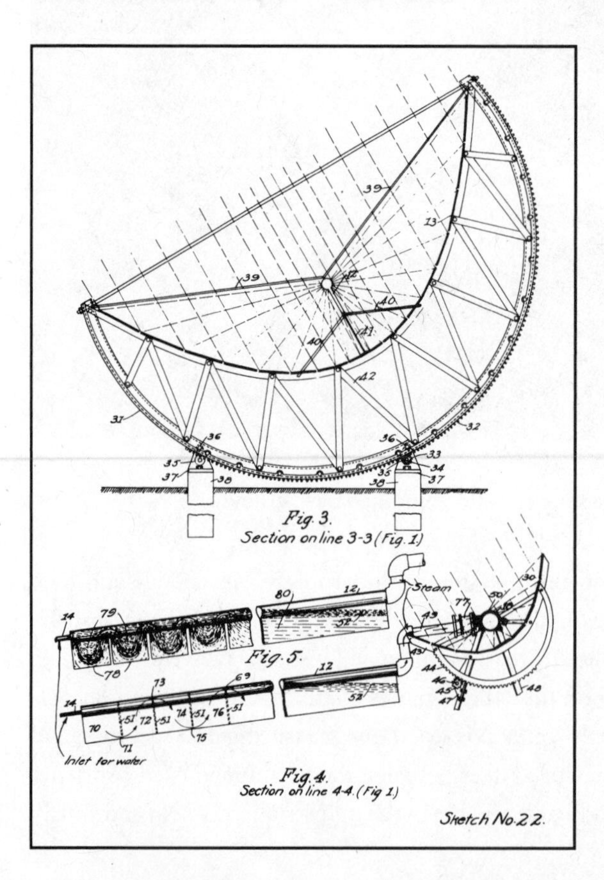

Figure 10.10. Patent drawing shows the thermostat that kept the reflector aligned with the sun throughout the day.

mechanism sprang into action and moved the reflector a fraction of an inch toward the west.

The Meadi plant could operate twenty-four hours a day. A large insulated tank, similar to the one used by Willsie and Boyle, held excess hot water for use at night or during overcast or rainy days. This enabled the engine to drive a conventional irrigation pump at all hours and in all weather, further increasing the efficiency of the plant.

Shuman set up a public demonstration of his sun-driven engine in late 1912. But the boiler reached temperatures too close to the melting point of the zinc pipes. Consequently, the metal began to sag until, according to one observer, the pipes "finally hung down limply like wet rags." The trial run had to be postponed while the zinc pipes were replaced with cast iron.

Figure 10.11. Workers inspect the Meadi plant.

By July 1913, the plant was again ready for testing. Such luminaries as Lord Horatio Herbert Kitchener, consul-general of Egypt, were invited to witness Egypt's first display of solar power. Gentlemen in pith helmets and Panama hats and women carrying parasols to protect their fair skin from the tropical sun watched the giant solar machine swing into action.[15] It produced more than 55 horsepower, enough to pump 6,000 gallons of water a minute. The absorbers collected 40 percent of the available solar energy, which was much better than at the Tacony plant. This machine far surpassed in performance previous solar engines, including Willsie and Boyle's plant in 1908 and Eneas's conical reflector engine in 1904.

Shuman's solar engine compared very favorably to a conventional coal-fed

Figure 10.12. Officials and dignitaries attend the grand opening of the Meadi plant in 1913.

Figure 10.13. Egyptian farm workers divert to irrigation canals the water that has been pumped by the solar plant.

plant. True, the solar plant still had an enormous ratio of collecting surface to horsepower produced — exceeding 200 square feet per horsepower. And the purchase price, at eighty-two hundred dollars, was double that of a conventional plant. Nevertheless, the costs of running the solar plant were so much less than the operating costs of a conventional plant that the financial outlook for solar power was extremely bright. With coal going for fifteen to forty dollars a ton in Egypt, the solar plant would save over two thousand dollars annually in fuel costs. In two years the extra investment in the sun plant would be paid off; in four years the entire purchase price would be met; and in subsequent years the plant would operate for practically nothing, with the exception of expenses for repairs and labor.[16]

Plans Fail

After devoting seven years of work to solar power and spending nearly a quarter of a million dollars, it seemed to Shuman that his earlier optimistic predictions were beginning to come true. He wrote in February of 1914: "Sun power is now a fact and no longer in the 'beautiful possibility' stage. . . . [It will have] a history something like aerial navigation. Up to twelve years ago it was a mere possibility and no practical man took it seriously. The Wrights made an 'actual record' flight and thereafter developments were more rapid. We have made an 'actual record' in sun power, and we also hope for quick developments."[17] Many others agreed, avidly supporting solar power. Some were former skeptics, such as those at *Scientific American*, who now praised Shuman's solar engine as "thoroughly practical in every way."[18]

Besides the world of science, western Europe's colonial powers, too, lauded Shuman's work and looked forward to the enormous economic benefits of using solar energy in underdeveloped Africa. Lord Kitchener offered the Sun Power Company a 30,000-acre cotton plantation in British Sudan on which to test solar-powered irrigation. The German government called a special session of the Reichstag to hear Shuman speak, an honor never before bestowed on an inventor. Speaking in the German he had learned as a boy, Shuman described the fantastic possibilities of solar power and showed movies of the Meadi plant at work. Duly impressed, the Germans offered Shuman two hundred thousand dollars in deutsche marks for construction of a sun plant in German Southwest Africa.[19] With such enthusiastic demonstrations of support, Shuman now expanded the scope of his plans. He hoped to build 20,250 square miles of reflectors in the Sahara, giving the world "in perpetuity the 270 million horsepower [201 gigawatts] per year required to equal all the fuel mined in 1909."[20]

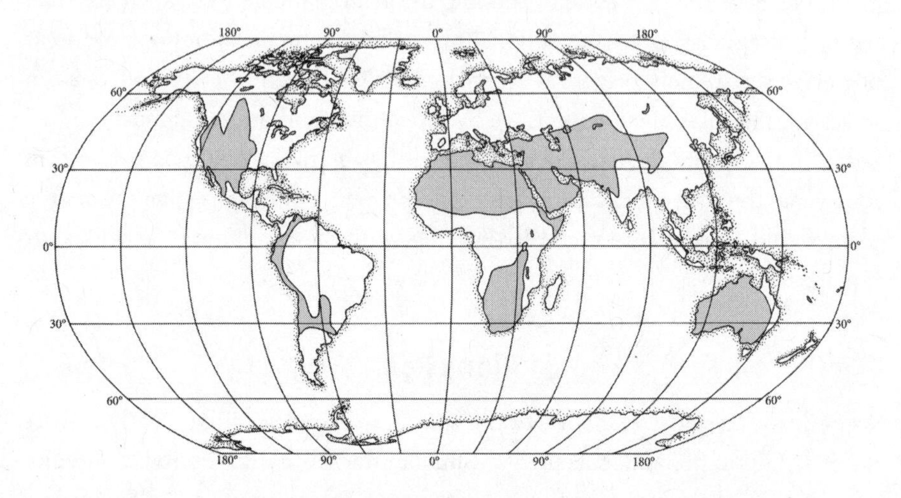

Figure 10.14. Shaded portions of the continents represent areas where Shuman believed the use of his solar plants would be economically and physically feasible.

But his grand dream disintegrated with the outbreak of World War I. The engineers running the Meadi plant left Africa to do war-related work in their respective homelands, as did Shuman, who returned to the United States. He died before the war's end.

Gone was the driving force behind large-scale solar development. Moreover, with the Germans in defeat and their African colonies taken over by the Allies, the promises made to the Sun Power Company were as worthless as the deutsche marks offered it. And the British too had lost interest in solar power. They began to turn toward a new form of energy to replace coal — oil. By 1919, the British had poured more than $20 million into the Anglo-Persian Oil Company. Soon afterward new reserves of oil and gas were discovered in many parts of the world — Southern California, Iraq, Venezuela, and Iran. These were almost all sunny areas where coal was difficult to obtain — areas targeted by Shuman, as well as by Mouchot and Ericsson before him, as prime locations for solar power plants. But with oil and gas selling at near giveaway prices, scientists, government officials, and businessmen became complacent again about the energy situation, and the prospects for sun power quickly declined.

PART III

SOLAR WATER HEATING

Figure 11.1. A Maryland gentleman enjoys a steaming hot bath provided by his Climax solar water heater.

[1891–1911]

11

The First Commercial Solar Water Heaters

The story of solar water heating begins in the nineteenth century, when Europeans and Americans began to bathe on a regular basis. During the 1800s, a heightened interest in personal hygiene, advances in technology, and greater material well-being all combined to increase the demands for hot water. Pasteur's germ theory of disease had underscored the need for frequent warm-water bathing; and with the introduction of iron plumbing and cheap manufactured soap, home hygiene became much easier than before. People also needed hot water for washing cotton clothes, which were rapidly replacing the woolens hitherto worn by everyone but the gentry.

Unfortunately, heating water remained a laborious and time-consuming task for the majority of Americans, who lived in small towns and rural areas without gas or electricity. They had to rely on the cookstoves for hot water. As one homesteader recalled, "You took just one bath a week, a Saturday night deal, because it was such hard work to heat water on the stove. You put the water in pots, pails, anything which would hold water and you could lift. It took a while for those old stoves to get going because the heat first had to penetrate through the heavy metal."[1]

Some people attached a water tank to the side of their stove, eliminating the need to crowd the top burners with pots and pans of water. The tank was made of cast iron and had a lid that lifted off to allow water to be poured in and then scooped out after being heated. Where there was enough water pressure, a more efficient method heated the water faster and did away with the burden of having to carry water from the pump or tap to the stove. Water circulated directly from the household pipes into metal coils looped through the firebox of the stove, and from there to a holding tank attached to its side. But even with this system the water took time to heat and, according to one old-timer, did not stay hot for very long: "Once you got the fire going really good, you'd have to wait about 15 or 20 minutes as the cold water heated up. The hot water would naturally rise up into the tank. And the holding tank was not insulated. That was a real problem because the water in the tank would be cold within an hour or so."[2]

Wherever water was heated — whether on top of, next to, or inside the stove — the job of starting the fire and keeping it hot was a chore. After the wood was chopped and brought in or a heavy hod of coal lifted, the fuel had to

Figure 11.2. A typical turn-of-the century water system. Water that had been warmed as it passed through pipes inside the stove was stored in the holding tank standing beside it.

be kindled and the fire periodically stoked. There were also the unpleasant side effects of smoke, ashes, dirt, and, in the case of coal, foul odors. In the winter, families endured such nuisances anyway as part of the price of helping keep the house warm. But in the summer, as one rural resident of the Santa Ynez Valley near Santa Barbara, California, exclaimed, "It was torture just to be in the house with the stove on! You nearly died from the heat."[3]

In large cities the fuel situation was a little better. There were gas heaters, which ran on "artificial" or "manufactured" gas made by baking coal in an airless environment; the gas was then distributed through pipes to consumers. Artificial gas had only one-half the heating capacity of natural gas, was not as clean burning, and left an oily residue. The most common type of gas heater was the "sidearm," so named because it was attached to the side of an uninsulated hot-water tank. The sidearm was not automatic. It had to be lit with a

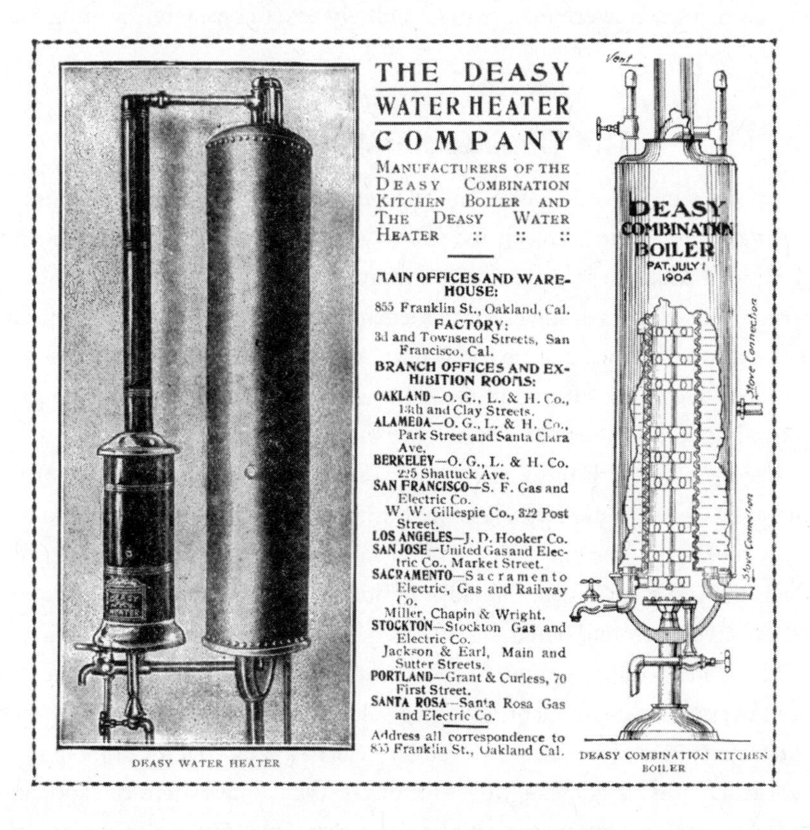

Figure 11.3. Typical gas water heater used in the late nineteenth and early twentieth centuries.

match. The water took a while to travel through the heating coils inside the side-arm and into the adjacent tank. And when the water got hot enough, "the tank would start jumpin' and you knew it was time to shut it off," said one plumber who had installed them. If you forgot, "you might get your hand scalded or get a face-full of steam if you opened the hot water faucet. There were times when [people] would split a tank. We had this one house where this woman started [the] side-arm up and went uptown and when she came back the back of the building was blowed off!"[4]

Besides being dangerous, these early gas heaters were too expensive for many families to use. The price of artificial gas around the turn of the century was about $1.60 per thousand cubic feet. Taking inflation into account, it cost more than twenty-nine times what a family paid in 2012 for a quantity of natural gas with comparable heating capacity.[5] And as exorbitant as gas prices were, electric rates were even worse; nobody even considered heating water with electricity.

The Climax Solar Water Heater

Fortunately, a much safer, easier, and cheaper way to heat water was discovered: metal water tanks painted black and placed where they would catch the most sun and avoid shade. These were the first solar water heaters on record, and they worked. Farmers harvesting crops in the American Midwest washed themselves after work with hot water from such tanks. A prospector in Nevada testified that sometimes late in the afternoon the water in the exposed tanks "would get so damned hot you'd have to add cold water to take a bath."[6]

The problem with these rudimentary solar heaters was not whether they could produce hot water but when and for how long. Even on clear, hot days, it usually took from morning to early afternoon for the water to get hot. And as soon as the sun went down, the tanks rapidly lost heat because they were bare and unprotected from the night air.

These shortcomings came to the attention of Clarence M. Kemp, a Baltimore inventor and manufacturer. Kemp sold the latest in home-heating equipment, including devices that produced artificial gas from coal for those living on large estates, and gas and coal stoves for the average homeowner. But appliances that ran on coal and gas weren't his only interest. Solar caught his eye as well.

In the fall of 1882 many newspapers and magazines carried the news about

Figure 11.4. The bare metal tank of this solar water heater was painted black for better solar absorption.

Figure 11.5. Interior view of Kemp's factory in Baltimore, Maryland.

Samuel Pierpont Langley using a solar hot box to heat water while climbing through the snow to America's highest peak, Mount Whitney, in extremely cold weather. Kemp, like many of his contemporaries, most likely read Langley's account. If Langley could boil water in below-freezing temperatures with a glass-covered box, Kemp thought, couldn't he improve solar water heating by placing water tanks in a similar glass-covered container under less trying conditions? In 1891 he patented a way to combine the old practice of exposing bare metal tanks to the sun with the scientific principle described by Langley, using the greenhouse effect to increase the tank's ability to collect and retain solar heat. Kemp called his invention the Climax, and it became the world's first commercial solar water heater.

Kemp sold the Climax in eight sizes. The most popular model was the smallest, a thirty-two-gallon heater that sold for $25 and measured 4½ feet long, 3 feet wide, and 1 foot deep. The largest heater held 700 gallons of water and had a price tag of $380. Every model contained four long, cylindrical water tanks made of heavy galvanized iron painted a dull black and connected to each other. They lay horizontally next to each other inside a pine box insulated with felt paper and covered by a sheet of glass. The box was usually installed on a

Figure 11.6. Clarence M. Kemp, inventor of the first commercial solar water heater, the Climax.

Figure 11.7. Advertisement for the Climax solar water heater. The 1891 price of twenty-five dollars was reduced the following year to fifteen dollars.

sloped roof or on brackets at an angle to a wall, so that the tanks lined up one above the other. The tanks, connected to each other, were completely filled with water, which was then heated by the sun.

To draw hot water from the tanks, a faucet in the bathroom or kitchen was opened. In a house with pressurized plumbing, cold water from the inlet pushed solar-heated water out of the tanks and down to the bathtub or sink. If the home had gravity-feed plumbing, opening the faucet drew hot water from the tanks. Cold water refilled the tanks from a small reservoir located above the heater. A float valve in this reservoir allowed it to refill. In either system, a drain allowed the tanks to be emptied before the onset of freezing weather, so that the water would not turn to ice and split the tanks.

Kemp advertised the Climax as "the acme of simplicity" compared with conventional heaters. Just turn on the faucet and "instantly comes the hot water," boasted the sales literature. Housewives could avoid the terrible heat of lighting the stove in the summer, and "gentlemen who occupy their residences alone

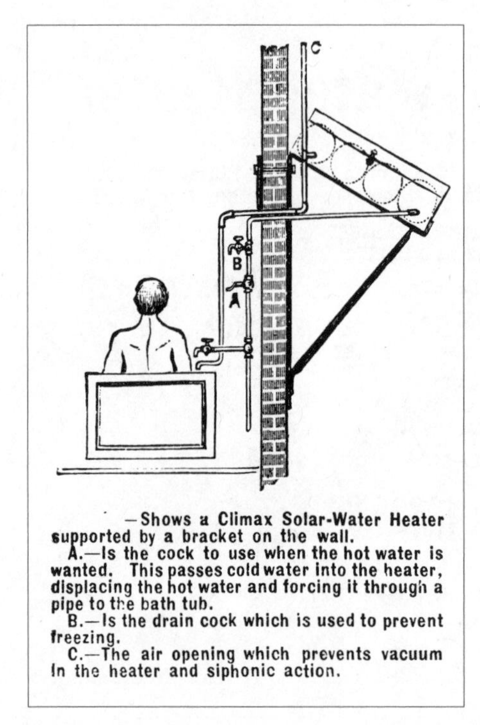

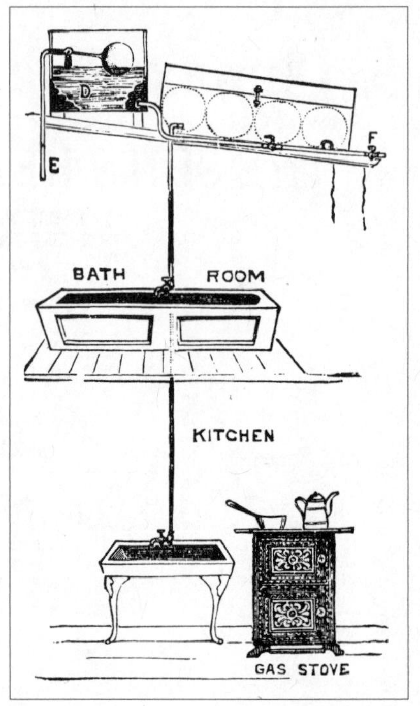

Figures 11.8 and 11.9. A Climax user's manual showed how to connect the water heater to either a pressurized system or a gravity-feed system.

during summer months, while their families [go] on holiday," advertising copy boasted, "can have the convenience of hot water without delay or attention."[7]

Solar Water Heating in California

In the Maryland area, Kemp claimed, the Climax could be used from the beginning of April until the end of October — producing water hotter than 100°F on sunny days, even during early spring and late fall, when daytime temperatures sometimes approached freezing. In areas of the country like California, the climate and fuel situation made the Climax even more attractive. Sunshine almost year-round meant free hot water most of the year, and extra savings because energy costs were high on the West Coast. California had to import coal at a price more than twice the national average, and artificial gas, too, was expensive. As one journalist wrote, it was essential for Californians to "take the asset of sunshine into full partnership. In this section of the country where soft coal sells

—Shows a swimming pool which can be supplied with hot water from "Our Heater."

—Shows another service it is qualified for.

Figure 11.10. The Climax Solar Water Heater Company recommended using the heater to warm water for, among other things, swimming pools and baptisms.

for $13 a ton (and the huge peaches bring only $12 a ton) a builder cannot afford to waste his sun rays."[8]

Two Pasadena businessmen, E. F. Brooks and W. H. Congers, recognized the potential market for solar water heaters in Southern California. In 1895 they paid Kemp $250 for the exclusive rights to manufacture and sell the Climax in California. Sales took off so quickly that just three years later, in 1898, Sarah Robbins was willing to pay Brooks and Congers ten times what they had paid Kemp for just the Southern California rights to the Climax. That same year Richard Stuart purchased the Northern California rights for $10,000.

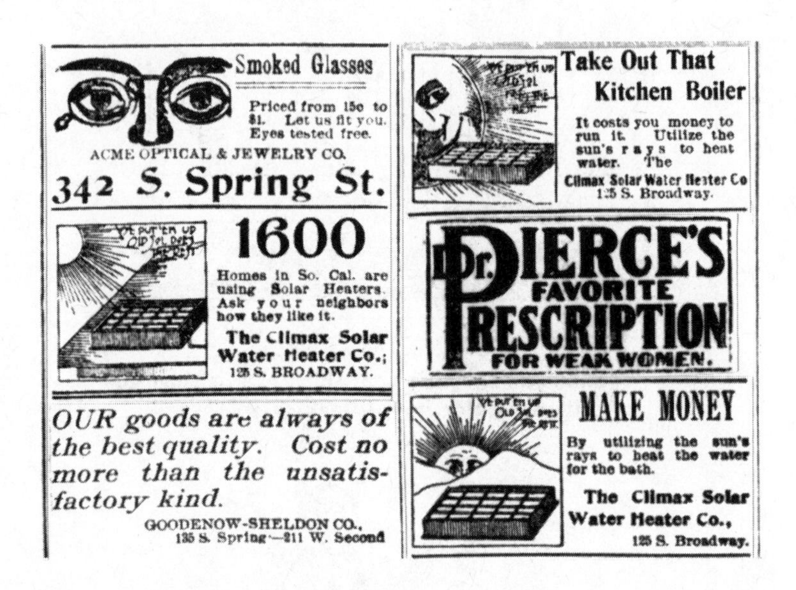

Figure 11.11. Three of the early advertisements for Climax solar water heaters that appeared in issues of the *Los Angeles Times*.

Climax installations spread from Pasadena to much of California and Arizona. By 1900 they topped the sixteen hundred mark in Southern California alone.[9] Economy was a prime lure of the Climax. For an investment of twenty-five dollars, the average homeowner saved about nine dollars a year on coal — and more if the homeowner would normally have used artificial gas for water heating. The Climax was also considered a wise choice by landlords — such as Samuel Stratton, who outfitted his six flats with solar heaters. The *Pasadena Daily Evening Star* called Stratton "a level-headed businessman who knows a good thing when he sees it."[10] One satisfied Climax household, the van Rossems, had their solar heater on the southwestern side of the roof of their house (which stood near the present site of the Rose Bowl). Walter van Rossem, who was a child at the time, recalled that solar heaters became so popular that he and the others in the neighborhood did not think of them as anything out of the ordinary. "Everybody had one," he said. "There was nothing uncommon about it at all. I can't remember a house on the block that was built at the time or soon after that that didn't have a solar heater. I don't think anybody would be thinking about paying for hot water, if they could get it from a solar heater."[11]

Figure 11.12. Four large Climax heaters atop this apartment in Southern California.

Figure 11.13. A Climax installation on the roof of the Pasadena home of Walter van Rossem, shown here sitting in his mom's lap.

The van Rossems were particularly strapped for money. As Walter pointed out, "We were pretty curtailed for expenses. Mother's inheritance only amounted to twenty dollars a month. The solar water heater was much more economical than buying wood or gasoline for the same purpose."

Van Rossem appreciated the Climax because he didn't have to fuel up the stove very often to heat water. "What the heck," he confessed, "I didn't like to chop wood any better than anybody else did!" The rest of the family also appreciated the solar heater, though there were a few drawbacks: "On an ordinary sunshiny day...by afternoon, my mother and our housekeeper would have enough hot water for baths and by evening there would be enough for us kids. Whether we had hot water the next morning depended on how much we used the night before. If we didn't use all the hot water up, it stayed fairly warm — enough to wash your hands and face."

As for laundry, van Rossem said the water was "hot enough for a small amount of washing, the things the women wore, but when we did the heavy washing, the stuff we kids wore like our overalls, we always had to boil water on the stove." Moreover, he noted, the seasons affected the amount of hot water available: "In the wintertime usually there were a couple of kettles sitting on top of the woodstove heating. They were used for dishes and a lot of things because the water in the solar heater never got as hot in the wintertime as it did in the summertime." Still, even on cloudy days "you'd be surprised how much it would heat up," van Rossem remarked.

The Walker Solar Water Heater

From the turn of the century until 1911, over a dozen inventors filed patents for improvements on the Climax. But only a few designs turned out to be technically and commercially successful. One of these was patented in the spring of 1898 by a Los Angeles contractor and real estate agent, Frank Walker. The Walker heater had only one or two cylindrical 30-gallon tanks. The tanks were set inside a glass-covered box, but the box did not protrude from the roof like the Climax — it fit inside the roof, and the glass cover lay flush with the rooftop. This arrangement afforded somewhat better insulation and looked less obtrusive.

But the major advantage of the Walker heater over the Climax was that it

Figure 11.14. Solar inventor Frank Walker served on the Los Angeles City Council at the beginning of the twentieth century.

was hooked into the conventional water-heating system to ensure hot water at all times. At night or during inclement weather, cold water from the bottom of the solar water tank ran down a pipe to a heating coil inside the wood or coal stove or gas heater. Afterward the heated water — which is less dense than cold water and rises naturally — flowed up through a second pipe leading to the top of the water tank. People found this method more convenient and cheaper because it was no longer necessary to have two sets of plumbing — one for the solar heater and one for the conventional heater.[12]

The Walker cost less than fifty dollars, including installation. While it cost more than a similar-sized Climax, many customers throughout Southern California willingly paid for the additional benefits.

The Improved Climax

In 1905 the rights to manufacture and sell the Climax in California were acquired by a branch of the Solar Motor Company — the firm founded by Aubrey Eneas, who designed and sold large solar concentrators to run pumps for farmers in the sun-rich but water-starved American West. Charles Haskell managed the Los Angeles headquarters of Eneas's business, which was listed as the Solar Heater Company.

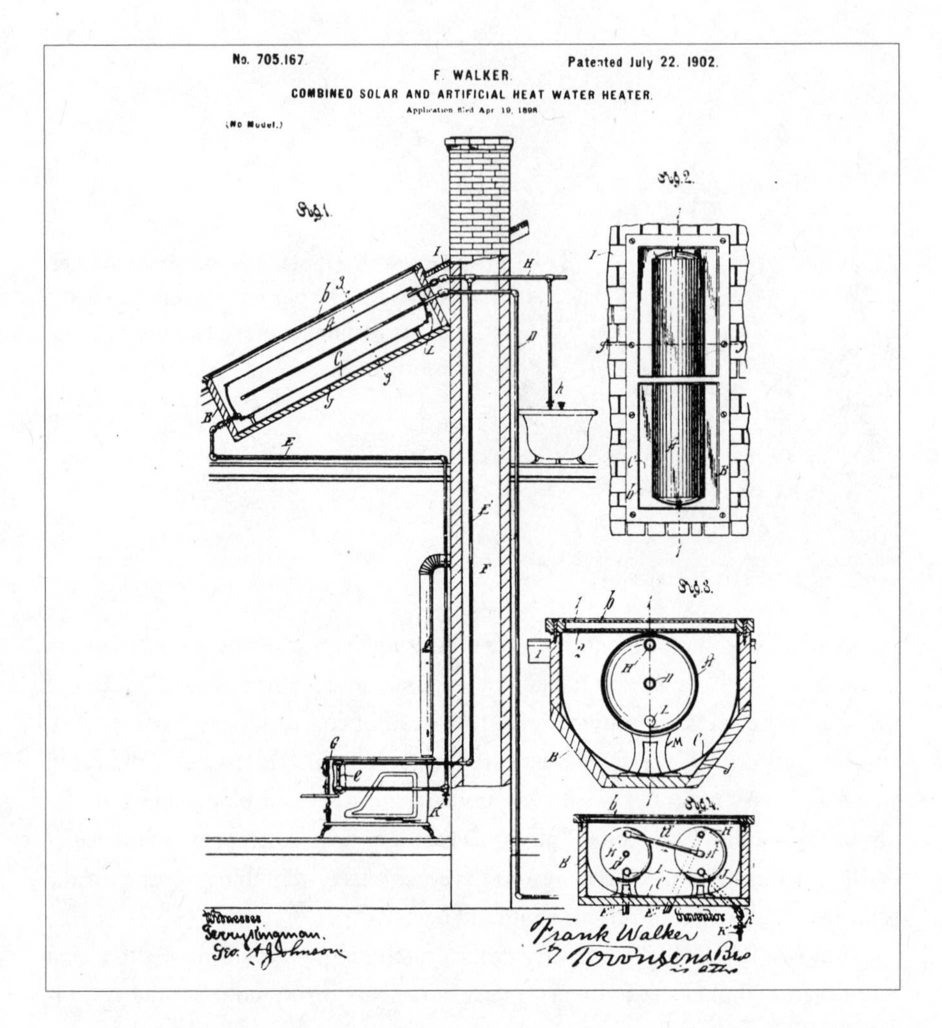

Figure 11.15. A patent drawing for the Walker solar heater shows how it would be connected to an auxiliary heat source — in this instance, the cookstove.

Haskell made a basic change in the design of the Climax water tanks. Noticing that it took many hours for the relatively deep body of water in the four cylindrical tanks to heat up, he decided to replace them with one large but shallow rectangular tank. It held the same total volume of water, but with less water per square foot the sun's heat penetrated more quickly and produced hot water earlier in the day. Like Walker's model, Haskell's was usually connected to a conventional water-heating system that took over during unfavorable weather.

THOUSANDS IN USE IN LOS ANGELES!
ASK YOUR NEIGHBOR ABOUT THEM

IMPROVED CLIMAX SOLAR HEATER

ABSOLUTELY RELIABLE

WHY let the sunshine go to waste, and your monev too, when at trifling expense you can put in your home an IMPROVED CLIMAX SOLAR HEATER that will furnish hot water from sunshine alone—winter and summer—for your bath, laundry, and all domestic purposes, without cost, damage or delay?
It can be connected with the range, furnace or gas heater to insure hot water on rainy days, and when so connected saves the expense of a kitchen boiler.
It insures a cool house during the hot season. Let us figure with you on your hot-water problem.

SOLAR HEATER COMPANY

342 New High St. - - - - - LOS ANGELES

Figure 11.16. Advertisement for the Improved Climax.

The Solar Heater Company called this updated model the Improved Climax. It was usually placed either on or in the roof, facing the direction with the best solar exposure. According to one of the company's installers, the Improved Climax worked well: "Even on a foggy day, the first one to use it would get warm water. But of course, on a sunny day it would be much hotter. Why, hell's bells! You'd have to use the cold with it because you couldn't stay under the shower with just the hot water turned on. It really got hot!"[13] Customers were just as laudatory. J. J. Backus, for example, a Los Angeles superintendent of buildings, wrote a testimonial that appeared in a 1907 issue of *Architect and Engineer of California*:

> I take great pleasure in saying that after a thorough trial extending over a year and a half, our solar heater continues to give just as much satisfaction as when first installed. I am ready to admit that [at first] we were unreasonably prejudiced against the heater, and feel that refusing to let you install one in my house for so long a time after you first approached me upon the subject, we lost a great deal of comfort and convenience. We are sure that every person having a [solar] heater will in a way become an advertising agent for your company, for so great will be his satisfaction that he cannot help talking about it.[14]

Figure 11.17. Downtown Los Angeles, 1900. Circles in the drawing indicate locations of solar water heaters in that area.

In Southern California and in many areas farther north, the Improved Climax and its predecessors, the Walker and the Climax, supplied large quantities of hot water for seven to eight months of the year — the Climax and Walker models heating water up to 120°F by late afternoon, and the Improved Climax reaching this temperature earlier in the day. But a serious defect hampered the effectiveness of these solar water heaters. While they lost heat less quickly than the early bare-tank heaters, their insulation consisted of only a pane of glass and a wooden box. The water did not remain hot for very long, especially on cloudy, cool days. Even under the best conditions, the water never stayed hot enough overnight to enable clothes to be washed in the morning. Kemp, Walker, and Haskell had brought the technology of solar water heaters a considerable distance in a decade and a half — but not far enough.

Figure 12.1. This home in Pomona Valley, California, had a Day and Night solar water-heater panel mounted on the eave, at the far right. Installation date: 1911.

[1909–1941]

12

Hot Water, Day and Night

In the summer of 1909, in a little outdoor shop in the Los Angeles suburb of Monrovia, an engineer named William J. Bailey began selling a solar water heater that eventually revolutionized the industry. It supplied solar-heated water not only while the sun was shining but also for hours after dark and the following morning as well — which gave it its name, the Day and Night.

Bailey had worked for Carnegie Steel in Pennsylvania before he moved west in 1908 to seek a cure for his tuberculosis. He soon discovered that his physician, Dr. Remington, had been experimenting with solar water heaters. To heat water faster and store the heat longer, Remington separated the solar heater into two units: a solar heat collector and a water storage tank. The collector consisted of a coiled pipe placed inside a glass-covered box that hung on the south wall of his house. The small volume of water in the pipe heated quickly. And instead of remaining outside where it would readily cool down at night or during bad weather, the hot water flowed through pipes to a conventional water tank in the kitchen.

Bailey adopted Remington's idea of a separate collector and storage tank. But a good deal of heat still escaped from the tank because it was made of bare metal. For better heat retention, Bailey insulated the tank — a new concept in water storage for solar heaters.[1] The average household-size tank made by Bailey held 60 gallons. It was encased in a wooden box, with powdered limestone between the box and tank at a thickness of 3½ to 9 inches along the sides and top. Bailey guaranteed that the water in this storage unit would not lose more than 1 °F per hour.

Figure 12.2. William J. Bailey, developer of the Day and Night solar water heater, stands next to the sign for his new solar company.

The Day and Night collector was better designed than earlier heaters. A key feature was the use of copper pipes that held only a small amount of water, as in Remington's model. But even more important, Bailey added a metal absorber plate to transmit the solar heat accumulated in the hot box to the water in the narrow pipes. The collector pipes and plate were enclosed in a glass-covered box measuring about 55 square feet. Only 4 inches deep, the box was lined with felt paper. Bailey put a large vertical cold-water inlet pipe made of iron at one end of the box, and a parallel hot-water outlet pipe at the other end. To these two headers he connected a series of smaller transverse pipes made of ⅝-inch copper tubing. The array of pipes was welded to a copper absorber plate at the bottom of the box, and both the plate and pipes were painted a dull black. Cold water entering the collector through the inlet pipe split into streams that flowed through the copper pipes. The streams heated up as the absorber plate conducted solar energy to the pipes. In Southern California the system produced hotter water than did previous heaters — 100 to 120°F on sunny winter days, and 115 to 150°F during the other nine months of the year. No pump was

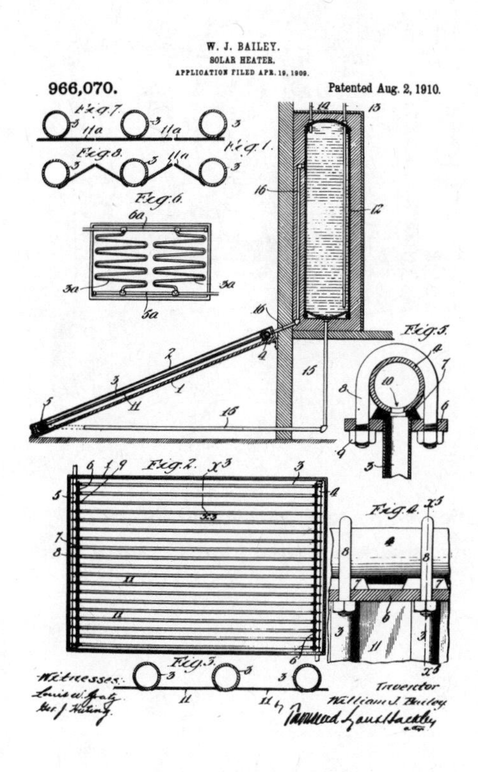

Figure 12.3. A patent drawing of the Day and Night solar water heater illustrates the system's novel design.

needed to circulate water between the collector and the storage tank. The Day and Night operated on the thermosiphon principle — hot water is lighter than cold and rises naturally. The storage tank was located above the collector so that cold water in the bottom of the tank would be pulled by gravity down a pipe to the collector inlet. After it passed up through the copper pipes, the heated water rose through the outlet pipe and into the top of the water tank. This influx of hot water forced more cold water out the bottom of the tank and down to the collector. The collector was usually installed on the south-facing slope of the roof. It was a heavy piece of equipment, and hoisting it into place was not easy. According to William Crandall, a Day and Night installer, the crew would tie a rope around the collector and "just by mean strength" pull it up on the roof on a

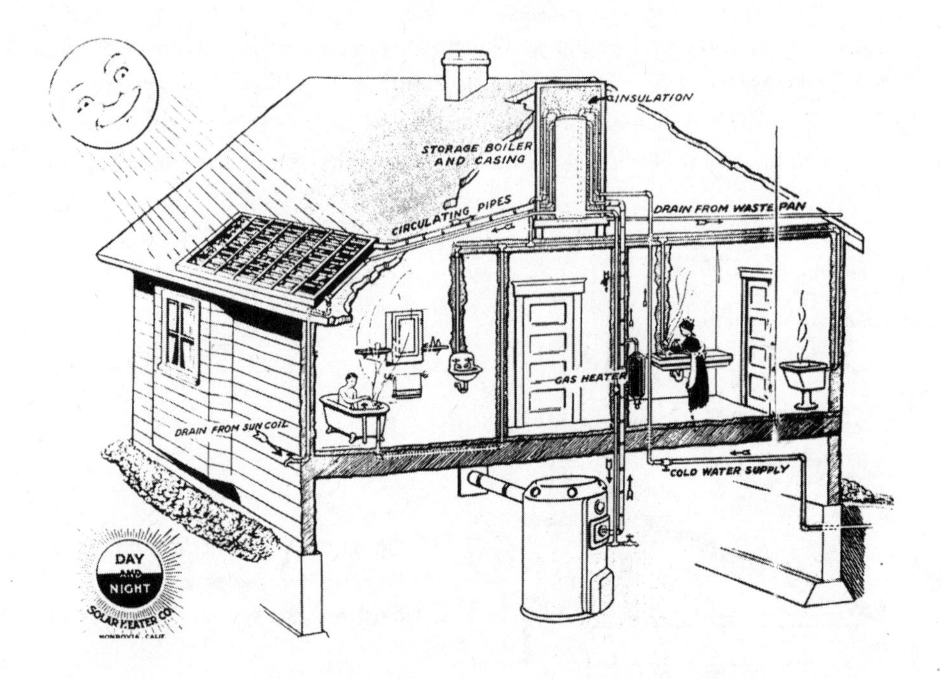

Figure 12.4. Typical Day and Night installation, with a rooftop collector and attic storage tank. A connection to the basement furnace allowed the latter to serve as an auxiliary water-heating source.

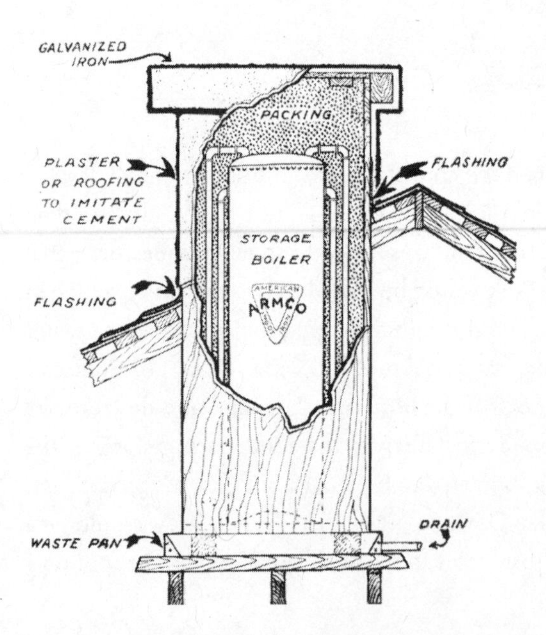

Figure 12.5. Cutaway view of the Day and Night storage tank.

Figure 12.6. A Day and Night collector sits atop the garage.

set of skids.[2] If putting the collector on the south side of a roof was impractical, they installed it as an awning or on the ground.

Installers usually placed the storage tank in the attic, making sure that the tank was higher than the collector. If the roof was too low, a hole cut in it allowed the tank to protrude above the roofline. The exposed section of the tank was then camouflaged with gray felt or roofing paper, cloth, and stucco so that it resembled a chimney.

To ensure plenty of hot water during periods of bad weather or heavy use, Bailey advised customers to provide an auxiliary heater. The Day and Night could be connected to a wood stove, gas heater, or coal furnace. In places where people could obtain and afford electricity, a small electric insert heater was sometimes placed in the storage tank. This heater turned on automatically when the water dropped below the desired temperature.

Day and Night Captures the Market

Newspapers hailed the Day and Night as the "ne plus ultra of solar heaters," because it could "heat water and keep it hot under conditions that would render most other heaters of little or no use."[3] Nevertheless, at $180, Bailey's invention faced stiff competition — a similar-sized Improved Climax cost much less. Before long, though, the Day and Night won over the buying public. Ned Arthur, one of the five original employees hired by Bailey in the early years of the business, elaborated: "It was always a battle; if someone was going to buy a solar heater, Climax was on the job and I was on the job, and we fought like tigers. But after we got a few heaters in, people saw the advantage of having water hot at night, and Climax was out."[4]

Unlike other solar water heaters, Bailey's product could supply hot water in the morning so that people did not have to wait until afternoon to do such chores as the laundry. "Many of our customers are reporting that they are putting out their entire washings early in the morning," one ad proclaimed. "Ask your neighbor if she can do this with her old-style heater."[5] Furthermore, the Day and Night's long-lasting hot water enabled city residents to use their auxiliary gas heaters less frequently. Whereas the Climax reduced gas consumption by 40 percent, the Day and Night reduced it by 75 percent. The demand for the Day and Night was also great in rural districts where no gas was available, because less coal or wood was needed.

The Day and Night soon began to edge out its rivals, but there was still a segment of the public that did not think any commercial solar heaters were worth buying. Paul Squibb, a rancher, spoke of the skepticism among some of his neighbors: "When they first began putting in these Day and Night jobs, old timers were giggling about how silly they were. They'd say, 'You couldn't get water any hotter than if you just stood a can full of water out in the sun.' " So there was always one poor guy ready to expose the scam, according to Squibb, "who'd say he'd put his hand in any water heated by the sun, and the poor guy got an awful roasting. He jerked his hand out before he lost his skin!"[6]

Demonstrations set up at fairs and at the Day and Night office gave others a chance to test the company's claims of "steaming hot water day and night." One advertisement challenged: "Step in some cloudy morning following a day of sunshine, hold your foot in the water from the heater for five minutes, and we will give you the heater. Cork legs are barred from the test." Such publicity stunts helped the Day and Night to catch on, and in 1911 — two years after Bailey had first opened up shop — the company was incorporated and moved to a larger plant.

The Sun Coil

Bailey's business continued to grow. But in January 1913, a freak cold spell that hit Southern California nearly killed the company. Temperatures dropped as low as 10°F in some parts of the area, the coldest it ever got there. The water

Figure 12.7. Employees of the Day and Night Solar Water Heater Company pose outside the factory, circa 1912.

Figure 12.8. An array of Day and Night Sun Coil collectors set up at a California county fair.

inside many Day and Night collectors froze, and the copper pipes "popped like popcorn all over the county," said Day and Night employee Ned Arthur. Bailey's son, William J. Bailey Jr., recalled that his father's telephone "rang all night long — irate customers were having problems with water coming through their ceilings. That sent him back to the drawing board."[7]

Bailey came up with a way to prevent such disasters. Instead of water, he used a nonfreezing liquid — usually a mixture of water and alcohol — that would be heated in the collector and then would travel to a coil inside the storage tank. The heat passed from the coil to the water in the tank, and the cooled liquid returned to the collector for another round. This method resembled the process in Willsie and Boyle's engine: the liquid that flowed through the collector gave off its heat to another liquid outside the collector without the two ever coming into direct contact.[8] Aside from taking care of the problem of freezing, there was another advantage to making the collector's circulation system independent of the hot water supply in the storage tank. In areas with very hard water, the collector could be filled with a mixture of alcohol and distilled water so that the tubing would not become encrusted with mineral deposits.

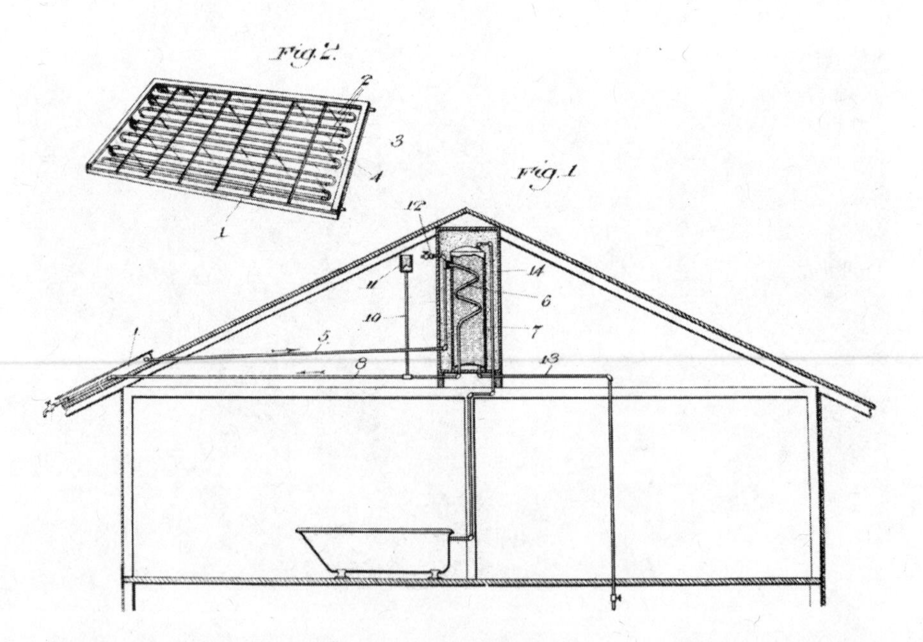

Figure 12.9. Patent drawing of the Sun Coil's nonfreezing solar collector. An antifreeze solution carried solar heat from the collector to a coiled heat exchanger inside the storage tank. The new collector is detailed at top left.

Bailey made some other improvements in the Day and Night collector at this time. He did away with the iron headers and transverse pipes, substituting what he dubbed the Sun Coil design. A series of parallel pipes running horizontally were connected to each other at alternate ends with U-shaped fittings, so that they formed one continuous length of tubing in a zigzag configuration. The tubing was then soldered to the copper absorber plate. This arrangement made the flow of water through the collector more efficient. The water entered at the bottom of the coil, and as it heated it rose up more quickly and out the top of the collector. Later Bailey switched from copper pipes for the coil to galvanized steel, which was more widely available.

The Industry Expands

Once Bailey had devised a circulation system that eliminated the danger of frozen water pipes, Day and Night's business flourished once again in Southern California. Sales began to take off in other parts of the American Southwest, too. Two brothers, C. M. Eye and H. A. Eye, bought the rights to the Day and Night for Arizona and New Mexico in 1913 and set up their headquarters in Phoenix. They wasted no time in marketing the heater. By the following year a reporter for *Arizona* magazine acknowledged that "the sight of the Sun Coil is becoming as familiar on Salt River Valley homes as in California, where they have been in general use for years."[9] Soon Day and Night heaters also spread to Hawaii. But in Northern California, solar water heaters were not as readily accepted. Ned Arthur described the problem: "It was awfully hard to break in. Everywhere I went... people would say, 'Oh well, they'll work in southern California but they won't work up here.' "[10]

Determined to prove the Day and Night's feasibility, Arthur arranged a demonstration in Palo Alto, near San Francisco. It was "right on the main street in the heart of town," he recounted. A sign attracted the attention of passersby with the message "Hot Water From California Sunshine — Try It!" An arrow pointed to a faucet connected to a Day and Night heater. It was a convincing test — painfully so, for "people would come by and scald their hands. I sold heaters right and left," said Arthur.[11] This marketing strategy worked equally well in dozens of rural towns in Northern California.

As Day and Night's reputation grew, the manufacturers of the Walker and Climax heaters were forced out of business. The solar water heater became

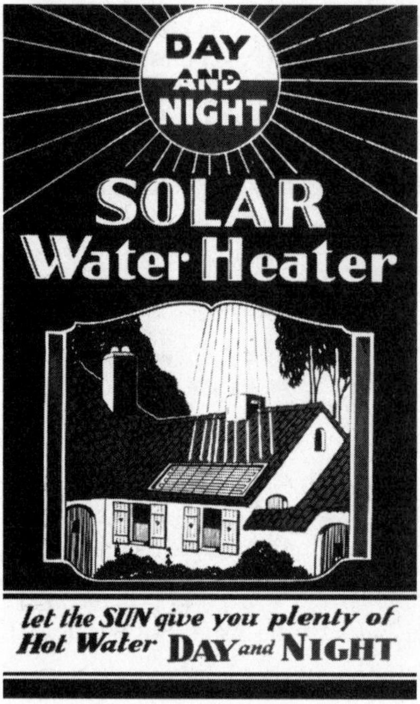

Figure 12.10. Day and Night sales brochure.
Figure 12.11. Advertisement for the Day and Night solar water heater in Arizona.

synonymous with Day and Night's name. Bailey's only remaining competition came from local plumbers and do-it-yourselfers. Because the basic design of the heater was not very complicated, "the man who was handy with tools and pipe wrenches could build his own," the trade magazine *Metal Worker, Plumber and Steam Fitter* commented in 1914.[12] One plumber in Ramona, California, sold collectors made of galvanized iron pipe coiled in a zigzag pattern on an absorber of black tar paper set inside a glass-covered box. A Santa Barbara plumber built a collector out of coiled pipe strapped to a copper absorber instead of welded to the plate as in the Day and Night model. People made storage tanks out of ordinary hot-water tanks, which they boxed in and insulated with any coarse, dry material, such as sawdust, ground cork, or rice hulls. Such heaters never had much impact on Day and Night's business, though.

During World War I, Bailey built his biggest unit ever, providing 500 gallons of hot water daily for soldiers at the Army Balloon School in Arcadia, California. Its success emboldened Bailey to write to the Department of War

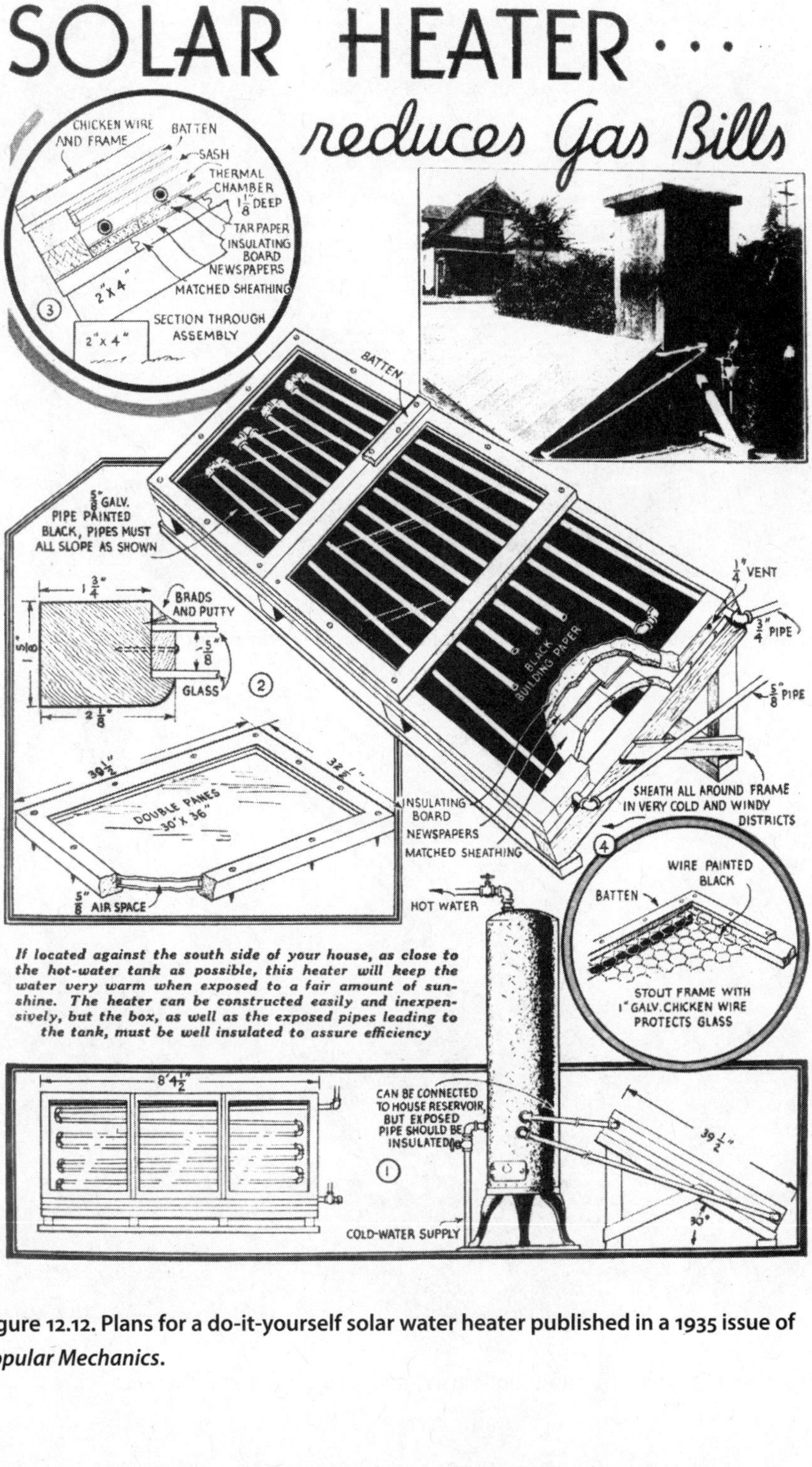
SOLAR HEATER . . .
reduces Gas Bills
CHICKEN WIRE AND FRAME
BATTEN
SASH
THERMAL CHAMBER 1 1/8" DEEP
TAR PAPER
INSULATING BOARD
NEWSPAPERS
2" X 4"
MATCHED SHEATHING
SECTION THROUGH ASSEMBLY
3
5/8" GALV. PIPE PAINTED BLACK, PIPES MUST ALL SLOPE AS SHOWN
1 3/4"
BRADS AND PUTTY
5/8"
GLASS
2 1/8"
2
39 1/2"
32 1/2"
DOUBLE PANES 30" X 36"
5/8" AIR SPACE
BLACK BUILDING PAPER
1/4" VENT
3/4" PIPE
5/8" PIPE
INSULATING BOARD
NEWSPAPERS
MATCHED SHEATHING
SHEATH ALL AROUND FRAME IN VERY COLD AND WINDY DISTRICTS
4
WIRE PAINTED BLACK
STOUT FRAME WITH 1" GALV. CHICKEN WIRE PROTECTS GLASS
HOT WATER
If located against the south side of your house, as close to the hot-water tank as possible, this heater will keep the water very warm when exposed to a fair amount of sunshine. The heater can be constructed easily and inexpensively, but the box, as well as the exposed pipes leading to the tank, must be well insulated to assure efficiency
8' 4 1/2"
CAN BE CONNECTED TO HOUSE RESERVOIR, BUT EXPOSED PIPE SHOULD BE INSULATED
1
39 1/2"
30°
COLD-WATER SUPPLY

Figure 12.12. Plans for a do-it-yourself solar water heater published in a 1935 issue of *Popular Mechanics*.

Figure 12.13. Five-hundred-gallon installation for the Army Balloon School.

Figure 12.14. Sixteen Day and Night installations in a single subdivision.

in Washington, DC, and say that the solar water heater was what every army camp in America needed. In the letter, he sent along test results and technical specifications to bolster his case. After a long wait, Bailey got a reply from a high-ranking officer. The board of investigators, the army man informed Bailey, concluded that heating water with the sun "was about as feasible as getting gold from gold fish."[13]

The discouraging news from Washington had no effect on business out west. By the end of World War I, over four thousand Day and Night heaters had been sold. In 1920 alone, over a thousand people bought Bailey's system.[14] But 1920 turned out to be the peak year for sales of solar heaters. Between 1920 and 1930, huge amounts of natural gas were discovered in the Los Angeles basin. Gas production soared and fuel prices plummeted. By 1927, consumers could get natural gas for about a fourth of what artificial gas cost in 1900. Networks of new pipelines brought cheap natural gas to urban and rural areas that formerly had no gas supplies.

Instead of trying to buck the trend toward gas, Bailey decided to capitalize on it. He began selling a Day and Night gas water heater that eliminated

Figure 12.15. Gas company ad promoting frequent baths to increase the use of natural gas.

the objectionable features of the old-fashioned sidearm. His gas heater used some of the techniques that had made the solar heater such a success. A copper heating element conducted gas heat to the water, just as the copper absorber transmitted solar heat to the liquid inside the solar-collector pipes. And the gas heater's storage tank was now insulated, as was the solar tank. Moreover, Bailey added a thermostat that automatically heated the water to the desired temperature. "No trouble, no fuss — simply turn the dial indicator," ran a Day and Night ad promoting its new gas heater. "All the hot water you need, heated quickly and kept hot in an insulated tank, constantly awaiting your needs."

The gas companies provided economic incentives to customers buying gas water heaters — one of many programs they initiated to encourage gas consumption. "They'd finance gas water heaters on a monthly basis or let you carry 'em for a year or two," a retired plumber related. "The gas company would do anything to get you buyin' from them."[15] In addition to easy terms, they offered cut-rate prices and free installation.

It was an unbeatable combination — cheap, accessible supplies of gas; the convenience of an automatic heater; and financial breaks that made purchasing a gas heater much easier on the pocketbook than paying for a solar heater. Solar water heaters were abandoned and new purchases of the Sun Coil dropped drastically. In 1926 Day and Night sold only 350 solar heaters; four years later, a meager 40 were installed. As Ned Arthur put it: "Whenever a gas main would run out into the country, our solar heater sales quit."[16] Bailey's old slogan of "Steaming hot water day and night" now meant water heated by gas, and his company became one of the largest producers of gas water heaters in the nation.

All in all, Bailey's company sold a total of about seven thousand heaters; it stopped manufacturing them at the beginning of World War II. The last

Figure 12.16. The Signal Hill oil field, a major oil and gas find in the Los Angeles Basin, helped put an end to the use of solar water heaters in the region.

production run was made in 1941, according to William J. Bailey Jr. The company would have sent solar water-heating technology more than halfway to Australia — but fate intervened:

> Pan American Airlines bought a big lot of them and had intended to ship them out to the South Pacific to put them on Canton Island. That was the time when Pan American flew the old Clipper Ship runs to Australia, and Canton Island was the stopover point. They wanted hot water there[,] and using solar was the only way they could get it. Those water heaters were on the dock in San Francisco, ready for shipment, when Pearl Harbor came along. They were never shipped.[17]

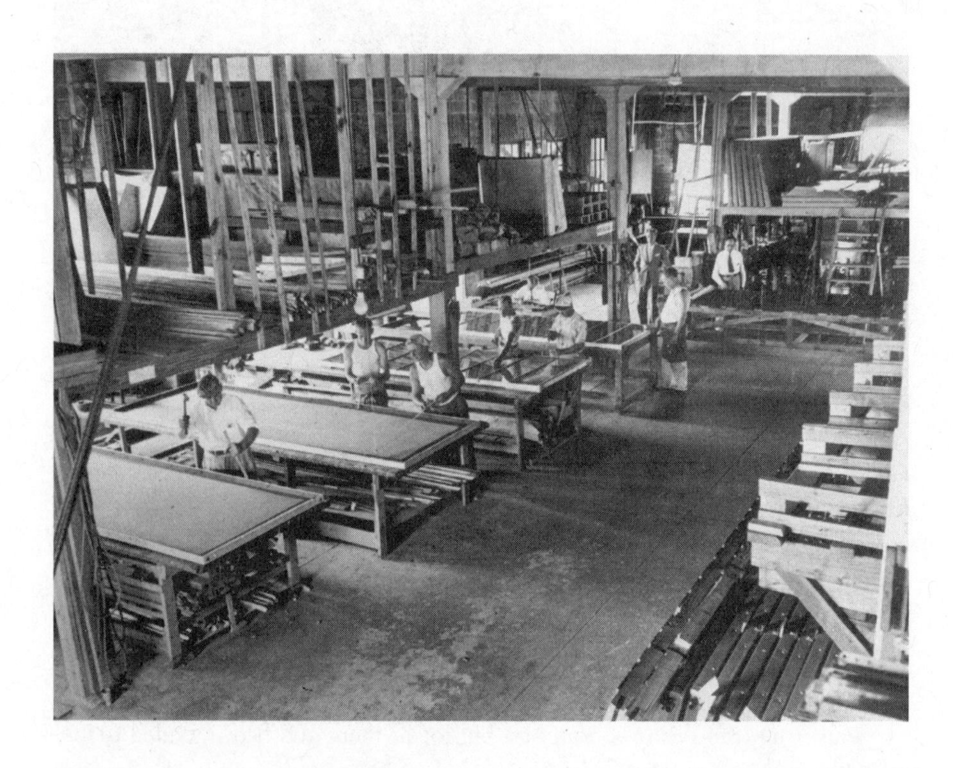

Figure 13.1. Workers assemble collectors at the Solar Water Heater Company in Miami.

13

A Flourishing Solar Industry

The 1920s may have marked the beginning of the end of solar water heating in California, but a new market was then opening up in southern Florida. Post–World War I Miami was a boomtown, with land speculators and real estate agents converting swamps into building sites for thousands of new hotels, apartments, and houses. The magnet of the sun and sea and the completion of a highway from New York to Florida brought vacationers and new residents to the area in droves. Miami's population almost tripled in five years — from twenty-nine thousand in 1920 to more than seventy-five thousand in 1925.

But newcomers to this tropical paradise found that there was no cheap way to obtain hot water. California's huge gas stocks did not ease the fuel shortage in Florida. Natural gas was unavailable there and artificial gas could not be supplied directly to homes and hotel rooms because it was difficult to lay pipelines through the subsurface marshland. Some people bought bottled artificial gas, which cost $4.60 for the equivalent heating capacity of 1,000 cubic feet of natural gas — about a third more expensive than artificial gas in California. The more common alternative was to use an electric water heater. The big stumbling block was the extremely high price of electricity in south Florida — $0.70 per kilowatt-hour, almost eight times its cost today, with inflation taken into account. Consequently, hot water was a luxury that many people did without.

H. M. "Bud" Carruthers, a wealthy builder who had come south from

Pennsylvania, knew of the great demand for an inexpensive way to heat water in the Miami area. On a trip to California he hit upon the answer — the Day and Night solar water heater. Carruthers purchased the Florida patent rights from Bailey for eight thousand dollars and an Oldsmobile touring car, and returned to Miami to set up the Solar Water Heater Company in 1923.[1]

Carruthers began manufacturing and selling a solar water heater exactly like the Sun Coil sold in California: a collector made of steel pipe soldered to a copper plate inside a glass-covered box, with a separate insulated storage tank. And the response? As Harold Heath, who later became the company's sales manager, put it: "Talk about a happy customer — they went nuts about this solar deal! People were living all this time taking cold baths or turning on electric [water] heaters. The cost of operating those things would eat 'em up alive — they used three thousand to four thousand watts, and the meter would just take off! In fact, some of our best customers were Florida Power and Light executives because they knew what it cost to heat water with electricity."[2] In contrast, a solar water heater saved so much money that it paid for itself in little more than two years.

People also liked the fact that solar heaters were automatic and appeared to be safe and reliable. In the words of one ad: "It can't stop working while the sun shines. It can't get out of fix, explode, or start a fire." By contrast, the electric heaters were dangerous. As Heath explained: "If you wanted a bath you'd turn the switch to high and wait fifteen or twenty minutes. Then you could take a bath. Nothing automatic about it. And if you got in your car and drove halfway to Palm Beach and somebody said, 'Did you turn the hot water heater off?' and you said, 'I don't know whether I did or not,' you'd turn around and head back home, because the damn thing would blow up." Consequently, "solar was real natural in Florida," said Heath.[3]

Carruthers did a tremendous business installing heaters on new houses springing up along the coast, and real estate agents and builders were eager to let prospective buyers know that their homes were equipped with solar heaters. L. J. Ursem, a local real estate agent, ran the following ad: "Don't fail to see this wonderful bungalow... [plus] garage with laundry and wash trap, solar water heater, Frigidaire, other fine features."[4] Carruthers's company also put heaters on many hotels and apartment buildings, using 14-by-4-foot bank of collectors connected to a 300-gallon storage tank. Before long "there seemed to be acres of solar collectors on apartment house roofs," commented William D. Munroe, a resident of Miami at the time.[5]

Figure 13.2. Exterior of the Solar Water Heater Company in Miami. Company trucks are carrying large bags of cork insulation to insure that the hot water in the storage tank remained warm.

Figure 13.3. Solar collectors atop a typical Miami bungalow, around 1924.

Because the Solar Water Heater Company owned the patent rights to the Sun Coil in Florida, there was little competition. Precise sales figures are not known, but Munroe observed that "the manufacture of solar water heaters was an established industry in southern Florida by the mid-twenties."[6] Carruthers built a factory occupying an entire city block, and in April 1925 the *Miami Herald* listed his company as one of the seven largest construction firms in Miami.[7]

Figure 13.4. Interior view of the new factory for the Solar Water Heater Company.

But the solar-heating industry was not destined to keep on climbing. In an ominous foreshadowing of what would soon happen to the country's economy as a whole, Miami's great building boom leveled off by early 1926 and completely collapsed that summer. Then a ravaging hurricane swept through the city in September. "Things were all broke to hell," said Heath.[8] Carruthers's business prospects were bleak and forced him to close up shop. He went back to Pennsylvania, entrusting his financial affairs in Miami to his associate, Charles F. Ewald.

The Duplex

Some five years later, around 1931, Ewald decided to try to crank up the solar water heater business again. Since the Depression had kept the construction industry in a slump, he figured he would gear sales to owners of existing homes. But first he wanted to take a closer look at the Sun Coil to see if its performance could be improved.

Ewald began experimenting with new designs. One problem he tackled had to do with the effects of Miami's humid weather. The wooden collector boxes and storage tank enclosures deteriorated quickly unless they were frequently repainted. To eliminate this expenditure of time and labor Ewald turned to all-metal construction, building the collector box out of galvanized sheet steel and the storage tank enclosure out of galvanized iron.

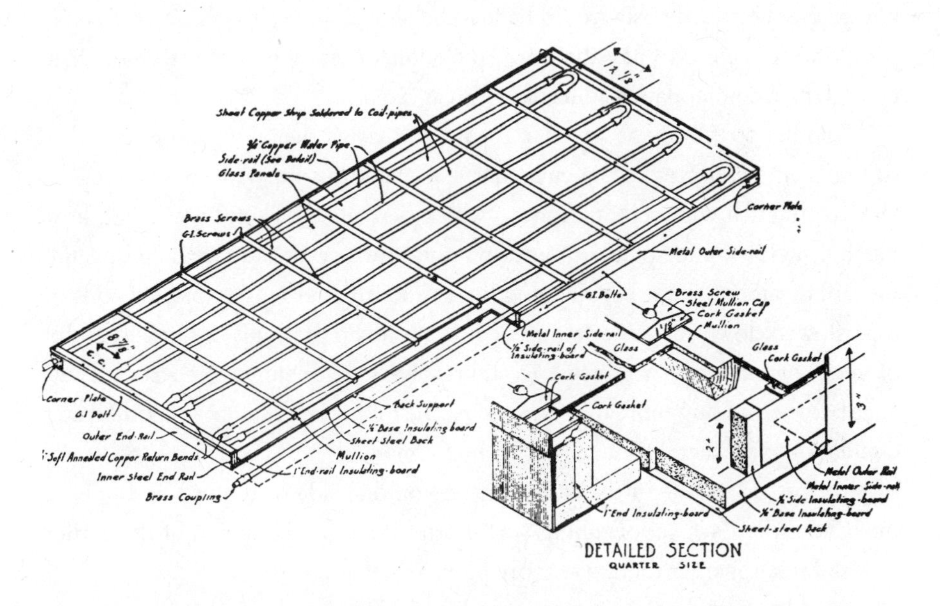

Figure 13.5. Construction details of Charles Ewald's all-metal solar collector, which used soft copper pipes to prevent ruptures caused by freezing.

Better heat collection was another goal, so he insulated the sides and bottom of the collector box. He also replaced the steel tubing with soft copper — a better conductor of heat and more resistant to ruptures from freezing. Bailey had found that hard copper coils would split in very cold weather, and Ewald reported that soft copper held up under test conditions as low as -10°F. The use of soft copper made it possible for heaters to be sold in places subject to occasional frosts without the special nonfreezing adaptation developed by Bailey. In 1958 an unusually severe cold wave in Miami verified the durability of soft copper — the coils suffered no damage, while nearly all the hard copper coils in other collectors burst.

Ewald also investigated ways to increase the efficiency of solar heat collection by rearranging the pipes inside the collector. After building several test models with different lengths of tubing and checking the water temperature in the storage tank, he concluded that 1½ feet of pipe should be used for every gallon of 140°F-water desired. He also discovered that putting in a lot more pipe in the same zigzag configuration would not be practical for a household-size collector. A bigger collector would require a larger copper absorber plate and glass cover, and the added costs would have to be passed on to the customer. Even if people were willing to pay the price, the roofs of many homes were not large enough to accommodate a collector of great size.

Another possibility occurred to Ewald — spacing the tubing closer together so that it fit a collector of the customary size. But this introduced a new problem — sluggish water circulation — which he realized when he watched how much time it took for pieces of cork floating in the water to circulate through the coil of tubing. The sharp-angled twists and turns of closely packed coils of pipe slowed down the water's rate of flow so much that only a small quantity of very hot water was produced. Ewald resolved the dilemma. He struck a balance between adding more tubing, spacing the tubing adequately for good water circulation, and keeping the collector box a manageable size. He put two coils of pipe instead of one in the collector, one on one side of the box and one on the other side. Each set of tubing was shorter than the single coil of the earlier model, but altogether there was more pipe.

Ewald patented his new design as the Duplex. Because it produced hotter water in greater volume, the Duplex became the company's main product. The Sun Coil continued to be manufactured as well, but a Duplex collector, the company claimed, would provide 20 percent more hot water at temperatures 20 to 30 degrees higher than did a similarly sized Sun Coil.

Another Ewald improvement was granulated cork used as insulation between the hot-water tank and its metal shell. "Because it was a waste product[,] there was no problem in getting it," according to Heath. "The cork was good insulation — it could retain the heat for 72 hours."[9] As an auxiliary, the electric water heaters that residents already owned were piped to the solar storage tank.

When installing a heater, the company found it important to determine the hot water needs of each household and to match the required tank size to the

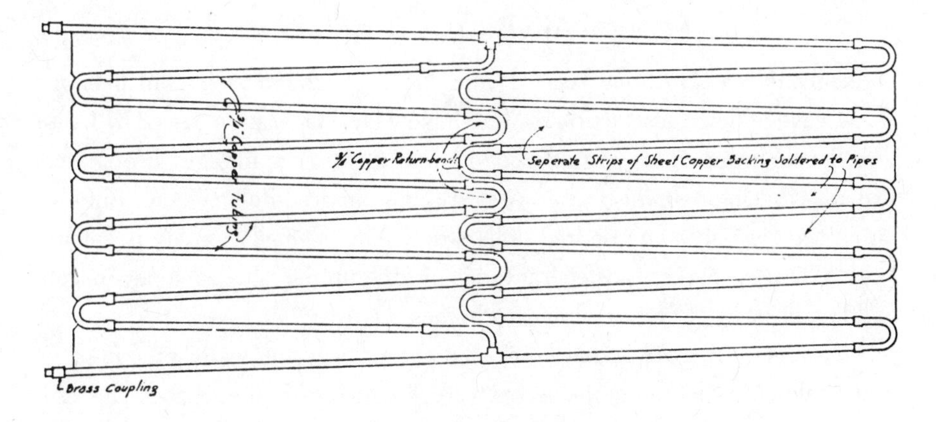

Figure 13.6. Piping pattern for the Duplex solar water heater.

Figure 13.7. Harold Heath, sales manager for the Solar Water Heater Company.

collector size. Heath explained that a collector sized too small for the tank produced only lukewarm water; and if too large, the water would get excessively hot, which was uneconomical. He described how a typical retrofit job was assessed:

> The first thing we would do was case the house, find out how many people lived there, and then we'd button them down. "Now don't mislead us," we'd tell 'em. "We're selling you a hot water service, not just a water heater, so we have to know the amount of hot water your family needs. How many in your family? How often do you do laundry?" [And so on.] Then we'd go from there to recommend the size of the system they needed.[10]

Homeowners were satisfied with such tailored installation service and with the performance of the Duplex or Sun Coil. Between 1932 and 1934 a growing number of retrofit installations revived the solar-heating business.

Federal Aid Boosts the Market

Suddenly, in 1935, the Solar Water Heater Company found itself amid an enormous wave of new construction in Miami. New Deal legislation passed by Congress late in the previous year offered low-interest home mortgages through the Federal Housing Administration. A thirty-five-hundred-dollar home could be purchased for only two hundred dollars down and a small monthly payment. With a hungry market and federal funds picking up the tab, developers began building projects of fifty or a hundred homes at a time.

However, the spate of new homes did not mean an automatic jump in solar heater sales. Most homes of the period were designed with sloping roofs, instead of the flat roofs of the 1920s that had made installation of the storage tank easy — it was simply bolted to the roof. Now a builder had to construct a platform in the attic on which to set the tank, and cut a hole in the roof through which it

Figure 13.8. Ground-mounted Duplex installation.

Figure 13.9. Workman installing solar water heaters in the roof of the laundry room in a Florida subdivision built during the late 1930s. Every home in this tract had its own solar water heater.

could protrude. Moreover, a solar heater cost about twice the price of an electric heater. Nevertheless, consumer pressure soon persuaded developers to go with solar. A strong advertising campaign and word-of-mouth publicity made customers aware of the fact that with solar water heaters they would save a substantial amount on their utility bills. Economy-minded homebuyers started confronting developers over the issue, as Heath related: "The buyers would ask, 'Where's the solar water heater?' If the home didn't have one, they'd say, 'Oh well, we're not interested.' It got to the point where developers couldn't sell a dang house unless it had a solar heater."[11]

Developers began to work together with the Solar Water Heater Company on installations in houses under construction. Usually the solar crew decided first where the tank should be placed. The building crew then built the platform and casing for the tank, and plumbers ran the necessary pipes to the nearest location in the attic. The solar installers came in with the storage tank, setting it in its enclosure. Next they generally installed the collector on the roof, but sometimes put it up as an awning over a south-facing window — which served the dual purpose of providing hot water and helping to shade the house in summer. Then they connected the tank and collector to the plumbing lines, and the building crew finished off the job by disguising the tank as a chimney.

Federal programs stimulated not only the new-housing market for solar heaters but the retrofit market as well. With an FHA home improvement loan a homeowner could buy a solar heater at 4 percent interest in installments of only six dollars a month, with no money down. With monthly payments lower than normal utility bills for an electric water heater, people started saving money right after buying a solar unit.[12]

The Solar Water Heater Company flourished, employing thirty-four workers to keep production rolling — three secretaries, five salespeople, eight installing crews of two persons each, and ten in the plant to make coils, collector boxes, and fittings. "We have no cause to complain about business," Ewald told the *Miami Herald* on August 5, 1935. "Only this week we have received from Pittsburgh our second carload of glass since March[,] and another large shipment of copper tubing is on the way, which is the third we have ordered this year."[13] Heath summed it up a little more colloquially: "We were selling heaters like bananas."[14]

Competitors Crowd the Solar Field

Until 1935 Ewald's company was the only major solar manufacturer in the Miami area. But as the demand for solar heaters began to multiply, so did the competition. Plumbers and roofers branched out into solar; companies in allied industries, such as U.S. Foundry — which worked with copper — also entered the field. Other firms, such as the Pan American Solar Heater Company and Beutel's Solar Heater Company, concentrated almost entirely on the solar heater business. In all, there were about ten sizable solar companies active in Florida by the late 1930s.

The Duplex patent still belonged exclusively to Ewald's Solar Water Heater Company, so it could not be copied by the new competitors. But this was not the case with the original Sun Coil, whose patent had expired by this time. Nearly everyone produced a facsimile, and rival companies promoted their products by giving them catchy brand names and emphasizing minor improvements in design. The Pan American Solar Heater Company advertised that its Hot Spot collector had flattened tubing rather than round pipe. They claimed this prevented shadows from being cast on the absorber plate.

Inevitably, some companies sought to reduce costs by cutting corners. A few soldered the copper plate to the coil of pipe only in spots rather than along its entire length, which resulted in poor heat conduction. A more common tactic was to eliminate the metal supports underneath the collector boxes. They laid the collector directly on the roof. But since the box was not watertight, rain seeped inside and leaked out the bottom. With no air circulation to permit evaporation of the water nor space for it to run off, water accumulated, rotted the roof and caused the collector box to rust. Other companies tried to capture the market by cutting down the size of the collectors. Unwary customers found out too late that these undersized collectors did not produce enough hot water.

People started to complain; they wrote to Washington, DC, since the FHA had financed most of the solar heaters. The government sent an investigator to Miami to check the veracity of their allegations. This induced the solar companies to form a voluntary association that adopted manufacturing standards — including proper guidelines for collector and tank sizes.

Such swift action from Washington and the solar industry restored public confidence in solar water heating, and sales climbed again. Estimates of the total number of installations made in the Miami area between 1935 and 1941 vary widely — from twenty-five thousand to sixty thousand. More than half the Miami population used solar-heated water by 1941, and 80 percent of the new homes built in Miami between 1937 and 1941 were solar equipped.[15] More than

Figure 13.10. Advertisement for the Solar Water Heater Company in 1941 reveals the number of cumulative customers that year.

five thousand solar heaters were installed on large structures, such as apartment houses, hotels, schools, hospitals, and factories, with more than half of them using 2,500-gallon storage tanks. The federal government purchased some of the largest solar-heating systems, putting them in the officers' quarters at the giant naval air station in Opa-Loka, outside Miami, as well as in the Edison and Dixie Court housing projects, which had a combined population of 530. In 1941, solar water heaters outsold conventional units in Miami by two to one.

Solar water heating also reached beyond the boundaries of Florida. In the late 1930s it spread to the Virgin Islands, where it was used in hospitals, and to Puerto Rico, Cuba, and Central America. To the north of Florida, public housing projects from Louisiana to Georgia began using solar water heaters. At one large project in Georgia, for example, solar heaters supplied 480 dwellings with

Figure 13.11. The solar water heater in Cuba. The sales brochure proclaims in Spanish, "Without electricity, without gas, without coal, without cost!"

a total of 35,000 gallons of hot water per day. After the first year of operation, one housing official reported that "solar water heating has proven to be most satisfactory, and while the initial cost is slightly more than for other types of hot water heating, the operation and maintenance costs are insignificant."[16]

The Postwar Solar Decline

World War II halted the burgeoning solar heater industry. There was a government freeze on the nonmilitary use of copper, one of the main components of solar water-heating systems. But many of Miami's companies returned to business after the war. Despite the comeback, however, Florida's solar business never regained its earlier momentum. Hot-water consumption began to leapfrog with the postwar baby boom and the advent of a newly affluent society that could afford dishwashers, washing machines, and similar home appliances. Families found that solar collectors installed on the basis of prewar calculations of hot water usage were now too small. In an attempt to meet this increased demand, several solar companies introduced a thermostatically controlled electrical resistance heater as an auxiliary.

Insufficient hot water was not the only reason for the growing dissatisfaction with solar. Water tanks began to burst during the late 1940s and early 1950s, about ten years after they had been installed. According to Heath, "The tank would generally go around two o'clock in the morning when the city water pressure was up."[17] Residents would awaken to rooms beneath the tank flooded with water. This badly tarnished solar energy's reputation of being a trouble-free way to heat water. The whole catastrophe could have been avoided if there had been better communication between the solar industry and the engineering community. As early as 1935 an article in an engineering journal explained that, to prevent corrosion, all parts of the solar system should be made out of the same kind of metal.[18] Unfortunately, most of the tanks were made of iron, and the pipes in the collectors were made of copper. The high-temperature water promoted an electrochemical reaction between the two metals, which gradually ate away the tank until it ruptured.

The sharp rise in the price of solar heaters also alienated consumers. From 1938 to 1948 the price of copper doubled, and by 1958 it had tripled. Labor costs also skyrocketed. An unskilled solar-industry worker who had earned

$0.25–$0.40 per hour in 1938 made about $1.10 by 1955. The localized nature of the industry prevented large investments in labor-saving machinery, and it also took many man-hours to do the on-site installation work. Such high labor and material costs made the price of solar water-heating systems shoot up from $125 in the 1930s to $350 in 1948 and $550 in 1958.

Meanwhile, electric water heaters became an economical and convenient alternative. Electric rates fell dramatically after the war, dropping to $0.40 a kilowatt-hour in 1948 and $0.30 in 1955. Many servicemen based in Florida during the war stayed on, and the subsequent growth in Miami's population allowed the power company to charge less per customer because its high capital investment in equipment could now be spread over a larger number of users. Florida Power and Light also pursued an aggressive campaign to increase electrical consumption, structuring rates to promote greater usage and offering free installation of electric water heaters. Demonstration heaters were displayed at its offices, and the businesses of the builders installing these and other electric appliances received promotional assistance from the utility.

The low initial cost of electric heaters also attracted customers. Large-scale production techniques had kept the price low — in the early 1950s an electric water heater cost only forty dollars more than in 1938. Because the payback time on a solar heater system had ballooned to eight years by 1955, many found it cheaper to buy an electric water heater instead. And those who had the unfortunate experience of a ruptured tank were reluctant to pay more money for a new tank that might need to be replaced in another ten years. They preferred to change over to electric heaters, especially since the models now being sold were automatic.

With solar energy no longer the bargain it had once been, and with electric water heating looking better and better, few people bought solar water heaters after the late 1950s. The industry became a service business — flushing out coils, fixing broken glass collector covers, and putting in new tanks for those who preferred to keep their solar water heaters because they were happy with what they were getting — free hot water. To this day one can find on the rooftops in Miami neighborhoods dating from the 1920s to the late 1930s hundreds of old installations, ranging from seventy to ninety years old, still in operation.

Figure 13.12. Day and Night–style solar water heaters installed atop a Florida apartment house in 1923 and still in operation today.

Figure 14.1. An advertisement for vinyl solar waters popular in Japan during the 1960s urged the Japanese to "shower with solar-heated water."

14

Solar Water Heating Worldwide, Part 1

Not all areas of the world enjoyed the cheap, abundant energy supplies that allowed Europe and America to forget about solar energy in the years after World War II. Many countries did not have easy access to the river of oil issuing from the Persian Gulf states and other oil-rich nations. Solar water-heating industries were established in some of these fuel-short regions — especially in areas with ample sunshine. Business flourished for manufacturers in Israel and Japan during the 1950s and 1960s. Solar water heaters were also successful in parts of South Africa and Australia.

An electricity shortage served as the impetus for the establishment of a solar water-heater industry in Israel. One cold winter day in 1951, the people of Israel got an unpleasant surprise. A newspaper headline announced, "Country Wide Electricity Curb Ordered." From the article people learned that no one could use electricity for cooking, baking, or heating between four thirty in the afternoon and eight in the evening. Nor could they run hot water from ten at night until six in the morning or install new electric water-heating appliances. Inspectors would be knocking on doors to enforce compliance.[1]

Despite such mandatory rationing, power shortages worsened over the spring. Water-pumping stations failed, leaving many without drinking water and threatening farmers' livelihoods. Officials contemplated closing down factories. A special Inter-Ministerial Committee set up by the Israeli cabinet to study ways of resolving the electricity crisis could only suggest continued rationing and the purchase of more generators to overcome the problem.

Frustrated by what he considered a lack of foresight by the committee, an Israeli engineer named Levi Yissar wrote a letter on June 20, 1951, to *Haaretz*, Israel's equivalent to the *New York Times*, which was published under the title "The Electricity Crisis and an Alternative Solution." Yissar complained that "no one has recommended an already existing energy source which our country has plenty of — the sun," adding, "surely we need to change from electrical energy to solar energy, at least to heat our water."[2]

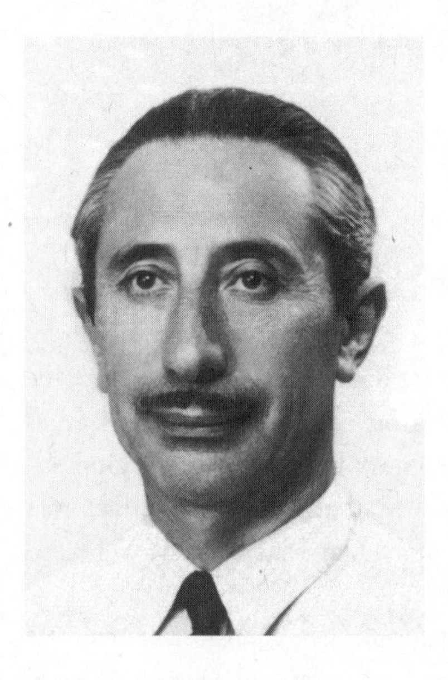

Figure 14.2. Levi Yissar, the civil engineer who introduced the solar water heater to Israel.

Yissar had first learned of solar energy's capability to heat water from his wife, Rina. Rina had needed hot water to bathe their baby boy, Gonen. The year was 1940 — a time when extreme scarcities of fuel oil plagued the area, which was then a part of the British Mandate of Palestine. Most people were forced to accept cold baths as yet another sacrifice to help found the Jewish state. But Rina, whom Gonen would later describe as a woman who "lacked formal technical education but had excellent common sense," refused to resign herself to this hardship.[3] Instead, like many rural folk in the United States years before, she took an old tank, painted it black, filled it with water, and left it out in the sun. After a few hours, she had enough hot water to give her baby a warm bath.

This simple demonstration of solar water heating fascinated Levi, a civil engineer. At Rina's urging, he began to look into the matter, but World War II and then the bitter battles of 1948 for Israeli independence forced him to put off his research. When a shaky peace finally arrived, Levi Yissar resolved to devote all of his efforts to harnessing the sun. He scanned the technical literature and found several professional accounts of solar water heating in California and Florida, as well as a study of solar collectors published by Hoyt Hottel, a professor of chemical engineering at MIT, and Byron Woertz, a graduate student at MIT, in 1941. He also attended an international conference on solar energy held at the Massachusetts Institute of Technology in 1950.[4]

Returning home, Yissar combined what he had learned about solar heat collection and storage with several of his own innovations and built a prototype solar water heater. Yissar did not pretend that he had invented the solar water heater, stating in an Israeli newspaper profiling him and his invention: "I don't want to give the impression that I have created something new. No way. This system has been used in other countries successfully, but no one in Israel has paid much attention to it."[5] Yissar's contribution was adjusting the solar water heater to Israel's special circumstances and climate. This meant a collector and storage tank resembling those of the Day and Night, which would provide all of one's hot-water needs year-round. To increase efficiency in summer, Yissar

Figure 14.3. Early Israeli solar water-heating installation at a kibbutz (communal farm).

inserted a dehumidifier in the collector box to reduce the moisture accumulation that would occur on the humid Israeli coastal plain, where most of the people lived. To supply all the hot-water needs of a family in winter, the tank was well insulated, and it was sized to hold enough hot water to last forty-eight hours, since during this season rain clouds rarely covered the sky for more than two days in a row. The collector could fill the tank with enough hot water for household tasks with only two hours of sunshine, an operating criterion necessitated by the fact that between December and March there is often cloudy weather.

Yissar's first efforts at promoting solar heating encountered stiff resistance. Despite his claim that solar energy could help conserve precious fuel reserves (all fuel in Israel was then imported), his colleagues remained skeptical. "Everyone laughed at me," he recalled. "No one believed that the sun could produce water hot enough for general household use."[6] But Yissar ignored his critics and soon built his first heater, which was installed at the home of a family living near Tel Aviv. They loved it. The family became one of the few in the Jewish state who had hot water day and night. "Was the family happy with the sun heater?" queried a newspaperman. "How can you ask such a question?" the lady of the house replied. "Look, my neighbor, who uses oil, can't get any. And those who

התחיל שיווק מכשיר
חימום מים בקרני השמש

**בימים אלה החלה חברת „נריה" בייצור והפ־
ת „מחמם-מים שמשי". מכשיר חדש זה עשוי ל־
פק מים חמים בכל שעות היממה — על ידי ניצול
רני השמש בלבד, לחימום וללא צורך בחשמל, או
דלק.**

הוצאה היחידה היא —
קעת היסוד" לרכישת המכ־
והדוד המחובר אליו. כ־
ר 220 לירות ישראליות ר־
פת צינורות מן הגג לדירה.
בהתחשב בעובדה, כי ההו־
לחימום מים בזרם חשמלי
זכמת אצל משפחות רבות
רות אחדות לחודש, יש לה־
כי המכשיר החדש יעורר
נינות.
פתיעה צורתו החיצונית ה־
ווסה" של המכשיר, כפי ש־
נן והותאם לתנאי הארץ על
המהנדס מר לוי ישר. ה־
נין באפשרויות גדולות ב־
ל האנרגיה של השמש לטו־
משק הארץ ומתמסר לענין
בכל לבו ונפשו. הדוד דומה
ר החשמלי הרגיל, אלא ש־
ולו 200 ליטר. לעומת 120

ל"י נוספות) אפשר, איפוא, ל־
הבטיח 100 ליטר מים חמים ל־
יום, גם בימות החורף.

מים חמים גם בחורף

שאלה רצינית היא, כמובן,
הבטחת מים חמים, בימי גשמים
ממושכים. ההתעננות נמשכת ל־
פעמים בארצנו שבוע, או שבו־
עיים ואין ספק, כי אי־אפשר
„לשמור" את החום במשך תקו־
פה כה ממושכת, כאשר השמש
„שובתת". ולעומת זאת, דווקא
בימים כאלה מעונין בעל המכ־
שיר במים חמים יותר מבשאר
ימות השנה. תשובת החברה ל־
שאלה זו היא, כי תספיק הופע־
תה של השמש לשעה־שעתיים
בלבד ושוב יגיעו המים לדר־
גת החום הרגילה.

Figure 14.4. Yissar standing by one of his first solar water heaters. The headline announced, "The Commercialization of Solar Water Heating Begins."

rely on electricity are not allowed to use it during the day and then don't have hot water left for the next morning. Besides, we have no bill to pay at the end of the month."[7] Soon twenty-five more solar water heaters were set up. Customers were impressed — the heater was built to last fifteen years and would pay for itself in only two. A popular Israeli newspaper predicted, "There will be much interest in this new approach."[8] Yissar obtained sufficient capital to establish the Ner-Yah [God's Eternal Flame] Company, Israel's first manufacturer of solar water heaters. One of his first customers was David Ben-Gurion, the founding father of Israel, who had a solar heater installed in his home.[9] Within a year Yissar's company had sold sixteen hundred.

Miromit Enters the Field

Eyeing Yissar's success, several formerly doubting colleagues decided to establish their own solar companies. Yissar's son, Gonen, claimed that "some of them got into the market with their own 'inventions' by violating my father's rights on his patents."[10] As in Florida, many of these new companies built substandard equipment in their scramble to gain a competitive advantage. This trend worried the Israeli government, because many of the heaters were being installed in government buildings — mostly housing projects for newly arrived immigrants. The government finally intervened and worked out a plan giving Miromit, Israel's largest metal fabricator, exclusive rights to Yissar's patent on the condition that it maintain high production standards. The government also awarded Miromit a license to use a special coating called a "selective surface" on the absorber plate. Invented by another Israeli, Harry Tabor, the Jewish state's leading solar advocate and scientist for many decades, this coating inhibited thermal radiation from the absorber and, as a result, cut heat losses from the collector by 30 percent.

Miromit also introduced several improvements of its own. The serpentine coil arrangement of the collector tubing was replaced by parallel piping. This change ensured better heat transfer from the absorber plate to the water and made it easier — and consequently cheaper — to flush out the pipes when they became clogged with mineral deposits. Israel's hard water was also the primary reason for not soldering the collector pipes to the absorber plate. Instead, the tubing fit tightly into grooves in the absorber. These pipes could be readily scrapped when they became too badly blocked by minerals, while the absorber plate could be reused. Miromit also made a collector box with a tighter seal to keep out the rain and dust.

Miromit's high-quality product helped upgrade the heaters marketed by other Israeli companies. According to Harry Tabor, "Keeping up with Miromit" became the song that many others in the industry sang. In fact, their competitors conformed so slavishly to every new Miromit development that when the color of the collector boxes was changed for purely aesthetic reasons, all the other companies followed suit. "Presumably," mused Tabor, "they assumed that there was some scientific reason for the change!"[11]

Figure 14.5. Rainier Sobotka, managing director of Miromit, with one of the company's high-quality solar water heaters.

Nearly fifty thousand solar water heaters were sold in Israel from 1957 to 1967. Miromit also ran a booming export business. According to the managing director, Rainier Sobotka, the company shipped heaters to "the Canary Islands, Reunion, and Honduras, to Trinidad and the Saint Vincent Islands in the Caribbean, to Manila and Singapore, to Iran, Turkey and Chile — all in all, to about 60 different countries" by 1961.[12]

Ironically, victory in the Six-Day War of June 1967 had a devastating effect on Israel's solar water-heater industry. The nation captured large oil fields on the Sinai peninsula — ending decades of fuel scarcity. When this oil began to flow toward Jerusalem, people stopped buying solar water heaters. Once again the siren call of fossil fuels led people away from the sun.

The Australian Government Lends a Hand

The government in Australia played an active role in getting solar water heating started there. Roger Morse, an engineer with the Commonwealth Scientific and Industrial Research Organisation, began to develop a government program in 1952. He and his colleagues developed a solar collector very similar to the model used at MIT, basing their design on the seminal study published by Hottel and Woertz in 1941. They also encouraged private industry to enter the field, and a number of firms soon began to manufacture and sell solar water heaters. Government advisers visited the factories and maintained a close rapport with the fledgling companies. But many of their early systems functioned poorly or did not work at all, because plumbers had difficulty installing these solar water heaters. To rectify this problem, Morse wrote an installation manual that became the industry's standard guide.[13]

Figure 14.6. Roger Morse, who introduced solar water heating to Australia.

Aside from providing technical advice, the government also developed a market for solar water heaters. In the midfifties, a government study recommended that all state buildings in tropical areas be equipped with solar water heaters.[14] These areas had the highest electric rates in Australia and usually enjoyed the most sunshine. Morse called this government attitude "the catalyst which gave encouragement to the industry."[15] The resulting demand helped two major solar firms to prosper — Solahart and Beasley Industries.

Solahart chose a completely new integral collector-tank configuration to facilitate easy installation on pitched roofs. The storage tank was mounted along the upper edge of the collector. Not only did this new arrangement eliminate the need for extensive piping, but it also removed the necessity of installing a heavy storage tank in the attic. Soon other firms adopted this modular system. Solahart, Beasley, and other firms slowly built up their market — primarily through word-of-mouth promotion by satisfied customers. Between 1958 and 1973, about forty thousand solar collectors were sold throughout Australia and its adjacent islands.

Figure 14.7. An early Solahart water heater installed in the Cocos Islands.

Problems in South Africa

Solar water heating met a different fate in South Africa. Lewis Rome, an English-trained engineer living in Johannesburg, founded the Economic Solar Water Heater Company in 1954. Rome had learned about solar heating from Austin Whillier, a fellow South African who had studied under Hoyt Hottel at MIT, and so it is not surprising that his collector resembled the MIT model.

Most of Rome's customers lived in rural areas, primarily in the northern Cape region and in Southwest Africa — then a mandate of South Africa. The high cost of shipping fuels to such remote areas made solar water heating economically attractive there. On the other hand, coal prices and electric rates in the country's urban centers were then among the lowest in the world. Even here, however, the payback time on a solar water heater was only seven years. Apparently that seemed too long a time for many city residents, for Rome's sales were poor in South Africa's cities.

Figure 14.8. Advertisement in Afrikaans for solar water heating in South Africa.

Some progressive citizens felt that cost should not hinder South Africa's full-scale use of solar water heaters. They urged the government to push solar energy to help alleviate pollution caused by coal burning, the primary fuel for generating electricity in South Africa. "For every solar heater installed where coal had been previously used," declared one engineer, "smog must become proportionately less."[16] But the government did not act. On the contrary, a bureaucratic measure killed solar energy's economic appeal even in rural areas. In 1961 the government doubled the rail rates for all manufactured goods and reclassified solar heaters so that they were subject to the highest transportation tariffs. The cost of shipping them from Johannesburg to outlying areas now outran production costs — boosting their list price beyond the reach of most people. The loss of this market forced Rome out of business that year.[17]

The Japanese Industry

Like the Romans, the Japanese have always loved their hot baths — especially farmers who came home covered with mud after spending long hours in the rice fields during the hot, muggy summers. They usually bathed in large wooden tubs heated by fire underneath; a sheet of metal beneath the tub protected it from the flames. Wood was abundant in the mountains, but in the low-lying districts people had only rice straw for fuel. During the Depression years of the

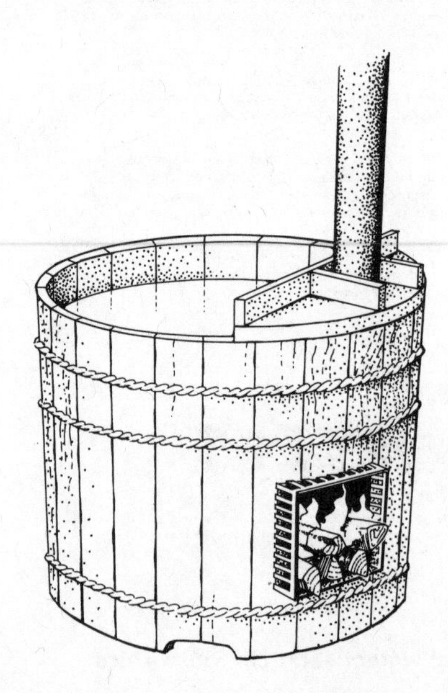

Figure 14.9. A wooden Japanese hot tub, which required burning scarce logs to heat it.

1930s, some farmers in fuel-short regions began to use the sun for heating their bathwater. However, none of these early attempts spread beyond the local level, despite Japan's ranking among the industrialized nations as "one of the most favored places for utilizing solar energy," according to a report presented to the United Nations.[18]

On a trip to the countryside during the 1940s, Sukeo Yamamoto saw one of these primitive heaters — a large bathtub filled with water whose top was covered by a sheet of glass. When set out in the sun early in the morning, it produced water hot enough for bathing by about two in the afternoon. Yamamoto was taken with the simplicity of this device, and two years after the war's end he designed Japan's first commercial solar water heater. It consisted of a rectangular wooden basin measuring 6 feet long, 3 feet wide and 6 inches deep. Glass covered the top, and a thin sheet of blackened metal lined the interior. The heater was usually mounted horizontally near the bathtub. In the morning a faucet was opened to fill the basin with water. At night the sun-heated water was emptied into the bathtub by another faucet. From late spring through the early fall, the 53-gallon heater could provide bathwater up to 140°F by late afternoon — the customary bath time. From November to March, when the low-lying

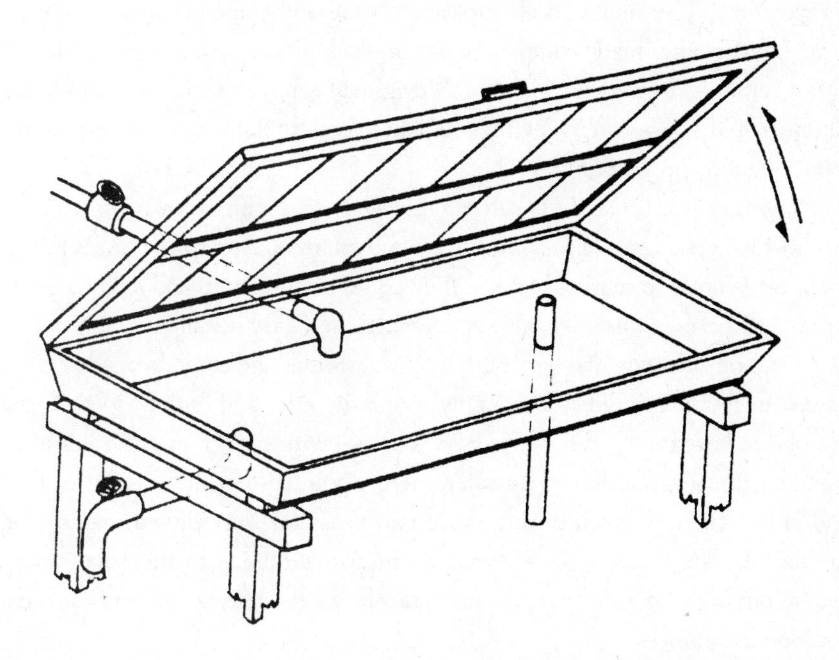

Figure 14.10. Japan's first solar water heater, invented by Sukeo Yamamoto in 1947.

sun cast its oblique rays on the heater, the water temperature reached only 70 or 80°F. Auxiliary heat was then used to raise the water another 30°F or so for a good steaming bath.[19]

The Kaneko-Kogyosho Company began mass-producing Yamamoto's heater in 1948. Farmers in the valley rice regions found it ideal — it was simple to operate, worked fairly well, and annually saved each family about 1½ tons of rice straw, which could then be used as cattle fodder or fertilizer. The heater cost only twenty dollars, and farmer's associations and government authorities helped the farmers finance the heaters with direct subsidies and low-interest loans. The solar heater quickly caught on and became popular in farming villages throughout the country. By 1955, over twenty thousand had been installed in Japan.[20]

Yamamoto's invention did have its drawbacks. Many found that the loose-fitting glass cover allowed too much dust and dirt to contaminate the water in the basin, and that algae often grew in the stagnant warm water. Competitors anxious to flourish in this growing market offered a completely sealed water heater that avoided these problems. Made of soft vinyl plastic, this heater was the same size as Yamamoto's but resembled an inflatable air mattress. Two models were available: one with clear plastic on top and black plastic on the bottom, and the other made completely out of black plastic. Both styles produced hot water about as well as Yamamoto's original heater. In winter a clear plastic canopy could also be put over either version to better retain the solar heat. Some left the canopy up year-round

The very low price of the soft vinyl heater — ranging from six to ten dollars — made it accessible to almost everybody. Urban department stores found them convenient to market because they could be folded up and sold in small cardboard boxes. Consumers also appreciated their easy installation. The vinyl heater sat on a flat wooden base built by the customer and could be readily connected to the household water supply by plastic inlet and outlet hoses. True, the plastic lasted only two years or so. But as Professor Ichimatsu Tanashita, a leading Japanese authority on solar energy, pointed out, the amount of fuel and labor that was saved during those two years far outweighed the cost of the heater. The Japanese public took an immediate liking to this solar water heater, buying 20,000 in 1958, the first year they were sold, 70,000 in 1959, and 170,000 in 1960.[21]

More and more companies were eager to take advantage of the growing

Figure 14.11. The vinyl "air mattress" solar water heater, introduced in the late 1950s.

Figure 14.12. Professor Ichimatsu Tanashita, an early Japanese advocate of and authority on solar water heating in Japan.

solar demand among Japan's farmers and the burgeoning middle class. Newcomers to the field looked for ways to enter the market, which until the late 1950s had been dominated by one firm. In 1960, a manufacturer came up with an improved solar water heater, one that was more solidly built so that it would last longer than the soft vinyl heater. It could also be placed on an incline facing toward the south to receive the sun's rays more directly in winter. Without knowing about early American solar water heaters, this Japanese manufacturer designed a device bearing a very close resemblance to the old Climax model. It was a glass-covered box containing several cylindrical water tanks made of blackened aluminum or copper. The heater held an average of 85 gallons and was usually installed on a slanted platform or on the south-sloping roof of the house. Connected to the city water lines, the heater reached its maximum temperature by midafternoon. Because of its increased heat collection in winter, it produced more hot water during the course of a year than earlier models had.

Figure 14.13. Advertisement for the tank-type solar water heater.

Before long, however, the aluminum tanks corroded and developed leaks. Manufacturers turned to more durable materials, such as glass, or stainless steel. Because city dwellers generally used more hot water than people in the countryside, they often coupled two or three water heaters together. At about fifty dollars each, this tank-type heater was a far more expensive proposition than the soft vinyl or wood-basin heaters. Nevertheless, they also sold extremely well; in 1960 alone over 15,000 of these heaters were sold. Sales increased to 50,000 the following year, and to more than 200,000 by 1963. Over the next five years, the combined sales by the five major manufacturers of tank-type heaters averaged more than 250,000 per year.[22] The soft vinyl heater continued to hold its own in rural districts, averaging over a quarter of a million sales annually during the same period. As a result, total sales of solar water heaters in Japan reached 3.7 million by 1969.

But 1966 turned out to be the peak year for sales. The advent of supertankers in the 1960s allowed the Japanese to import huge quantities of oil more cheaply than ever before. Inexpensive and plentiful fossil fuels drew many customers away from solar heaters. Rural electrification gave farmers another alternative to rice straw for water heating, and many took advantage of low nighttime electric rates. In addition, an increasing number of middle-class families preferred a domestic water-heating system tied into the central house-heating system — a new feature in urban Japanese homes and apartments.

The tank-type solar water heaters experienced many of the same problems that had plagued the early Climax heaters. None of them had insulated storage tanks, so all lost large quantities of heat at night. Furthermore, the relatively large volumes of water in the tanks did not get really hot until midafternoon or later. To eliminate these shortcomings, Professor Tanashita had introduced a system with a collector and separate insulated storage tank as early as 1948. But few could afford its high price in postwar Japan, and the simpler versions remained popular. Of course, the hundreds of thousands of solar water heaters that had been installed continued functioning. Impressed by the number of pre-1973 solar installations on rooftops in Japan, American solar pioneer George Löf stated at the 1974 Japanese/United States Symposium on Solar Energy Systems: "It is a bit like carrying coals to Newcastle [when speaking about solar energy to] a Japanese group, because the lead certainly had been in that country as measured by the number of solar water heaters in practical use at that date."[23]

Figure 15.1. Airmen use a solar desalinator on a life raft.

[1943–]

15

Saving Airmen with the Sun

Sometimes it takes an event that moves the hearts and minds of a nation to inspire a scientist to come up with a major discovery. In the case of solar scientist Dr. Maria Telkes, the downing and rescue in 1943 of ace Eddie Rickenbacker, the pilot who had beaten the infamous Red Baron in World War I, led to the invention of the first practical emergency solar desalinator. Temporarily lost at sea and adrift on two life rafts, having bailed from their B-17, Rickenbacker and his crew nearly died of thirst before rain miraculously fell to provide them drinking water. They wrung their soaked clothes into buckets. Now they had water but had to ration it — one-fifth of a cup for each person per day — for who knew how long they would remain adrift. Yet for one of the crew, their new source of water was no help, because he had drunk too much from the ocean. On the twenty-fourth day, rescue came. Their ordeal, the pain, the suffering, and the heroism filled Americans with awe and pride, so necessary for the morale of a nation in the midst of war.

"It was the experience of Eddie Rickenbacker drifting on his little life raft, almost dying of thirst," Telkes recalled, that made her "realize the necessity of a source of fresh water for those men" and led her to drop other work in search of a practical, compact solar desalinating kit so future downed airmen didn't have to endure a similar or worse fate.[1]

The design Telkes came up with adhered to the principles discovered by Saussure and utilized by Kemp and Bailey. Called a solar or sun still, it consisted of inflatable plastic material that was folded in a portable watertight can. When inflated, the still took the form of a 24-inch clear plastic sphere — very much like a balloon — through which sunlight entered from morning to late afternoon. A black cloth in the form of a pad bisected the sphere to maintain its shape, absorb the salt water, and collect and emit solar heat. The invention was of such singular importance to the military that the commander of the Seventh Naval District, which included Miami, lent Telkes all the facilities necessary to test the new invention. Southern Florida provided real-world testing conditions, as its latitude and climate resembled the tropics, where, if successful, Telkes's invention would primarily be of use.

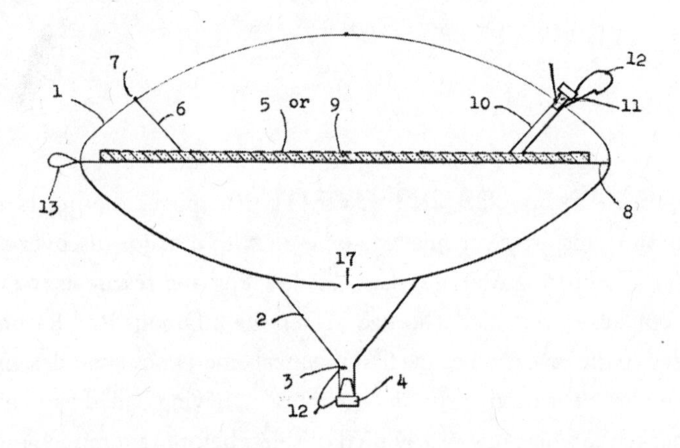

1. Transparent envelope
2. Distilled water bag
3. Distilled water outlet
4. Plug
5. Evaporator pad (or bag)
6. Pad (or bag) strap
7. Pad (or bag) support attachment
8. Pad (or bag) support
9. Bag transmitting water vaport,but not liquid water
10. Filling tube for sea water,with funnel
11. Plug for funnel
12. Plug attachment
13. Towing loop
14. Inflatable tube
15. Inflation orifice
16. Plug or closure
17. Valve

SOLAR STILL FOR LIFE RAFT

Figure 15.2. Design of the solar desalinator illustrating how it worked.

Figures 15.3–15.5. Step-by-step directions for use of the solar desalinator by airmen on a life raft.

In February 1944, a crew of seven launched their raft through the surf and headed out to sea, where the sailors inflated desalinating devices designed by Telkes, filled them with seawater, and dropped them overboard, testing them over a period of five days. Attached to the raft by lines, they looked like jellyfish bobbing to the rhythm of the incoming swells. They desalinated the water inside over a four-hour period. Telkes wrote, "When in actual use on a life raft, the operation naturally would be carried out during the entire day."[2]

The black cloth absorbed the seawater. As the sphere heated up, the water evaporated. The resulting vapor rose to the top of the plastic — just as steam does from a heated teakettle — leaving the pad saturated with the salt and allied minerals. Cooled by the outside air and seawater, the vapor condensed as droplets of freshwater trickled down into a reservoir attached at the bottom for those onboard to drink.[3] The results were heartening. The tests demonstrated that each still could provide 2 pints of drinking water per day, and that

Figure 15.6. Chicago artists play with converted desalinators.

the average salt concentration stayed within safety limits set by the Naval Medical Research Institute.[4] Furthermore, no matter how turbulent the sea became, Telkes reported, "the distillers skimmed over the waves[,]... never covered by water." Just as important was the still's simplicity, according to the inventor: "Any person who read the instructions could operate the distiller[,]...[with] tests 10 miles off shore indicat[ing] that distillers could be easily operated."[5]

Manufacture of the solar still couldn't have come at a more propitious time. As World War II continued, the military came to see rescuing its personnel as a major priority, pointing out that "every pilot forced down and every service man in the water is a well trained man needed to shorten the war. Every man lost at sea means another man will have to be trained for months, perhaps years."[6] The navy and coast guard took note of this concern, quickly adding new equipment in the quest to save the lives of those downed, with the solar still becoming standard equipment, since "drinking water," in the words of the military, "is the most essential thing" for those stranded in Pacific waters.

Rickenbacker himself praised the new device. He told an audience, "The story of my own experience as a survivor would have been stripped of much of its aura of stark tragedy had it occurred a year later than it did. Perhaps then, instead of 'seven came through' it would have been eight. We would have not known thirst.... The solar still, which uses the energy of the sun to produce more than a pint of water a day, would have obviated that."[7]

When the war ended, DuPage Plastics, the manufacturer of the solar still, discovered they could only sell gutted spheres. They became quite the rage as huge balls for endless entertainment at get-togethers and parties. *Life* magazine, which had run the dramatic story of Rickenbacker and his crew of castaways just three years before, picked up on the fad. Its headline for a 1946 feature read, "Life Tries Out a New Toy: Chicago Artists Play with Giant Plastic Balls Reconverted from War Use."[8] The magazine noted that "the new plaything is derived from one of the war's most ingenious inventions, the solar still. Adrift at sea, a man would put sea water in the still and let the sun's heat distill the undrinkable salt water into drinkable fresh water. DuPage[, which] was caught at war's end with a lot of solar-still machinery and material, has bounced into peace with ridiculous ease." The inventor herself acknowledged the change of circumstances, writing on her copy of the *Life* article, "The war ends: Solar Stills are not needed anymore."[9]

PART IV

SOLAR HOUSE HEATING

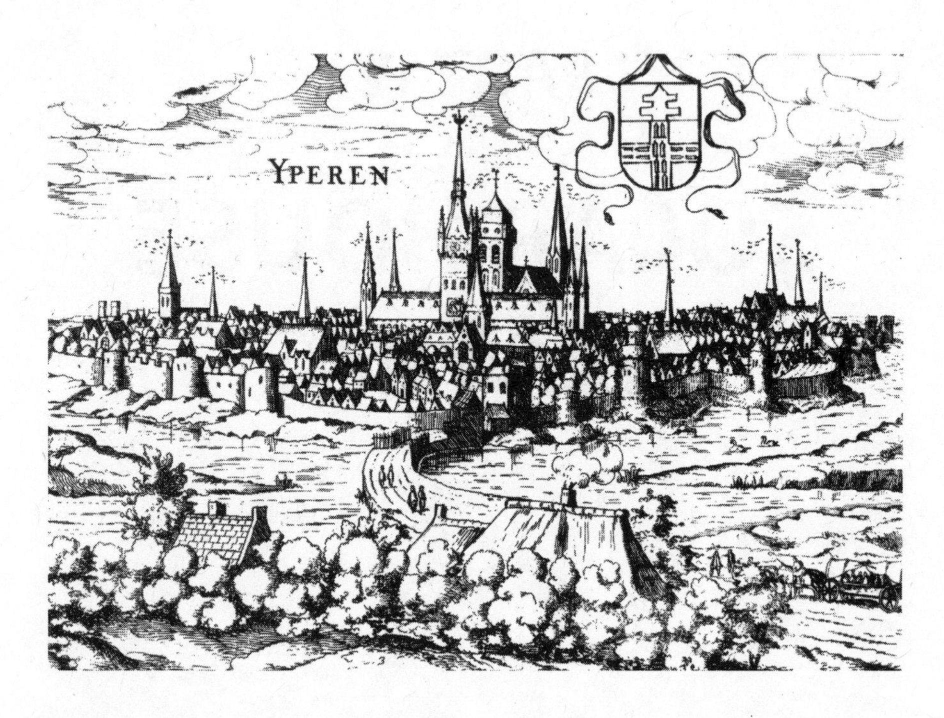

Figure 16.1. A view of a typical medieval city, which suffered from an overall lack of access to sunlight.

[1807–1850]

16

Solar Building during the Enlightenment

Solar architecture persisted in southern Europe and western Asia Minor from classical times through the nineteenth century. While excavating the ancient solar city of Priene, in what is now Turkey, for example, Theodor Wiegand observed that those living near the ruins built as their ancient neighbors had. The facades in villages, farms, and cities that the archaeologist visited during the digs almost always faced south.[1]

Rural people in old England also knew that for winter warmth a southern exposure was best. To keep their hogs happy and healthy during colder months, didactic farm poet Thomas Tusser advised farmers to "warmly enclose [the sty], all saving the mouth, / And that to stand open, and full to the south."[2] In parts of the Rhine Valley and in Switzerland, the main living areas of traditional houses were situated in the south corner of the house. All their windows faced southeast and southwest, guaranteeing that, in winter, sunlight would enter the main rooms throughout the day.[3]

When the classical influence in building reached northern Europe in the eighteenth century, however, architects copied the outward forms of the architecture, such as columns and porticoes, but neglected the solar principles that made them beautiful *and* useful, missing an opportunity for an assist from the sun in winter. The early-nineteenth-century English architect Humphry Repton, one of the few architects to recognize the irony of this misuse of classical architecture, remarked, "I have frequently smiled at the incongruity of Grecian architecture applied to buildings in this country, whenever I have passed the beautiful Corinthian portico to the north of the Mansion House.... Such a portico towards the north, is a striking instance of the false application of a beautiful model."[4]

Faust's Book

Cities of medieval northern Europe, too, were built without planning, leaving residences "confined, squashed, stifled in dark, crowded" conditions.[5] Dr. Bernhard Christoph Faust, court physician to the nobility of Bückeburg, decided to ameliorate the situation by writing a book on how to construct houses, apartments, villages, market towns, and cities to maximize solar heat gain during the colder months and to minimize solar heat gain during the warmest months.

Figure 16.2. Bernhard Christoph Faust.

Faust's book, the first ever written solely about solar architecture, got its start with just a single engraving of a dream, a plan for an ideal solar city — a *sonnenstadt*.[6] Far away from the madding crowd, Faust, physician to Dowager Princess Juliana of Schaumburgh-Lippe of Bückeburg — a town so small that from the center you can still walk to its farthest limit in less than ten minutes — sent a copy of the plan to the well-known architect Gustav Vorherr. The drawing so intrigued Vorherr, who aspired to high places in the construction section of the Bavarian state civil service, that he encouraged Faust to produce more on the subject.

When Faust took up residency at Bückeburg, he found himself housed outside the castle grounds in a large three-story house of wood and plaster.

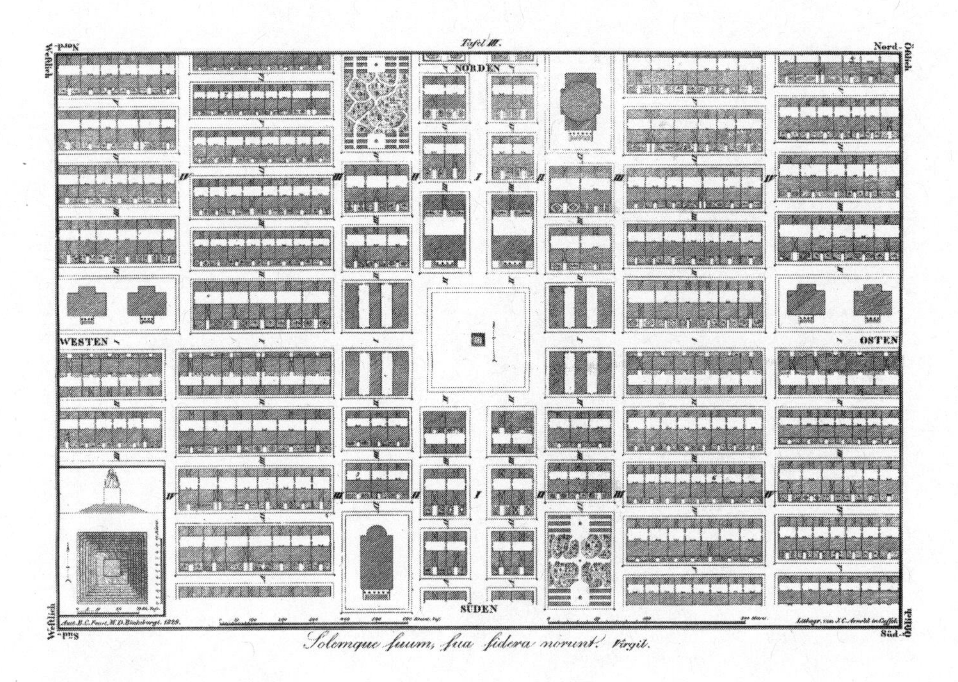

Figure 16.3. Faust's plan for a solar city.

But what a house! Built in 1649, it had a facade filled with windows all facing south-southwest that allowed the sun to march into the house beginning in early autumn and then flood all the rooms in winter. Warmed by so much sunshine, Faust wrote that the sun "lit me to think — yes — thank you sun for the idea: All houses should be directed toward the sun, all of humanity should live in sunlight." For decades he thought and obsessed on the notion that his sole purpose of living was "to enlarge upon this idea" for humanity to understand. "I was thinking and working on the subject for over 34 years" before writing a word about it, because, he confessed, "I did not know what was true and right, much less could have I proven what is true and right."[7]

Faust and Vorherr discussed through letters all the solar ideas that were swimming in the doctor's mind. Meanwhile, Vorherr continued his rise in the state bureaucracy and finally attained his life's goal of becoming royal Bavarian building officer for the court of Joseph Maximilian, king of Bavaria. There he started, with the blessings of the king, the *Monthly Journal for Building and Land Improvement* (Monatsblatt für Bauwesen und Landesverschönerung in Bayern)

Figure 16.4. Gustav Vorherr.

Figure 16.5. Faust's house in Bückeburg.

and the Königlichen Baugewerksschule, a school in which architects could learn the techniques of building construction.

Back in Bückeburg, Faust struggled to find the proper orientation: "whether buildings in our northern hemisphere should be directed towards south-southeast." He chastised himself for his previous erroneous thinking: "What a mistake! And yet that is how I thought in the beginning! Or toward the south? — The last one! Towards the sun at midday!"[8] he exclaimed in a fragment that begins his groundbreaking book — *All Buildings of Men Should Face towards the Midday Sun* (Zur Sonne nach Mittag sollten alle Häuser der Menschen gerichtet seyn) — which would have a powerful influence on practical building in Bavaria and beyond in the first half of the nineteenth century. "And with my heart and soul," he explained to the reader, "I turned my eyes to the sky, to the sun," and went to work, researching and writing feverishly. Faust believed that what he would one day publish would outrank any other piece of literature, even the Bible, because, as he said, "The sole aim of life is the correct orientation of buildings to the midday sun. Everything else fades compared to the sun and its

Figure 16.6. TItle page of Vorherr's monthly architectural journal.

benefits — to receive the sun in its greatest abundance, the most important gift that God gave to man and animal."[9]

Significant as his task was, he put the work on hold when he heard that a terrible fire had engulfed the city of Hof. "Oh, let Hof arise from its ashes toward the sun," he wrote to Vorherr, and immediately tasked his skilled copyist with the job of making a facsimile of a drawing meant for the book. The drawing would show the way streets should be laid out and the manner buildings ought to be placed for the new sun-oriented city of Hof. Here he had the chance to carry out his ideas and let the world see whether they worked or not. But despite his and Vorherr's efforts, the city fathers decided to rebuild as they always had built in times past.[10]

Faust's Sonnenbaulehre [Solar-Building Principles]

Fragments of what one day would become his book arrived in Munich in 1822 for Vorherr's critical comments. The bits and pieces he received excited Vorherr to such a degree that he wrote, "We hope [publication] happens soon," because this book "should be considered fundamental for the badly needed reform of building methods. May it soon be completed and distributed throughout the entire world!"[11]

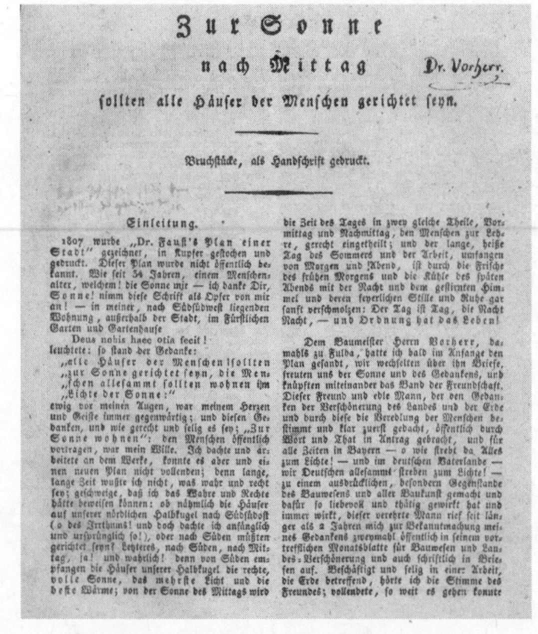

Zur Sonne
nach Mittag
sollten alle Häuser der Menschen gerichtet seyn.

Dr. Vorherr.

Bruchstücke, als Handschrift gedruckt.

Einleitung.

1807 wurde „Dr. Faust's Plan einer Stadt" gezeichner, in Kupfer gestochen und gedruckt. Dieser Plan wurde nicht öffentlich bekannt. Wie seit 34 Jahren, einem Menschenalter, welchem! die Sonne mir — ich danke Dir, Sonne! nimm diese Schrift als Opfer von mir an! — in meiner, nach Südsüdwest liegenden Wohnung, außerhalb der Stadt, im Fürstlichen Garten und Gartenhause

Deus nobis haec otia fecit!

leuchtete: so stand der Gedanke:

„alle Häuser der Menschen sollten
„zur Sonne gerichtet seyn, die Menschen allesammt sollten wohnen im
„Lichte der Sonne:"

ewig vor meinen Augen, war meinem Herzen und Geiste immer gegenwärtig; und diesen Gedanken, und wie gerecht und selig es sey; „Zur Sonne wohnen": den Menschen öffentlich vortragen, war mein Wille. Ich dachte und arbeitete an dem Werke, konnte es aber und einen neuen Plan nicht vollenden; denn lange, lange Zeit wußte ich nicht, was wahr und recht sey; geschweige, daß ich das Wahre und Rechte hätte beweisen können: ob nähmlich die Häuser auf unserer nördlichen Halbkugel nach Südsüdost (o des Irrthums! und doch dachte ich anfänglich und ursprünglich so!), oder nach Süden müßten gerichtet seyn? Letzteres, nach Süden, nach Mittag, ja! und wahrlich! denn von Süden empfangen die Häuser unserer Halbkugel die rechte, volle Sonne, das mehrste Licht und die beste Wärme; von der Sonne des Mittags wird die Zeit des Tages in zwey gleiche Theile, Vormittag und Nachmittag, den Menschen zur Lehre, gerecht eingetheilt; und der lange, heiße Tag des Sommers und der Arbeit, umfangen von Morgen und Abend, ist durch die Frische des frühen Morgens und die Kühle des späten Abends mit der Nacht und dem gestirnten Himmel und deren feyerlichen Stille und Ruhe gar sanft verschmolzen: Der Tag ist Tag, die Nacht Nacht, — und Ordnung hat das Leben!

Dem Baumeister Herrn Vorherr, damahls zu Fulda, hatte ich bald im Anfange den Plan gesandt, wir wechselten über ihn Briefe, freuten uns der Sonne und des Gedankens, und knüpften miteinander das Band der Freundschaft. Dieser Freund und edle Mann, der den Gedanken der Verschönerung des Landes und der Erde und durch diese die Veredlung der Menschen bestimmt und klar zuerst gedacht, öffentlich durch Wort und That in Antrag gebracht, und für alle Zeiten in Bayern — o wie strebt da Alles zum Lichte! — und im deutschen Vaterlande — wir Deutschen allesammt streben zum Lichte! — zu einem ausdrücklichen, besondern Gegenstande des Bauwesens und aller Baukunst gemacht und dafür so liebevoll und thätig gewirkt hat und immer wirkt, dieser verehrte Mann rief seit länger als 2 Jahren mich zur Bekanntmachung meines Gedankens zweymahl öffentlich in seinem vortrefflichen Monatsblatte für Bauwesen und Landes-Verschönerung und auch schriftlich in Briefen auf. Beschäftigt und selig in einer Arbeit, die Erde betreffend, hörte ich die Stimme des Freundes; vollendete, so weit es gehen konnte

Figure 16.7. Title page of Faust's seminal work on solar architecture. Note Vorherr's signature, indicating that this copy belonged to Faust's patron.

Faust came to realize as his ideas developed that the major challenge was homeowners' seemingly contradictory demand that the same structure "be cool in summer and yet warm in winter." This apparent paradox, which he described as the "torment of all builders," had to be addressed for his ideas to be applicable in the world he wished to transform.[12] The hard data he needed were difficult to find. The lack of "accurate observations on sunlight, its power, duration, and brightness in many countries in the inhabited part of the earth" at first stymied his research.[13] Delving deeper, he finally found the necessary information to develop tables that accurately displayed the changing angles of the sun and the time the sun spends at different points of the sky throughout the year at different latitudes.[14] Studying this information led Faust to his great discovery: the

Nördliche Halbkugel der Erde	§. 21. Tafel I. A. Die Sonne beleuchtet aus der Mitte ihrer Scheibe die vier Seiten oder Wände der rechtwinkligen, orientirten Häuser nachstehende Stunden lang: I. Im Winter-Halbjahre:				§. 22. Folgerungen aus nebenstehender Tafel, A. (alle Zahlen bedeuten hier, wie in der Tafel, Sonnenstunden)	
Breite-Grade	Oestliche S.	Südliche S.	Westliche S.	Nördliche S.	Die Oestliche und die Westliche Seite der Häuser hat der Sonnenstunden im Sommer-Halbjahre mehr, als im Winter-Halbjahre:	Die Südliche Seite der Häuser hat der Sonnenstunden mehr:
	A.	B.	C.	D.	N. (=E–A; =G–C.)	O (= B – F)
30°	970	1940	970	0	255	im Wint. 430, als i. S.
35	946	1892	946	0	304	: : 242, : : :
40	917	1834	917	0	363	: : 84, : : :
45	886	1772	886	0	476 1/2	im Som. 53, als i. W.
50	846	1692	846	0	609	: : 208, : : :
55	799	1598	799	0	606	: : 362, : : :
60	733	1466	733	0	712	: : 544, : : :

	II. Im Sommer-Halbjahre:				Jedes Paar Seiten (die Oe. u. We. die Süd. u. Nö.) hat im Sommer:	Die Südl. Seite hat im Sommer der Sonnenstunden mehr, als d. Oe. u. W.	Die Südl. Seite hat im Sommer der Sonnenstunden mehr, als die Nördl.	Jede, die Oe. u. d. W. hat im Sommer mehr, als die Nördliche
	E.	F.	G.	H.	P. *)	Q.	R.	S.
30°	1225	1510	1225	940	2450	285	570	285
35	1250	1650	1250	850	2500	400	800	400
40	1280	1750	1280	810	2560	470	940	470
45	1312 1/2	1825	1312 1/2	800	2625	512 1/2	1025	512 1/2
50	1355	1900	1355	810	2710	545	1090	545
55	1405	1960	1405	850	2810	555	1110	555
60	1475	2010	1475	940	2950	535	1070	535

Figure 16.8. Faust compiled tables like this one, which showed the hours of sunlight falling on the eastern, southern, western, and northern sides of a house during the winter half of the year (top left) and during the summer half of the year (bottom left). He concluded that the east and west walls receive many more hours of direct sun in summer than in winter at all latitudes, and that the south receives more hours of sun in winter than either the east or west side.

sonnenbaulehre (solar-building principles) that provided the underlying science for building homes with interiors that would be comfortable year-round, simply by adapting their design to the changing angles of the sun.

As the ancients knew millennia before Faust, measurements of the sun's angles show certain unchanging facts: At midday the sunbeams fall upon the southern side at a narrower or wider angle depending on the position of the sun in the sky. In summer, when the sun is highest in the sky, its rays merely glance upon a south-facing vertical wall with little effect. The angle of the winter sun is just the reverse — it lies low in the sky and its sunbeams focus on that wall for most of the day. "In other words," he concluded, "the vertical south-facing wall in winter receives [many more] times the amount of sunlight than the same wall during summer, guaranteeing that the south-side of a house is not so hot during summer nor [as] cold during winter" as the other sides of the house.[15] In contrast, "the west side of houses in summertime receives full exposure to the sun throughout the afternoon. Worse, the declining sun after midday faces the west wall at lower and lower angles so that its rays enter ever deeper into the interior, making us hotter and hotter until we want to choke."[16]

Calculating the time the sun spends around the cardinal points of the compass throughout the year also added to Faust's solar architectural theory. Using Munich as his example, he showed that the sun at the winter solstice remains in a south-facing room almost the entire day — and so is capable of helping to heat the house — while east or west rooms receive almost no sun, and the sun they do receive shines into them for a very short time at that. In summer, Faust came to learn, while the hours of sunlight coming from the south do not exceed its winter duration, the time the sun spends in the east and west doubles.

Since we cannot change the ways of the sun, Faust concluded, "human beings should build, orient, and arrange their houses" adaptively. For this reason, he advised, build yourselves "a house on the southward side. And on this side should man and wife gratefully and wholeheartedly receive heaven's highest gift and the highest perfection in comparison with houses to the east, west and north — the low, the imperfect."[17]

Living with Sun in Mind

Faust's interest in solar architecture made sense from his medical perspective. A physician to nobles, he also ministered to the desperately poor. He saw firsthand the dank hovels they existed in, which, the doctor remarked, were "no better than human stables."[18] Many of the complaints he attended to — colds, coughs, aching bones, chronic fever — could be attributed, Faust noted, to living in housing that was shut off from the sun's heat and light, and fresh air. Mold and disease flourished in damp, cold rooms, while in interiors exposed to the midday sun, "the flame of life breathes effortlessly and burns calmly, nourishing body and soul."[19] In his professional observation, "The warmth nourishes, the cold wastes [life] away."[20] Living in insulated and properly oriented buildings, people "need little or no heating indoors even in the coldest winters," he argued, because houses "warm up very easily, requiring the smallest amount of fuel,"[21] creating the ideal condition that any conscientious health practitioner would prescribe.

"Living with the sun" would also alleviate the worst aspects of poverty. In the world Faust hoped to build, "the human body would have to give up less heat and, in the process, waste less energy. Naturally warmer circumstances would immensely help the poor since they could survive more healthfully given the meager nourishment and coverings they could afford."[22] The rich had the means to escape the worst of winter by traveling south; and creating through architecture a warmer, sunnier indoor climate would permit the poor to do the same while staying at home. It would be, Faust wrote, "as if people had moved" to a more southerly clime. Furthermore, the rich, by building properly, would not have to make their seasonal migration, he believed. And in this realizable utopia, "the rich and the poor within their houses would equally enjoy the golden light and beneficial warmth" of the sun, since "the sun shines for all."[23]

Just as important, in Faust's opinion, solar houses saved fuel for the simple reason that, in colder climates, heating homes that were poorly designed and built "waste[s] horrible amounts." He reminded readers that "the more coal burned, the less remains" for future use. In contrast, heating by the sun does not follow the rule of diminishment by consumption. With solar energy as our heat source, "we could also save so many forests... [therefore] what a blessing will buildings oriented to the sun become," he declared.[24]

Putting Faust's Ideas to Work

Faust, though, had no illusions about the challenge his crusade faced. "The crooked streets of [cities throughout Europe,] where no sun shines on their hovels," demonstrated these cities' shortcomings. Faust thought that builders in northern Europe had for the longest time, to paraphrase Aeschylus as he did, emulated "anthills," by burying "those they build for" in cavelike darkness.[25]

In a country where almost every house was oriented in an arbitrary direction, where the majority lived "in dark, crowded medieval towns," Faust and Vorherr had their work cut out for them.[26]

Vorherr was, however, optimistic in light of the changing times brought on by the American and French revolutions. He believed that the old way of building was passé, representing a "dark, repressed mindset," and that "Dr. Faust's marvelous Solar Architecture theory" embodied "the spirit of our age — the spirit of enlightenment and liberty."[27]

Faust credited Vorherr for the completion of his book, saying, "He urged me for more than two years to publish my ideas. I heard his voice urging me on and so I finished the work that I had started."[28] Furthermore, without Vorherr's connections as a well-placed architect in the Bavarian state government, and without his embrace of and zealous promotion of Faust's work, the ideas in *All Buildings of Men Should Face towards the Sun at Midday* would have never found their way beyond Bückeburg. Vorherr played every card in his deck — the influence he wielded as a high-ranking official, as head of the state-run school of building arts, and as publisher of the *Monthly Journal for Building and Land Improvement* — to gain support for Faust's plan to apply solar principles to the built environment.

At the building school, which was attended by more than twenty-five hundred architectural students from everywhere in Europe, Vorherr integrated into all projects the principles of building in relation to the sun. During the winter of 1824, for example, the students in the master planning class were assigned to complete a design for a solar suburb in Munich. One of Vorherr's lesson plans asked the top student of each graduating class to write an essay on the topic "What advantages are there for correctly orienting a building toward the midday sun[,] and what must builders do so that such an orientation becomes universal?"[29]

Two similar schools propagating Faust's teachings were established in other areas of Germany, and former students of Vorherr's from Greece and Russia established institutions of their own in Athens and in Saint Petersburg to foster the spread of Faust's ideas in their countries. Vorherr also publicized Faust's work through his monthly journal, which he sent to leading statesmen throughout the world, including America's sixth president, John Quincy Adams. That an 1824 issue of the *London Times* would carry commentary on Faust's work — praising his promotion of "the embellishment and comfort of habitations of men, by placing the principal front of their dwellings towards the south"

— provides a singular indication of how well Vorherr had spread his hero's ideas beyond Bavaria.[30]

Bavaria's most respected technologist, Anton Camerloher, the royal Bavarian engineer first class, learned of Faust's solar building principles through Vorherr. After submitting these strategies to rigorous scientific analysis, he declared them "well founded" and enthusiastically joined Vorherr in his fight for their implementation in construction throughout the region.[31] Camerloher's opinion on the Faustian doctrine greatly influenced King Joseph Maximilian to mandate implementation of Faust's teachings in the construction of all new public and communal buildings in Upper Bavaria. Several years later the Bavarian government published the basics of Faust's solar building principles with the intent of guaranteeing that "all districts, police, and building departments in the Isarkreis [Upper Bavaria] will give these architectural ideas special attention and support."[32] Other German states, such as Hessen and Prussia, followed suit.

Two years later his son Ludwig succeeded Joseph Maximilian to the throne. Ancient Greek and Roman architecture fascinated Ludwig. Friends of Vorherr let the royal architect know of the new king's consuming interest in classical building. They also informed Vorherr of the publication of recently translated classical works that supported the *sonnenbaulehre* of Faust. Vorherr wasted no time publishing excerpts from Socrates and other classical writers regarding solar architecture, and he posed the rhetorical question, "How much time has to pass until the principles taught by the Greek sage will finally be adhered to?"[33]

The appearance of Vorherr's article on solar construction in antiquity coincided with Ludwig's decision to have Vorherr design his new palace. Vorherr saw to it that the palace's principal side, which had multiple large windows, would look to the sun at noon.[34] New schoolhouses were given a similar design during the construction boom that followed, sparked by growing support for education throughout the Continent. The Bavarian Royal Ministry of the Interior distributed blueprints created by Vorherr to help builders comply with the royal mandate for the construction of solar-oriented school buildings. The royal Bavarian government of the Lower Danube District stated on January 3, 1829, that the first consideration in building a schoolhouse should be "that it had a healthy and welcoming location, with the front facing the midday."[35] Vorherr learned of the success of the mandate through a letter from a colleague, who advised him that "in many villages [of Upper Bavaria] there are presently many school buildings that have been built oriented to the midday sun."[36] Documents in the Munich City Archive verify the construction of schoolhouses according

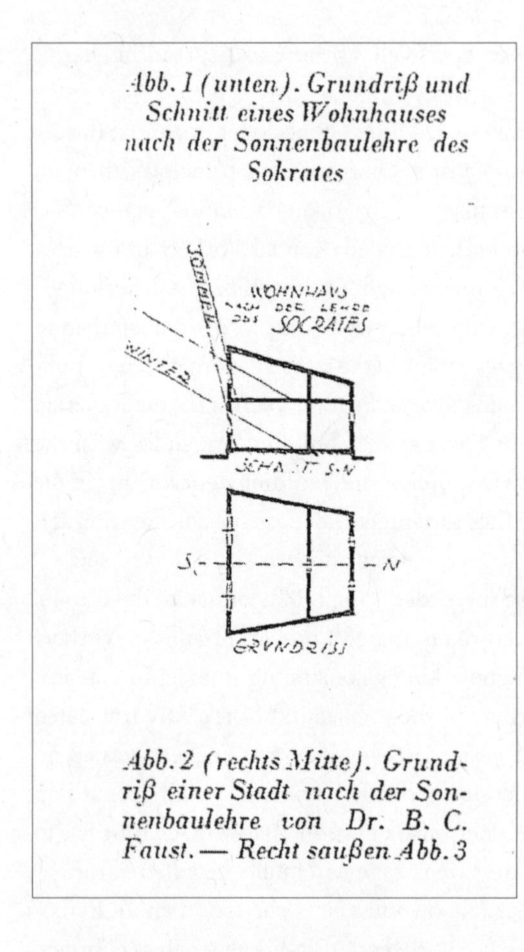

Figure 16.9. Sketch by Vorherr illustrating Socrates's architectural ideas as articulated in his dialogue recorded by Xenophon.

to Faust's solar principles in the Bavarian towns of Arnbach, Grossdingharting, Nussdorf, Surberg, and Vilsbiburg.[37]

Vorherr considered the solar orientation of schoolhouses exceedingly important. He even had a model of a schoolhouse built according to Faust's doctrine and displayed at his architectural trade school in Munich. He considered schools "as points of light" for the communities they served.

Solar Buildings Rising from the Ashes

Private homes were being built in great numbers, too, primarily to replace old ones, which were constantly lost to fire. The combination of wood houses and stables full of straw, together with the ubiquitous use of open flames such as candles, lanterns, and hearths, made the burning of multiple structures a common

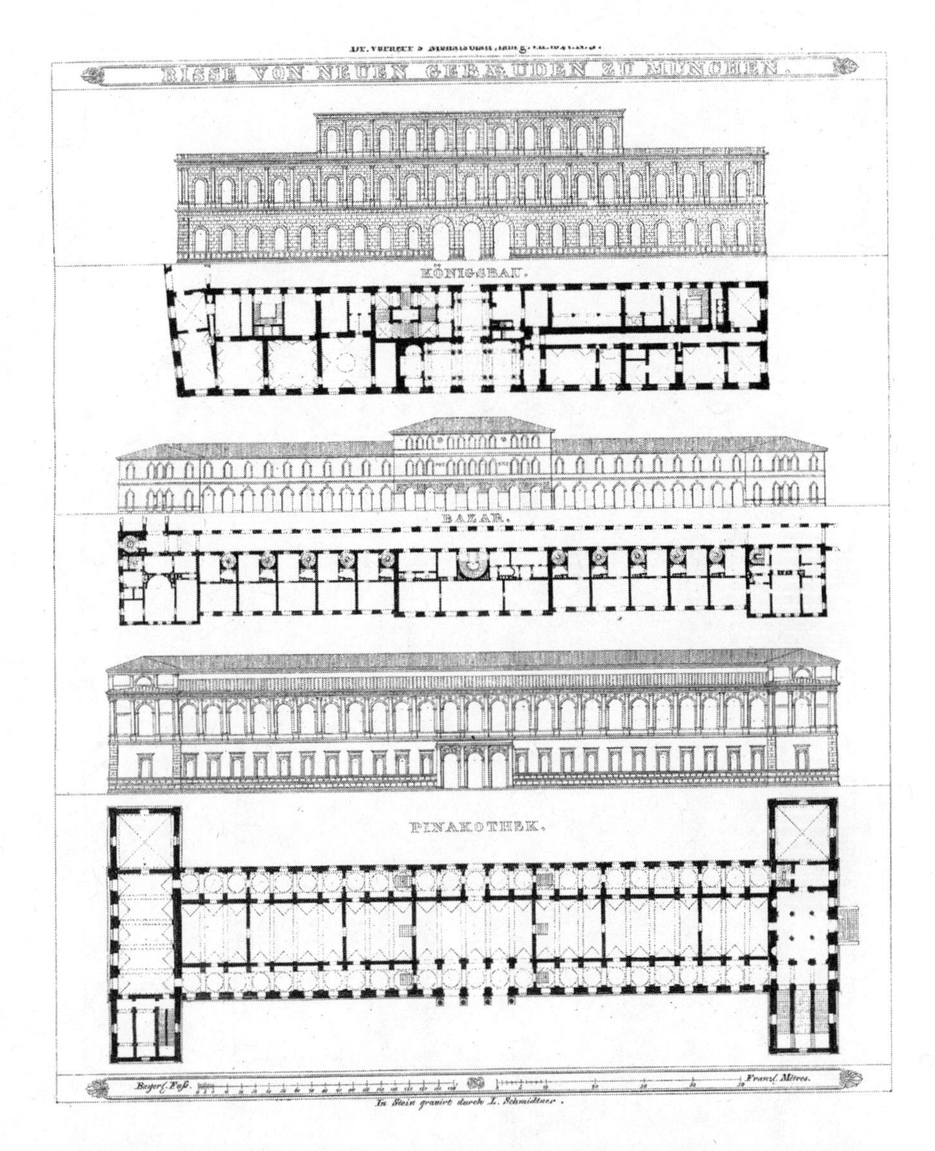

Figure 16.10. Vorherr's plan for the palace of the Bavarian king, today known as the Sun Palace and a popular tourist attraction in Munich.

experience in nineteenth-century life. Such powerful "flames write down [a] verdict," in Faust's opinion: "don't build in the same old fashion!"[38] Use brick, use adobe, instead.

Two municipalities lost to fire were reconstructed in line with Faust's tenets — Schwaboisen in Bavaria and Palotsay in Hungary. Out of the seventy-six

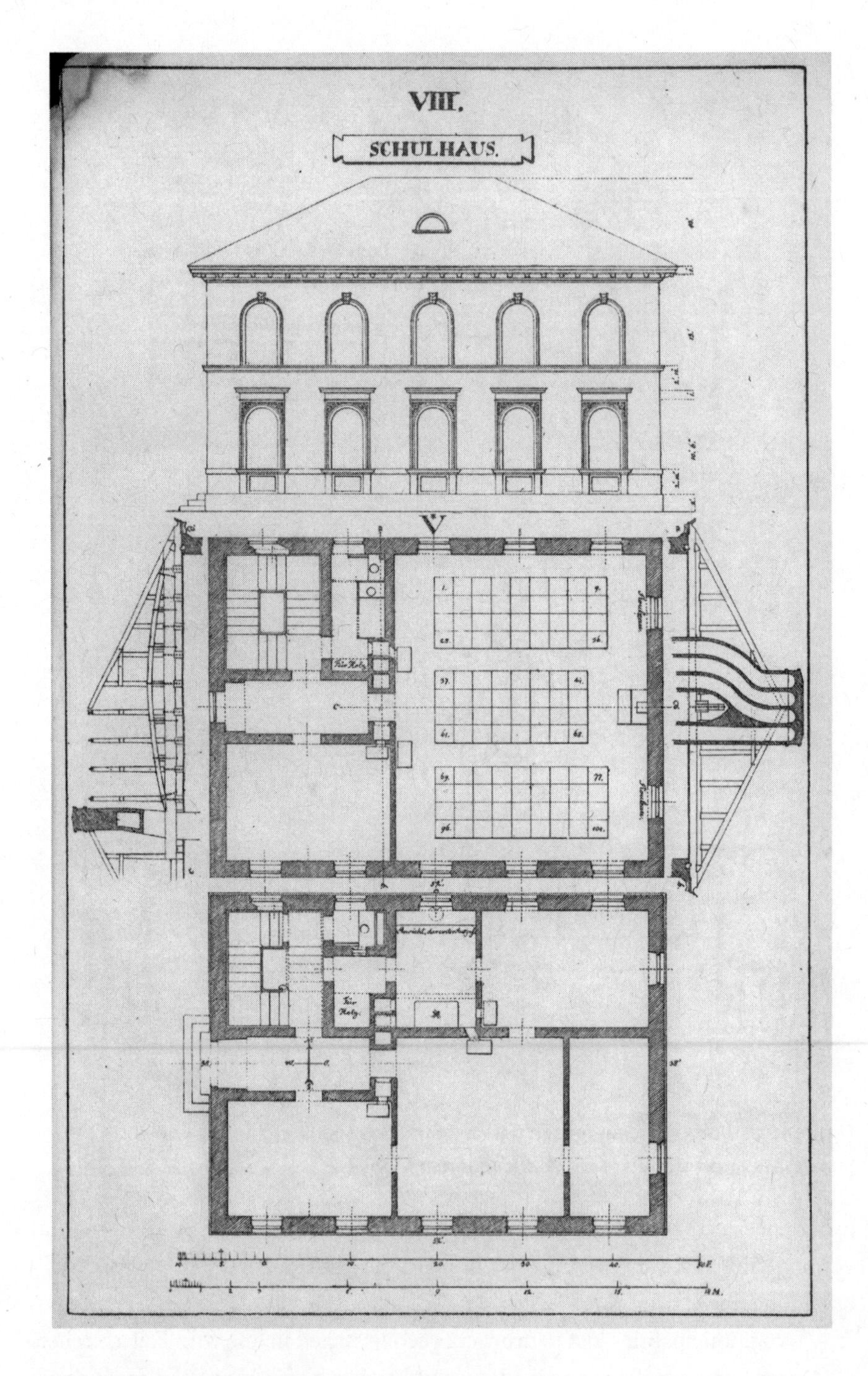

Figure 16.11. Vorherr's plans for a solar schoolhouse.

Figure 16.12. Fires were common in nineteenth-century cities and towns.

homes at Schwaboisen, flames destroyed fifty-three. During rebuilding under Vorherr's direction, the town straightened a portion of its main street so that it ran perfectly east-west, as specified by Faust, giving houses along the road better access to the midday sun. In the case of Palotsay, its ruler, Baron Palotsay, on his own initiative, rebuilt with Faust's ideas in mind.[39]

Faust's Solar City — Sonnenstadt — Realized

In an inscription to Vorherr by Faust on the cover of his second work, *Brief Discussion Regarding the Building of Houses and Cities towards the Sun* (Andeutungen über das Bauen der Häuser und Städte zur Sonne), the author wrote that he hoped his writings might "help to advance and increase solar architecture throughout the world; may it be taught in every home; may it become an ideal for humanity to use the sun."[40] While Faust did not see the fulfillment of this grandiose goal in his lifetime, the technologist Anton Camerloher credited that "charming old gentleman [Faust] with enlightening and inspiring us to direct our gaze steadily at the sun."[41]

Faust died before the largest project based on his work was realized. It began in the 1830s, when Frederick William IV, emperor of Prussia, embraced solar architecture. In 1834 he saw to it that his government sent out a circular to builders throughout the Prussian Empire outlining Faust's ideas and recommending their implementation. Charles-Henri Junod, then the chief engineer

Figure 16.13. Charles-Henri Junod.

for La Chaux-de-Fonds, the watchmaking capital of the world and part of Prussia at the time, read the circular and decided to adopt Faust's solar-oriented city plan as the basis for reconstruction of the part of the city that had earlier burnt down. It took another twenty years, though, before Faust's *sonnenstadt* took form as the new quarter of La Chaux-de-Fonds. Given its perpendicular streets, with the main avenues running east-west, it calls to mind the ancient city of Olynthus. Three to six buildings, each containing three flats, line every block and realize the Faustian ideal of sunlight for all. The egalitarian vision of its developers — to provide reasonably priced housing for watchmakers and their families — fit well with Faust's principles.[42]

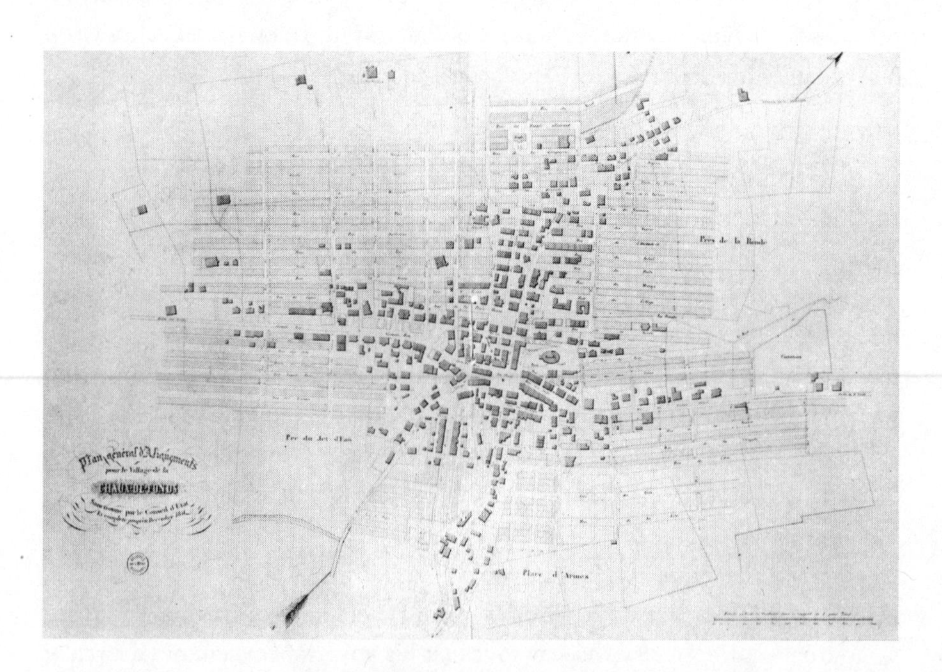

Figure 16.14. Charles-Henri Junod's plan for the new city of La Chaux-de-Fonds based on Faust's ideal solar-city plan of 1807.

Figure 16.15. The new solar housing district of La Chaux-de-Fonds, built according to Faust's plans.

In 2009 the United Nations chose La Chaux-de-Fonds as a World Heritage Site. The selection was made, according to the World Heritage Site web page, because of the " 'rationalist' principles . . . adopted[,] which addressed the relationship between living conditions and 'health.' A town plan was developed in 1835 designed by one of Pestalozzi's pupils (Charles-Henri Junod) and inspired by an ideal town called 'Sonnenstadt,' planned in 1824 by a Dr. Bernhard Christoph Faust. Features included having most houses facing onto small gardens receiving the midday sun."[43]

This monument to Faust's dream resonates with his exclamation written more than 150 years before the city's selection as a World Heritage Site: "Oh people, face your houses toward the midday sun to give yourselves and your children and their children until the tenth generation the warmth, life, power, joy and blessings of the sun."[44]

Figure 17.1. A typical nineteenth-century English slum.

17

Solar Architecture in Europe after Faust and Vorherr

Believing that Britain was more advanced than Germany in most matters, Vorherr wrote, "How far ahead of us the British are.... They can serve as role models for all peoples." Seeing Britain in this light, the Bavarian believed that the island was fertile ground on which to promote solar architecture, and so he went to England precisely for that purpose.[1] Ironically, just fifteen years later, another German, Friedrich Engels, visited Manchester, England, to see capitalism's ugliest manifestation: the Industrial Revolution's first manufacturing center. In this city, massive numbers of squalid slums were built to house the proletarian army filling its factories, an army who traded medieval hovels for modern squalor. Seeing such wretched conditions, Engels was shocked by how "little space a human being can move [about in], how little air — and such air! — he can breathe...and yet live." "How is it possible," he asked, "under such conditions, for the lower class to be healthy?"[2]

Other burgeoning British industrial cities followed Manchester's horrid pattern. Tenements in Edinburgh were "often so close together," a journalist noted in an 1843 article, "that persons may step from the window to that of the house opposite."[3] On the first floor of one such structure lived ten families. A single room, measuring 11 feet by 6 feet, was home to a family of four. The "walls were rough and broken," a visitor recollected. "The only light came in from the open door, which let in unwelcome smells and sound. It was abominable.... It was nearly dark at noon, even with the door open.... In such circumstances it was...impossible to be cleanly, impossible to be healthy."[4] Dickens's description of London's slums in *Bleak House* shows that in London the poor fared no better.

Social Change through Solar Architecture

One of the loudest voices advocating better housing for the working class belonged to John Ruskin, the eminent Victorian art critic and social critic. In lectures and writings he told his pious and wealthy audience: "Neither new-moon keeping, nor Sabbath-keeping, nor fasting, nor praying will in anywise help" the blighted cities such as Edinburgh "to stand in judgment higher than Gomorrah."[5] Salvation for these urban wastelands could only come, according to Ruskin, through good works — passing humane housing laws, fixing up slums, and, ultimately, building better structures. Ruskin foresaw the day when enlightened captains of industry or the state would sponsor construction programs that would provide the working class with freshly aired cottages overlooking gardens and orchards.

Communities along the lines Ruskin suggested soon came about, providing improved living conditions for the poor. They offered housing with plenty of sunlight, sufficient space, and adequate sanitation. Port Sunlight, one of the earliest of these projects, was built by Lever Brothers for workers in one of its soap

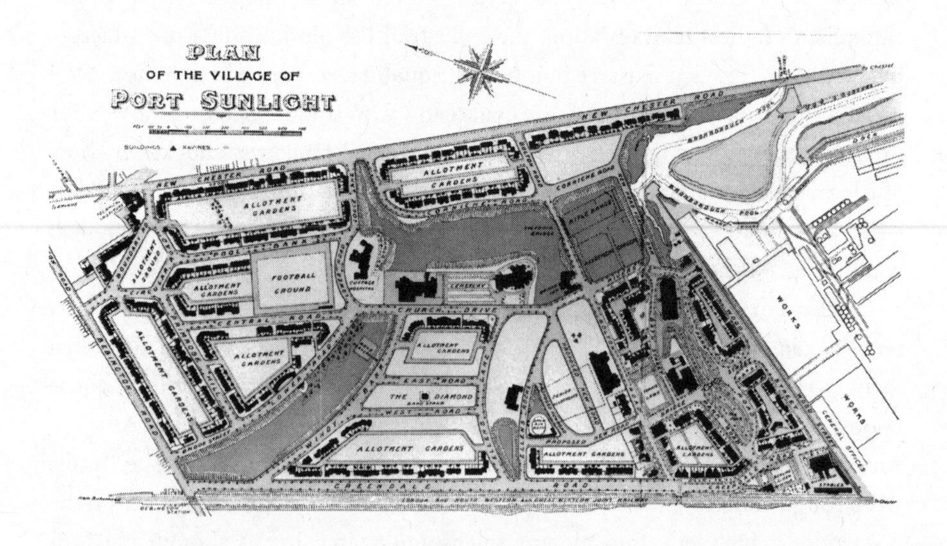

Figure 17.2. Street plan of the working-class community of Port Sunlight.

factories. Compared with the sordid tenements of nearby Liverpool, Port Sunlight was a godsend. And it was accurately named. As the developers boasted, "The worship of sunshine is characteristic of every building in the village." The rows of houses were surrounded by wide areas of open space. According to one observer, "Roads varying in width between 40 and 120 feet separate the various blocks, so that air and light can penetrate [the homes] freely on all sides."[6]

As planned working-class communities became popular in England, architects and planners emphasized the importance of access to sunshine. "Into as many more rooms as possible let the sun come in," urged Raymond Unwin, an early-twentieth-century urban planner and reformer. Unwin believed that every house "should turn its face to the sun, whence comes light, sweetness and health."[7]

To help architects maximize a structure's solar exposure, Unwin compiled information on the sun's annual movement with respect to the earth. Through his studies he reached the same conclusion that the Chinese, Greeks, and Faust had come to earlier: "In taking the whole year round there can be no doubt that an aspect south or slightly west of south may be considered the most desirable for dwelling rooms." Whenever possible, he followed these guidelines in the design of homes for working-class communities. If facing south meant designing the house's main rooms to face the backyard instead of the street, Unwin would not hesitate to do so. He called this having "its front behind." For a quad where land expenses had forced the builder to go to more than two stories, Unwin suggested keeping the southern side "lower to allow the sun to get well into the court," just as Socrates had suggested thousands of years before.[8]

The crowding of the poor in medieval cities and towns had stimulated Faust to come up with a more humanitarian approach to architecture. Similarly, the slums of late-nineteenth- and early-twentieth-century European cities led architects on the Continent to embrace new ideas in building that would ameliorate these horrid conditions. According to Sigfried Giedion, the leading architectural theorist of the period, healthful "housing for lower-income classes was in the foreground" of every discussion.[9] The dominance of this subject among the more influential architects in Europe became apparent at the first International Congresses of Modern Architecture in 1928.

The Congresses got their start when the patroness of the new architecture, Helen de Mandrot, informed Giedion that she wished him to bring together the outstanding architects of their day at her castle in Switzerland. To those attending, the old ways of building offered no remedy to the growing misery caused by haphazard construction and overcrowding. The onus lay, they believed, on an architecture that focused on form and not function. At the end of the 1928 conference, this select group of twenty-six issued a declaration — their renunciation of "an established aestheticism." They stated that in its place the driving force would be functionality.[10] "Maximize accessibility for all to enjoy fresh air, light, and sun" became the modern architects' antidote to the scourge of Europe's slums.[11]

One architect commented a few years later that although everyone at Mandrot's castle thought the conference's participants had hit on a novel and revolutionary track, probably no one realized that Faust and Vorherr had done just that — ameliorate horrid living conditions by "building to maximize air, light and sun [more than] 100 years" before the Swiss meeting "through the work of Dr. B. C. Faust of Bückeburg and the Royal Bavarian building officer Dr. Gustav Vorherr of Munich."[12]

Germany: Once Again the Center of Solar Architecture

Post–World War I Germany provided the most fertile ground on the Continent for these new ideas in building, as the old ways seemed irrelevant to the new problems faced. "Spring of 1919 [in Germany]: revolution, runaway inflation, and housing shortages," recalled Ferdinand Kramer, an architectural student at the time. "For [us] at the Munich Institute of Technology, there were lectures on Renaissance palaces, courses in Gothic arch construction, and exercises in drawing Greek moldings. What good was such knowledge?"[13] Modern functional architecture, in contrast, addressed these acute issues. So in Germany, more than anywhere else, young architects started to look at the practical concerns of construction, such as making use of the sun's heat by means of design strategies.

In Germany, conserving coal that would be used in house heating was particularly important because the Allies occupied the country's Ruhr district — Germany's source for most of its coal. For this reason alone, according to the noted architect Marcel Breuer, a pioneer in solar design during this period, saving fuel with the sun was a major goal of the German housing movement.[14] Wilhelm von Moltke, then an architectural student, who would later become a professor of urban planning at Harvard, added that "sun orientation was important because there is not much sunshine in Germany during winter, and, therefore, it was very much a concern to capture what little of it we could."[15]

The political environment during the 1920s also prompted the development of solar housing. Left-leaning Social Democrats took power, and many German city officials sought to thwart land speculation by purchasing large tracts that were then selling at depressed postwar prices. These public properties became the sites of vast housing projects for the war-torn nation. This building explosion was aided by the efforts of cooperatives — groups of people who pooled their resources to buy land and build low-cost, high-quality housing. Sympathetic local governments often aided these housing cooperatives. Whether the land was owned by the government or by cooperatives, architects were often able to implement their newfound solar ideas.

Where Is the Sun in Winter?

Because of the acute housing shortage in Germany, large apartment complexes were favored over individual homes. Traditionally, throughout most of Europe large apartments lined the perimeter of a block, facing in all four directions. However, long narrow buildings several stories high were now being built in parallel rows. Walter Gropius, a leading exponent of this architecture, explained that "parallel rows of apartments have a great advantage over the old blocks, in that all apartments can have an equally favorable orientation."[16] But a problem arose here: architects throughout Germany began to wonder, in which direction should these buildings face for best access to the winter sun?

Most chose an east and west orientation. This was the best arrangement with regard to land economy. Called the Zeilenbau (row house) plan, this east-west orientation became the pattern for numerous long, narrow apartment complexes that went up in many parts of Germany. Ernst May designed probably the greatest number in Germany, for the city of Frankfurt. The rows of four-story buildings were erected far enough apart that no apartment blocked another's sunlight. The majority of rows ran north-south, and each apartment was only two rooms deep. The living room and balcony usually faced west, and a bedroom looked toward the east; theoretically, half the main rooms received the morning sun and the other half got the evening sun.

One architectural critic called the Zeilenbau "the ultimate in heliotropic housing, [where] *licht, luft, und sonnschein* [light, air, and sunshine] became so important that other site design factors became ignored."[17] The solar aspect excited most of the world's architectural community when these structures went up. Lewis Mumford, writing in 1933, reflected this enthusiasm: "Above all, Zeilenbau permits the orientation of the whole community for a maximum

Figure 17.3. Zeilenbau structure running north-south, facing east-west.

amount of sunlight. In every other type of plan, a certain number of rooms will face north, but in Zeilenbau, when a correct orientation is established, every room and every apartment share equally in the advantage."[18]

Not all architectural critics agreed. Adolf Behne, a respected German architectural writer, wrote sarcastically that the orientation of Zeilenbau apartments "would be correct... if the sun traveled back and forth between east and west" throughout the year. But it does not, he thundered in his critique: "And there lies the failure, since Zeilenbau architects 'boycott the south.'" As Behne saw it, they would have done far better to locate "the living areas... in the south [where] the solar exposure is... far more intensive [in winter] than that from east and west."[19]

Just as Behne had suggested, the Zeilenbau plan did not achieve the expected results. As Faust had shown almost a century before, east and west windows receive only miniscule amounts of sunlight on winter days, while in summer just the opposite occurs, heating the rooms in the wrong season. The Zeilenbau debacle further proved the nineteenth-century physician was right when he denigrated housing oriented to the east and west as "the low, the imperfect."

How did these renowned architects err so badly? The trouble was that none of the Zeilenbau architects had done the meticulous research that Faust had done before starting construction. Marcel Breuer, another invitee at the 1928 congress, gave one explanation, stating that when it came to determining the orientation of building, "most architects made their own rules."[20] Walter Schwagenscheidt, one of Germany's leading architects, proved Breuer's point with his comment, "[The] sun [like] butterflies cannot be expressed in numbers and diagrams."[21]

Hannes Meyer, Heir to Faust's Legacy

The Swiss German architect Hannes Meyer, another attendee at the 1928 conference, who as a young architect had studied under Raymond Unwin in England, could not have disagreed more with Schwagenscheidt's approach. No other architect in Europe at the time embraced functionality with more scientific precision than Meyer. Functionality in architecture meant to Meyer designing a house that satisfied the basic needs of those living in it. The dire straits of his time pushed Meyer into forgetting about frills. He built for survival. In Germany, where Meyer worked and lived, 7 million were jobless. Their families lacked basic amenities. Under such harrowing conditions, a contemporary, Teo Otto, who later enjoyed an illustrious career as a theater designer, observed, "No wonder that Hannes Meyer, driven by... the economic misery of his time, gave priority to the important, the useful, the purposeful, over good taste, beauty, the aesthetic."[22]

The idea of functionality prompted Meyer to build the sort of home that acted as a warm haven despite the cold German climate, without running up a huge fuel bill. An architect wishing to build a useful house, according to Meyer, must design its body "to be a solar accumulator."[23] "Maximal use of the sun" in winter, as far as he was concerned, was "the best plan for construction."[24]

While Meyer's rhetoric about designing structures to collect solar heat did not set him apart from those who failed miserably in its execution, his way of achieving the goal did. He believed, no differently than Faust, that success lay in obtaining the latest scientific knowledge about solar radiation. He called such data "co-determinate factors for architecture."[25] Without them, Meyer noted, the architect will surely fail to create a functional house. To gather such knowledge, he studied at Dr. Carl Dorno's Swiss Meteorological Institute, which the international scientific community rated at the time as the foremost research center in the world for solar heating and lighting.

After becoming well versed in the science of solar energy, Meyer went on to design a very large and impressive solar-oriented construction project in 1928. He presented his plan when the Federation of German Trade Unions held an architectural competition to design a building to house its school near Berlin, where workers would live while studying for leadership roles in Germany's trade union movement. Meyer's design won over those submitted by some of the most prominent German architects of the period. One of the features that impressed the judges was the arrangement of the sixty-room dormitory. It was "governed by the intention of obtaining maximum exposure to the [winter] sun for all sixty rooms," which would house 120 worker-students, Meyer explained to the architectural jury.[26]

Meyer faced all the dormitory windows to the southwest. Worker-students attending the school in summer would find their rooms almost free from unwanted solar intrusion when they returned from their seminars. Sunlight at that time of the year hardly entered the rooms. During winter, while the worker-students were in class, the afternoon sun streamed in. Returning to the dormitories in the evening, they joyfully found that the warming rays of the sun had made their down bedding fresh and fluffy. During spring and fall, too, the sun swept into the rooms. Those occupying the rooms in March and early April welcomed the incoming solar heat, because it would still be chilly during these months. Fall never heats up that much in Berlin, so the temperature gain did not irritate residents as it might in other parts of the world.

The pleasing way the windows bring in the vista of the nearby forest and lake integrates the building with its natural surroundings. Looking out from their dormitories after classes must have allowed the students to unwind after a day full of class work. The structure, which still stands today, serves as a perfect example for architects who are learning the art of constructing a true green building that combines pleasurable living conditions with functionality.

The success of the Berlin building brought Meyer an even larger commission. The city of Dessau contracted with Meyer to design five apartment buildings to house workers from a nearby factory. He completed the project in 1930. Just as Faust had suggested, Meyer faced the living area of each flat south to "get as much sun as possible" in winter. He cleverly placed on the north side of each apartment the less-used rooms, such as closets, bathrooms, storage areas, kitchen, and laundry room, as Faust had recommended, using their position to isolate the most-lived-in area of the apartment from the bad weather that blows in from the north.[27]

Figure 17.4 (above). Hannes Meyer's structure built in 1928 to instruct future leaders in the left-wing trade union movement.

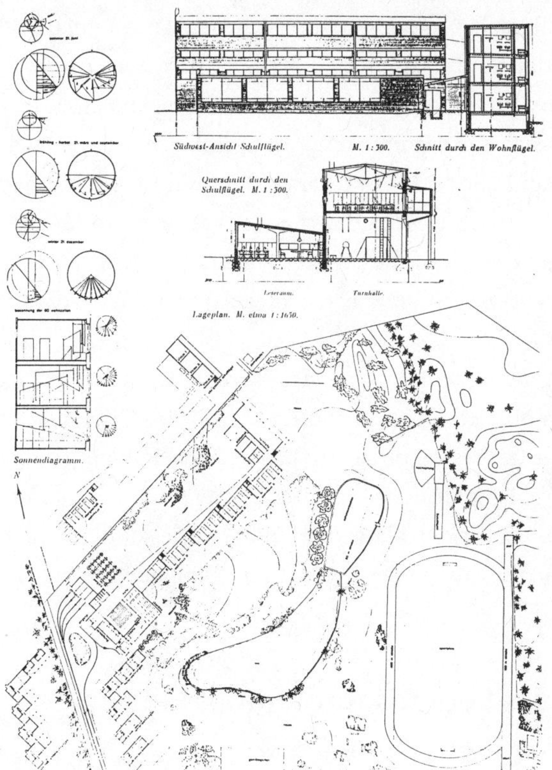

Figure 17.5 (left). Sun-angle studies at the top left corner show the course of the sun in summer, at the equinoxes, and in winter. These studies also showed how the orientation worked to keep the sun out of the building in summer and in winter.

Meyer got his chance to pass his ideas to some of the finest future architects in the world when he became the director of the prestigious Bauhaus. Here many of the greatest modern practitioners in architecture, art, music, technology, and the theater did their work and taught. Meyer let future students know that he did not believe art had any place in architecture, and that they should look at it as solely a scientific study. Under Meyer's direction, architectural students "took part in exercises...in design, orientation, the calculation of sunlight and its direction and position," explained one of his students.[28]

Attending lectures and working inside one of the Bauhaus's buildings gave the would-be architects firsthand experience in choosing between the merits of functionalism and aesthetics in building design. Walter Gropius, former director of the Bauhaus and a critic of Meyer's functionalist approach, had given no thought to the well-being of those working and studying inside the structure he had built. Its appearance to the outside world overshadowed every other consideration. Those attending classes in the typography and printing area "understood the aesthetic effect, created by the glass cube...set proud above the supporting storey of pillars," recalled Philipp Tolziner, an architectural student who had studied under both Gropius and Meyer. "Its rooms [though] became 'sweat boxes'...almost unusable...for teaching, working and especially drafting[,]...because of the way the gigantic areas of glass let in" the unwelcome solar heat in summer and allowed it to accumulate inside. One hot day spent roasting in Gropius's "sweat boxes" sufficed to convert most to Meyer's approach of designing for the comfort of the people inside rather than for some artistic notion. It "determined our position," Tolziner exclaimed. "We were for Hannes Meyer!"[29]

Others, too, followed Meyer's example. Renowned architect Hugo Häring, who also attended that groundbreaking Swiss conference, stated, "There is no doubt anymore that we, in the interest of obtaining proper solar radiation, must choose a south orientation." Häring designed rows of single-story solar homes running east-west whose main rooms had large windows on the south side and very few windows on the north. In summer, retractable awnings kept out the high summer sun.[30]

Solar Architecture Spreads throughout Europe

News concerning solar architecture in Germany spread. People all over the continent clamored for sun and light in construction and town planning. An

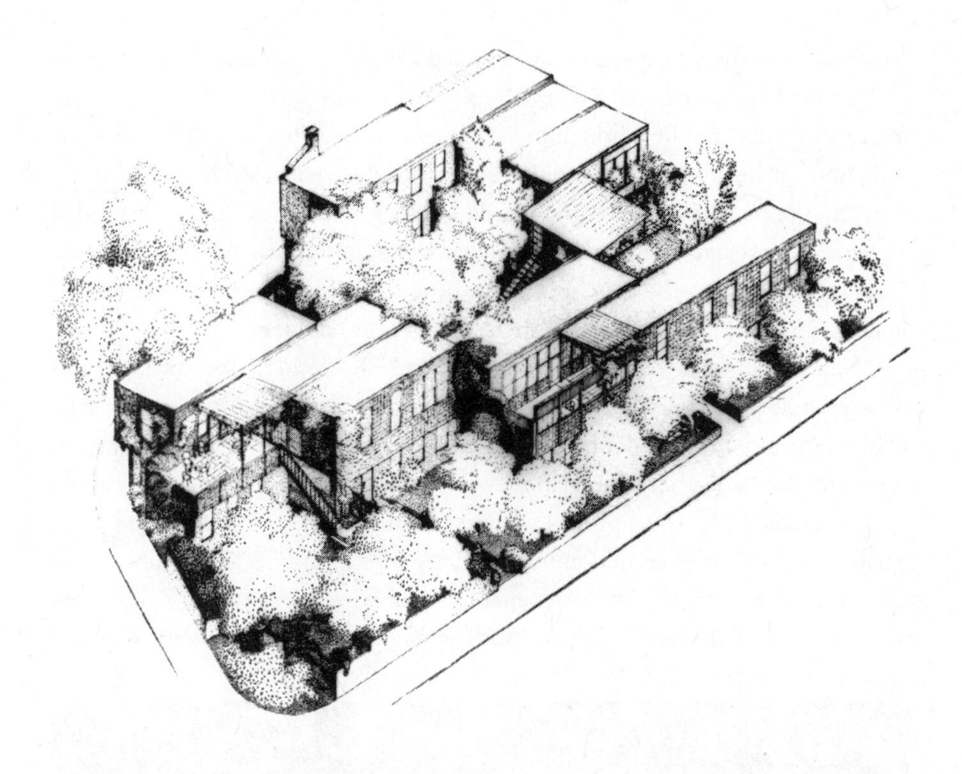

Figures 17.6 and 17.7. Two solar housing projects by Hugo Häring.

international panel of experts on insolation and lighting convened at the League of Nations headquarters in Geneva as a consequence of the public's growing interest in solar building principles. The panel found that the effects of solar radiation on heating, cooling, and ventilation were considerable. One study cited by the panel showed that a building which opened to the north needed 17 percent more heat during the winter than did a similar structure facing south. Such findings led to the conclusion that proper siting could go a long way to holding down heating and ventilating costs for householders.[31]

Nations represented by the panel included the Netherlands, Sweden, and Switzerland, countries where solar developments recently had sprung up. One of the largest and most sophisticated examples was the Swiss community of Neubuhl, now a district of Zurich. Seven young architects organized Neubuhl as a cooperative housing project. The two hundred apartments ranged from small bachelor residences to family dwellings with six rooms. These units were apportioned among thirty-three separate structures perched on a mountain slope. Almost all the buildings faced south or slightly southeast and were spaced

Figure 17.8. The Swiss solar community of Neubuhl, near Zurich, as seen from the southwest. Almost all the apartment buildings face slightly east of south.

far enough apart so that no building blocked another's solar access during winter. Every unit received the same number of hours of sunlight in winter.[32]

Glass spanned the entire south wall of each living room as well as a good portion of the south bedroom wall. The kitchen, stairwell, bathroom, and pantry occupied the north and northwest parts of the apartment. Consequently, the kitchen remained cool in the summer, an advantage when the stove was in use. Retractable wooden shades and canvas awnings were rolled out to prevent the summer sunlight from entering the south-facing rooms. The considerable decrease in fuel consumption in winter was great enough that it surprised the project's designers.[33]

Unfortunately, despite all the progress, solar building techniques fell victim to the "culture wars" that erupted in the late 1920s in Germany and pitted liberals and socialists against Nazis. The entire modern architectural movement enraged Nazi stalwarts. It betrayed Teutonic culture, they charged. Consequently, when Hitler gained control of Germany in 1933, the practitioners of modern architecture fell from favor. Their opportunities for work ended, and modern architectural magazines that had flourished in the late 1920s and early 1930s withered and died.

Figure 17.9. Architecture during Nazi rule once again embraced a medieval ideal, one that architects from Faust to Meyer had tried to eradicate.

Figure 18.1. Close-up view of the dwellings at the solar city of Acoma, New Mexico.

[1200–1912]

18

Solar Heating in Early America

American solar architecture began with indigenous peoples. Some highly sophisticated solar communities were built by the Pueblo Indian tribes of the American Southwest. During the eleventh and twelfth centuries CE, the Anasazi Indians built a number of large community structures — some were south-facing cliff dwellings, and others were situated out on open plateaus — that display a remarkable sensitivity to the sun's daily and seasonal movements. Such well-known ruins as Long House at Mesa Verde and Pueblo Bonito in northern New Mexico were built during this classic period of Anasazi culture.

Like other ancient cultures, the Anasazi had a keen sense of the changing location of the sun throughout the year. Instead of using gnomons, Native Americans in the Southwest commonly relied on geophysical formations on the horizon as markers for the changing seasons. For example, to ascertain the coming of the winter solstice, an observer can watch from a specified point the sun getting nearer every day to a natural sandstone column. When the sun rises directly behind the column, it signals the arrival of the shortest day of the year. "Sun watching," according to noted astronomer Michael Zeilik, "was critical for keeping the calendar" of the Anasazi people.[1]

One of the most sophisticated examples of Anasazi solar architecture, no doubt springing from the culture's rich knowledge gained from sun watching, is the "sky city" of Acoma. Like the Greek city of Olynthus, Acoma was built atop a plateau. It has three long rows of dwelling units running east to west. Each dwelling unit has two or three tiers placed in a way that allows every residence full exposure to the winter sun. Most doors and windows open to the south, and the walls are built of adobe. The sun strikes these heat-absorbing south walls much more directly in winter than in summer. And the horizontal roofs of each tier, built of straw and adobe layered over pine timbers and branches, insulate the interior rooms from the high, hot summer sun.[2]

Figure 18.2. Distant view of Acoma.

A study conducted by Professor Ralph Knowles of the University of Southern California proves just how well these Pueblo dwellings are suited to their climate.[3] In winter, over one-third of the sun's heat reaches the interior, while in summer only one-quarter makes it inside the rooms. Even more surprising is the orderly town plan that guarantees all residents full, equal access to the sun's heat. Continuously inhabited longer than any other community in North America, Acoma has likely more than fulfilled the expectations of its ancient designers.

The Spanish colonists, too, often built according to a solar plan when they settled in the American Southwest. Their dwellings took the form of single homes not very different from those common in many areas of Spain. The typical Spanish Colonial villa consisted of an adobe building oriented east-west with the main rooms facing south. The south wall absorbed sunlight during the day, and in the evening this heat was released from the adobe wall into the home. Shutters on the windows helped keep the heat inside the house at night. In summer, eaves sheltered the interior from the high, intense sun.[4] Spanish Colonial architecture did not survive the later influx of Yankees from the eastern United States, however. These immigrants, whose ancestral roots were in England and northern Europe, did not understand how well adapted the adobes were to their environment. They often replaced these structures with wood houses better suited to the New England climate.

Figure 18.3. An early California family in front of their Spanish adobe.

But settlers in the eastern United States were not ignorant of climatic considerations in their own building designs. In New England, the colonists often built "saltbox" houses facing toward the sun and away from the cold winds. These structures had two south-facing, windowed stories in front — where most of the rooms were placed — and only one story to the rear of the building. The long roof sloped steeply down from the high front to the lower back side, providing protection from the winter winds. Many of these saltbox houses also had a lattice overhang — a pergola — protruding from the south facade above the doors and windows. Deciduous vines growing over the pergola afforded summer shade but dropped their leaves in winter, allowing sunlight to pass through.[5]

Figure 18.4. Saltbox house with trellis.

In the mid- to late nineteenth century, solar building became a lost art in the United States. From the early 1860s to the late 1870s, nearly 5 million people poured onto America's shores — the vast majority of them ending up in eastern cities like New York, Philadelphia, Boston, and Baltimore. As the East Coast became increasingly urbanized, the commonsense designs developed by rural residents were all but forgotten. Where abundant open space had once allowed easy solar orientation, the arbitrarily planned street grids now made such orientation difficult. As in Europe, cities on the Eastern Seaboard became crowded with multistory tenements inhabited mainly by impoverished workers. Conditions in these cities were as bad as in the filthy, cramped, working-class quarters of European towns. In the 1870s, however, some of the more well-to-do began to desert the urban centers and settle in the surrounding countryside. But suburban housing was not much better. Architects like Bruce Price criticized the acres of "hideous structures" around New York and Philadelphia — suburban homes that were built shoddily, oriented randomly, and lacking adequate ventilation. The critics recalled a time when people built climate-sensitive homes, such as the Colonial saltbox. In response to this trend, Price and a group of his colleagues began to design houses that remained naturally cool in summer and warm in winter. "The heat of the summer demands shady porches," Price declared, and "the cold of winter[,]...sunny rooms."[6] In his houses, he placed louvers so that summer breezes could cool the rooms; they also had low ceilings and well-insulated walls to minimize winter heat loss.

Urban Solar Access

The work begun by Pierce and others did not spread to the cities. As in Europe, land speculation put a premium on urban space, and the resulting high-density developments made solar exposure difficult. The housing crunch continued to tighten as the population skyrocketed. Some 9 million immigrants landed between 1881 and 1900. With more and more people clamoring for housing in major cities, architects began building vertically.

Around the turn of the century the problem of solar access came to the attention of William Atkinson, a reform-minded Boston architect. The city's population had more than tripled in the previous fifty years, prompting a move to construct taller and taller buildings. Atkinson saw that "the sky scraper enjoys an advantage of light...at the expense of lower and more ancient buildings." He drew diagrams showing how high-rises shaded the surrounding buildings, and presented his findings to the Boston city council in 1904. He convinced them that it was essential to guarantee access to the sun, and legislation was soon passed restricting the height of new buildings in Boston.

Atkinson had first become curious about ways to maximize solar exposure in 1894, when he was working on the design of hospital buildings. He found no rational criteria for the proper orientation of hospitals. So he decided to investigate how much sunlight penetrates windows facing in different directions. His research took him to the Harvard Observatory, where he learned, as so many had before him, the changes in the angle of the sun over the year and how to use that knowledge for the natural heating and cooling of buildings.

One thing that no one had ever measured was how much solar heat each window orientation might trap in summer and winter. Atkinson did this by building a device he called a "sun box," which was similar to Saussure's hot box — except that Atkinson's box simulated a room and window. The sun box had an inner wooden shell measuring 1 foot square that was covered on one side, perpendicularly to the ground, by a single sheet of glass. The other sides were surrounded by insulation, an air space, and an outer box lacking a covering.

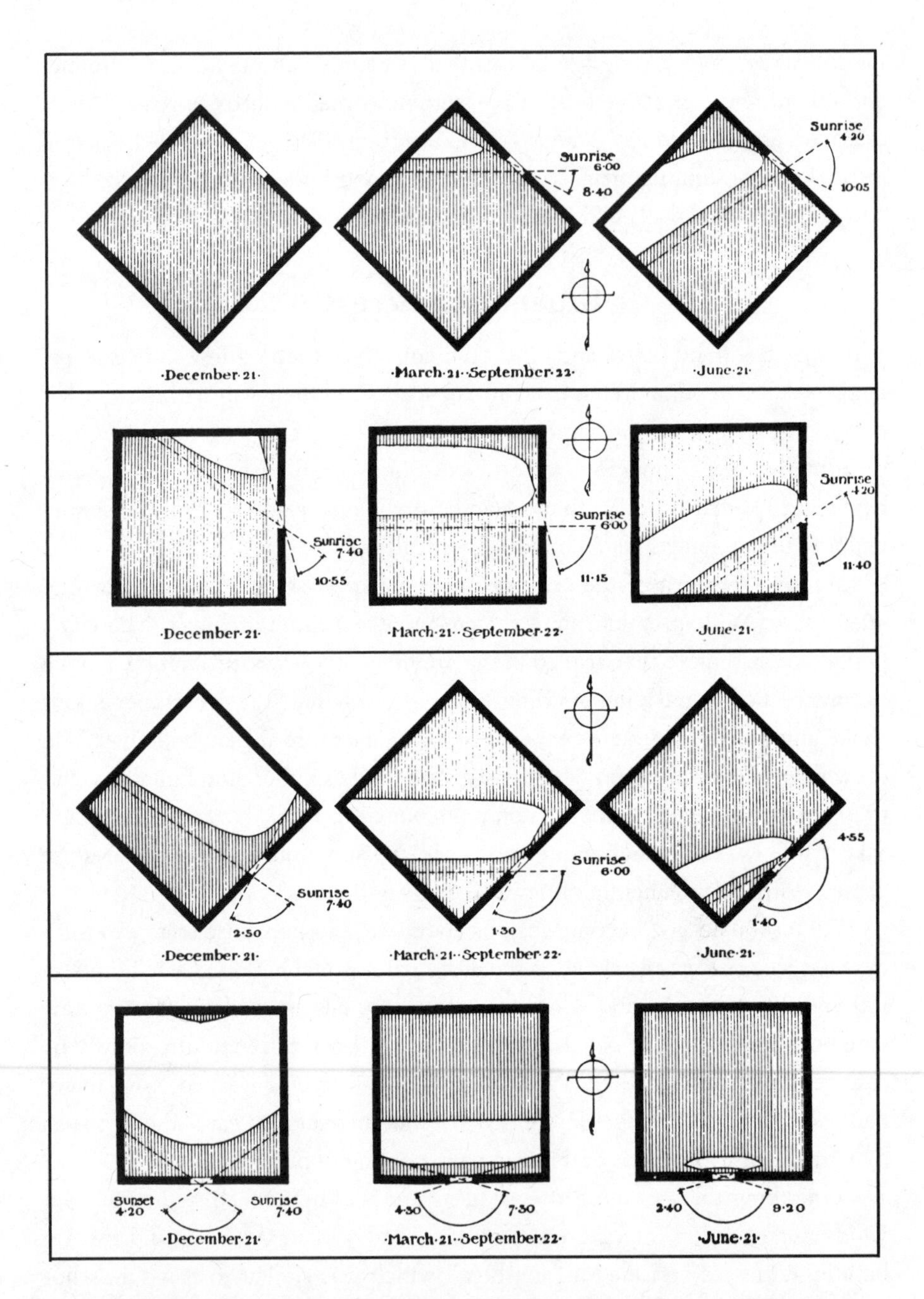

Figure 18.5. Window studies showing solar penetration into rooms at different times of the year for different orientations — northeast, east (west in the afternoon), southeast, and south.

Atkinson built two sun boxes and exposed them to the direct rays of the sun. Between the months of June and December, he changed their positions periodically and collected data for orientations toward the west, southwest, south, southeast, and east.

Figure 18.6. Sun boxes used by William Atkinson to mimic windows with different orientations.

The summer tests confirmed his hypothesis that east and west windows admitted too much summer sunlight. On June 21, for example, an east-facing box reached more than 120°F by eight o'clock in the morning — 52° hotter than a similar south-facing box and 42° hotter than the temperature outside. By afternoon, a box facing toward the west registered over 144°F — 44° more than the south-facing box and 66° above the ambient air temperature. It was no wonder people suffered in rooms oriented in this fashion during summer. The most spectacular results, however, occurred on December 22. The temperature of the south-facing box rose to 114°F, when it was only 24° outside, leading Atkinson to conclude that if houses are properly situated, "the sun's rays are not of indifferent value in the heating of our houses."

In 1912, Atkinson built what he called a "sun house" for Samuel Cabot, a

Figure 18.7. Atkinson's sun house on the property of Samuel Cabot.

wealthy Boston aristocrat, to determine more precisely the heating potential of the sun. Actually a small shed, the building stood next to Cabot's spacious mansion. It was long and shallow, with glass spanning the entire south face so that the sun's rays could fully penetrate the interior. The other three walls and the roof had generous amounts of insulation to help retain the trapped solar heat. Atkinson reported that despite freezing weather outdoors, "a temperature of 100°F and over has been frequently attained within this building...entirely from the warmth of the sun's rays." The dramatic success of the sun house led him to exclaim, "Every dwelling may be converted into a sun box!"[7]

Optimistic about the potential of solar heat for buildings, Atkinson published these findings in 1912 in his aptly titled book *The Orientation of Buildings; or, Planning for Sunlight*. John Wiley and Sons printed a thousand copies. Though brief, *The Orientation of Buildings* is a gold mine of information, rich in visuals and data. The book also contains extensive sun-box performance records, providing temperature readings for the differently positioned boxes from June through December at various times of the day. However, few, if any, of his contemporaries made use of the material presented in the book. Atkinson's ideas on solar architecture had to wait for their realization by the next generation of architects and builders.

Figure 19.1. Green's Ready-Built Homes of Rockford, Illinois, developed this prefabricated solar home for the postwar housing market.

[1931–1950s]

19

An American Revival

For two decades after Atkinson published his book, new ideas in solar architecture were scarce in the United States. But progress made in Europe during the years after World War I helped to stimulate a new wave of interest there. A consensus on optimal orientation had yet to be reached, but reports that aimed to finally resolve the issue began to appear in the 1930s. One of them, produced by the Royal Institute of British Architects, sparked new interest across the Atlantic. In its introduction the report observed, "During the last few years an extraordinary and even revolutionary change has taken place in all countries in the general appreciation of the values of fresh air and light, particularly sunshine."[1] The authors lamented the existence of only "a mere smattering of data, and that of little or no practical use," and noted that "such meager information as is to be found entails a prohibitive amount of labour before it can be used to solve even simple problems in practical design."[2] They advised that the report was "an attempt to supply the information which shall offer hope of adoption in busy drawing rooms."[3]

In their pursuit of more precise measurements, the report's authors made use of a recently invented machine called the heliodon. Consisting of two main, independent parts, the heliodon had a spotlight that mimicked the sun when an operator moved it up or down a vertical bar to simulate the months and seasons, which were marked off along the bar. And it had a model house that sat on a horizontal table, separated by a specific distance from the spotlight and vertical bar. This table could be tipped to an angle corresponding to latitude and rotated to depict the changing hours of the day. With the spotlight projecting onto the model, the operator could determine changes in solar penetration of a building for different orientations throughout the day and year at the desired latitude. The data that most surprised the English involved east, west, and south orientations. The heliodon showed that while sunlight would pour into rooms facing east or west at the summer solstice, very little came in at the winter solstice. In contrast, the sun barely entered the south-facing room in summer but flooded it in winter.[4] Here was incontrovertible evidence showing the failure of the Zeilenbau plan, with its east-west orientation for buildings, on the one hand, and confirming the building ideas of Socrates, Faust, Meyer, and Atkinson on the other.

The heliodon soon found its way into the studios of American urban planners like Henry Wright of Columbia University. In his influential book, *Rehousing Urban America*, Wright advocated using information derived from this

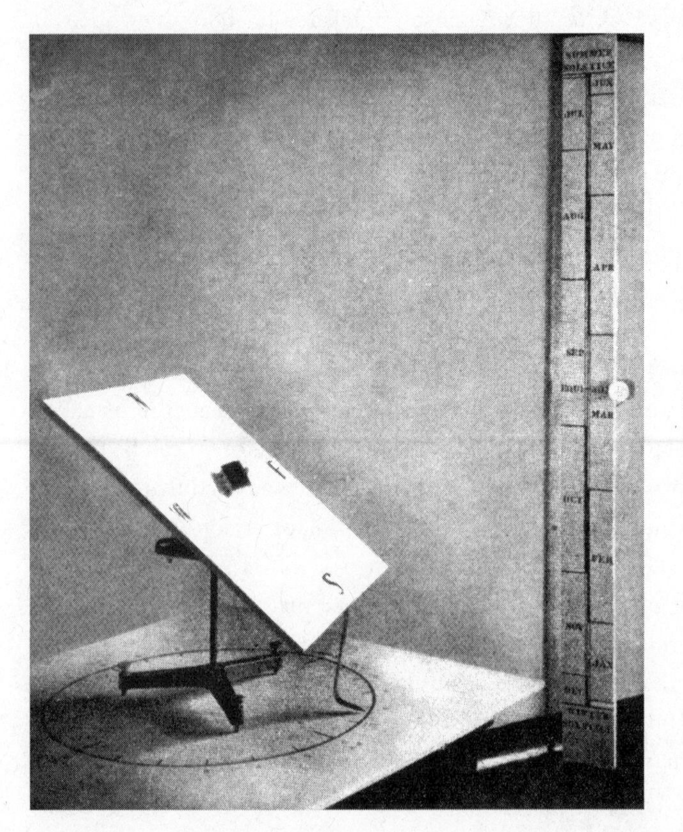

Figure 19.2. The heliodon.

machine to determine how to best take advantage of the sun's heat. His book circulated in manuscript form to many academic architects in 1933 and was published in 1935. Wright's conclusions about orientation were similar to those that the Royal Institute of British Architects had outlined in their report.

Another piece of empirical evidence that supported the preference for a southern orientation appeared in 1934. While experimenting with heat-absorbing glass, invented to keep buildings cool in summer, the American Society of Heating and Ventilating Engineers (ASHVE) unexpectedly made an important discovery. They built two identical test structures, one with a south window made of heat-absorbing glass and the other with a south window made of ordinary glass. The building with the ordinary window, which transmitted the most winter sunshine, took 9 percent less electricity to heat. Astonished, the president of the society remarked: "It has long been the custom to allow for extra heating on the north sides of buildings...because of the cold winds from the north. I think we are coming to realize that the provision of extra heat on the north side is not so much due to this effect as to the absence of sun heat!"[5]

Bundling the data gained by the use of the heliodon and the study by ASHVE into a meaningful whole, researchers at Yale University's Pierce Laboratory computed the amount of sun heat absorbed during the changing seasons by walls oriented in different directions. A southern orientation was found to be the best for winter heating and summer cooling.

From these studies it seemed that to best collect solar heat during winter would require a great expanse of glass on the south side of a house. Conventional wisdom suggested that all the heat captured by such windows would pour out of the house through the glass once the sun went down. But calculations by Columbia University graduate student Harry Fagin in the 1930s demonstrated the contrary. Fagin reported in his master's thesis that if brick or plaster, instead of glass, covered the entire south side, the heating bill would be far greater.[6]

A Boom Begins in Chicago

George Fred Keck, one of America's top architects, was the first in America to dare to translate the conclusions of the British and American sun studies into house design. Without access to institutional funding, however, Keck could only try out these proven facts on the homes he built for private clients. For this reason he had to be cautious. Many still held, despite the Columbia University study, that large windows would result in catastrophic heat losses during winter. Even Keck remained skeptical, writing to a colleague: "So little is known about heat losses and heating bills [for those] living under normal conditions that I have hesitated to increase the glass size for fear of large fuel bills for an owner."[7] Consequently, his approach was gradual. "Each year we would build a small house for somebody," he recalled, and "each year we tried to orient it [toward the midday sun] and open more and more glass to the south."

Keck periodically visited the homes he designed, recording their fuel consumption and the owners' reactions. He found that the more glass he used on the south wall, the more solar heat the house gained. Buttressing his convictions, greenhouse owners in the area told him that their furnaces were shut down from dawn to dusk — even on the coldest days — if the sun was shining brightly.

Figure 19.3. George Fred Keck.

The hottest days also concerned Keck, as he wanted those living in the houses he designed to stay comfortably cool in summer. He regretted that the roof of one of the first south-facing houses, built for William Fricker in Wisconsin, was painted black, because he felt it ought to have been white to reflect rather than absorb the high summer sun beating down. However, the placement of aluminized venetian blinds on the outside of south-facing windows did help to reflect the summer sun. As another strategy to keep the house cool, Keck advised the owners to open the windows at night and close up the house during the hottest hours of the day.[8]

Results from Fricker's and other houses were encouraging. That Fricker's house had never heated up to more than 78 degrees during the previous, very hot, midwestern summer was sweet news to Keck's ears. Keck also found validation of his design ideas in the owner's extremely low heating bills.[9]

Figure 19.4. The Fricker House.

The discovery and refinement of double-pane insulating glass during the mid- to late 1930s by the Libbey-Owens-Ford Glass Company also gave Keck's work a great boost, because this glass would retain more heat in his solar-oriented homes than the old single-pane type did. These new windows were made of two sheets of glass hermetically sealed together with a ½-inch air space between them. They cut 50 percent of the heat losses normally experienced with conventional window glass.

After seven years of experimentation, Keck finally felt confident enough to expose the entire south side of a house to the sun. His opportunity came in 1940, when he designed a house for an old friend, Howard Sloan, a Chicago real estate developer.[10] Sloan and his family truly enjoyed the house. He found the sun "a real aid in heating." "Day after day [the interior temperature] reaches 75 and 76 [degrees], relieving the furnace for 8 to 10 hours even in extremely cold weather," Sloan wrote in the architectural journal *Pencil Points*.[11] Sloan liked the house so much that he decided to develop a subdivision featuring Keck's bold solar-design ideas. He explained how he promoted this new type of architecture: "The house was opened to the public in September as the Solar House. On one Sunday we had 1,700 visitors. The demand of the public was such that I subdivided 10 acres into 38 lots and opened it in April, 1941. [Although] Hitler was overrunning countries in Europe, customers were becoming jittery, [and] prices were going up, houses sold faster than we could build them."[12]

Figure 19.5. The first Keck house with the entire south side walled with glass.

Figure 19.6. Two south-facing homes in Sloan's Solar Park, the first solar-oriented subdivision in the United States.

Figure 19.7. The Duncan House.

Putting Solar Homes to the Test

The Sloan project greatly boosted Keck's ideas in the architectural world. Still, not everyone agreed that solar homes used less energy. Keck compared such skepticism with doubts about the efficacy of motor cars years before: "I am personally old enough to remember when the first two came to our town up in Wisconsin and got stuck or had flat tires; and a crowd would develop on the street, and I would hear my elders say, 'Give me a horse and wagon anytime.' This is the position we find ourselves in at present," he complained.[13] The Chicago architect pushed for the testing of a house designed to his solar criteria to validate his claims.

Keck eventually persuaded Libbey-Owens-Ford, developers of the double-pane window glass used in all of Keck's homes, to sponsor a yearlong study of solar heating. Tests were conducted by the Illinois Institute of Technology under the guidance of Professor James C. Peebles and William Knopf, a graduate student.

The two engineers tested the solar performance of a typical solar home built by Keck belonging to Dr. and Mrs. Hugh Duncan of Flossmoor, Illinois. Knopf and Peebles wanted to study conditions in a real-life situation rather than set up an artificial laboratory experiment. In this house, they found that even on the coldest days those living in it had to open doors and windows to cool down the house when the sun was shining. Such periodic ventilation made it difficult to calculate how much solar heat had actually collected inside the house. The radiant heating system used was partially responsible. Because it took a long while for the heat from hot water to conduct through the massive floor slab, it continued to be released long after the thermostat had shut off the heating system. This added heat made it uncomfortably warm inside the house. And when the windows were opened to cool the house down, more heat often escaped than was desirable. Under such conditions, Knopf and Peebles reached the following conclusion: "A house of similar design equipped with a heating system better adapted to fully utilize the available solar heat input should show a substantially lower heating cost than that recorded in this test."[14]

Other problems that made the test results unclear were the random opening of doors by residents entering and leaving the house, and the loss of heat through cracks around doors and windows, which had not been caulked as stipulated. The editors of *Architectural Forum*, who analyzed the institute's report, estimated that the sun provided between 7 and 18 percent of the home's heat, leading the journal to conclude in its headline, "Survey proves large windows properly oriented save fuel even in rigorous climates."[15]

Although the researchers were unable to determine the precise solar heat gain of the Duncan house, it was apparent that a solar house with fewer cracks to cause heat leakage and with a more compatible heating system would collect a substantial amount of heat from the sun. And there was another energy-saving dividend: the large south windows created an ambience of total illumination. The editors of *Architectural Forum* reported, "At no time during the daylight hours did they require artificial light," which translated into reduced utility bills.[16] This reduction was not calculated in the study by Knopf and Peebles, since the engineers had focused solely on the heating aspect of solar energy. In fact, the effect of natural daylight in the Duncan house could have accounted for the apparent confusion over the measured heat gain from windows and the surprisingly low heating expenses, amounting to one-third less than the engineers had anticipated. Sloan explained why: "We never feel as chilly in a well lighted room as in a dreary one. Studies have been made showing that 67 degrees in a well daylighted home is as comfortable as 70 in a poorly lighted one."[17]

In a subsequent test, the local utility rigorously calculated the amount of daylighting in a Keck-designed school compared with one that was not solar oriented. Results showed that at the latter school, students sitting in the last row received a light intensity of 10 foot-candles, while those farthest from the windows in the solar school received six times as much. These results led one writer on the subject of solar houses to remark, "If solar architecture had no other advantages at all, it would more than justify itself through the daylight it pours into rooms."[18]

One piece of news from the Duncan house aroused national interest: on a frigid morning when outdoor temperatures dropped to -17°F and never rose higher than -5°, the house interior warmed up so quickly from the heat of the sun that by 8:30 in the morning the thermostat, set at 72°, shut off the furnace, which didn't go back on until midevening. Reporters and camera crews rushed to the Duncan house, and almost every news outlet in the nation picked up the story. *Newsweek*, for example, stated, "One way for Americans to hedge against future fuel shortages would be to build more solar homes like that of Mr. and Mrs. Hugh Duncan."[19] Journalist Ralph Wallace's report of Keck's work found its way to *Reader's Digest*. In it he praised the success of the solar house as "the most exciting architectural news in decades." Wallace told millions of readers that the monthly amount of heat from the sun pouring into the interior of a solar home equaled the burning of a ton of coal. "In effect," Wallace wrote, "the sun becomes your private coal mine."[20]

To satisfy public interest, Libbey-Owens-Ford published a short booklet on solar houses. Earl Aiken, the company's liaison to Keck, wrote to the architect: "I find we are hopelessly swamped with requests. We could get rid of more than 150,000 in a flash."[21] The demand for this superficial publication whetted Keck's appetite for writing a more substantial work, "a 'high class' 'how to' book," according to Keck, that would include chapter headings like "Fitting the House to the Climate" and "Orientation and Solar Principles."[22]

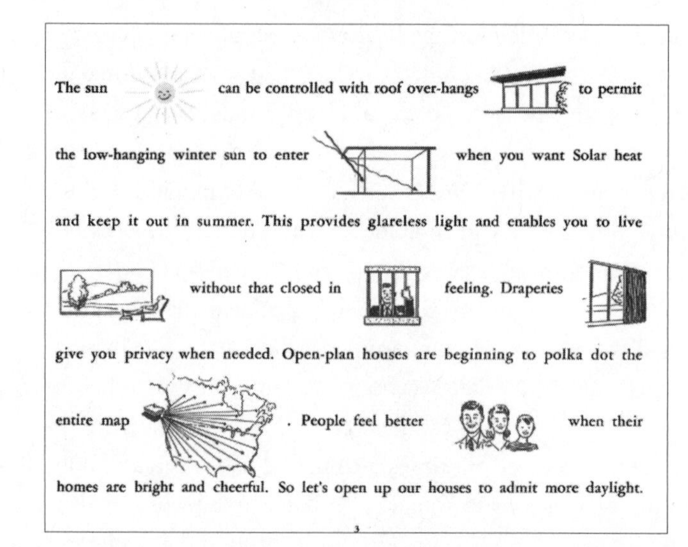
The sun can be controlled with roof over-hangs to permit the low-hanging winter sun to enter when you want Solar heat and keep it out in summer. This provides glareless light and enables you to live without that closed in feeling. Draperies give you privacy when needed. Open-plan houses are beginning to polka dot the entire map. People feel better when their homes are bright and cheerful. So let's open up our houses to admit more daylight.

3

Figure 19.8. Page from a short booklet published by Libbey-Owens-Ford popularizing Keck's work.

Although Keck never completed the book, fragments of intended chapters remain. They present Keck's architectural vision. First of all they reveal that he wanted to establish a truly American house free from its ancestral ties to Europe, just as Faust had wished to sever the medieval constrictions binding nineteenth-century architecture. "For houses to become completely indigenous," Keck wrote, "they must be designed for the climate in which they are to be placed; they must take the best advantage of natural phenomena [that is] possible." In most of the United States this meant "it would be very desirable to design [a house] in such a way that in the summer time, when we have too much solar radiation, no sun strikes the house," and so that in winter, when it is cold, "as much sun could strike and come into the house as possible." He added, "[This] is not very difficult to do, if we know enough" about the yearly changes in the sun's position in relation to the earth. When we realize that "the sun is always south of us, high in the sky in summer and low on the horizon in winter, we can plan to take advantage of these facts."[23]

Insulation materials would also help keep out the cold in winter and the heat in summer. Snow, too, assists in winter, especially, as Keck observed, "a good layer on the roof, if the roof is comparatively flat, [because it] forms added insulation[;] and snow around the north side of the house also insulates the house from the biting cold north winds." Keck concluded by noting that "it may be wise to dig in and allow the snow to heap around the house, just as many wild fowl dig themselves into the snow banks in winter to keep warm."[24]

A house that has been planned to take full advantage of the climate offers many benefits, according to Keck's jottings. Aside from the economic gain from lowering utility bills, "there is also the advantage of additional comfort, convenience, and cheerfulness." In Keck's view, "there is no heat comparable to solar radiant heat[;] throughout the winter the house has the feel of spring in it when the sun shines." In fact, Keck believed, just as Faust had so many years before, that

> such a fundamental concept in planning room arrangements cannot be wrong[,] for it is based upon taking the greatest possible advantage of the natural phenomena at any given site. Here is a sensible, straight forward approach to the problem of planning which has long since been recognized in certain other types of structures[,] such as chicken barns and structures for housing animals in the country. We seem to have overlooked the fact that a structure that might house a healthy chicken and make her lay an extra egg might apply to human beings.[25]

Figure 19.9. Sun entering the living room of a Keck-designed house in winter provides warmth and light for both the woman of the house and her dog.

Solar Building Takes Off

Keck never got beyond writing the initial chapter synopses. When he had first proposed the book, the outcome of World War II remained in doubt, which meant that most resources were allocated for the war effort, leaving little for domestic projects such as housing. Instead of making glass for house windows,

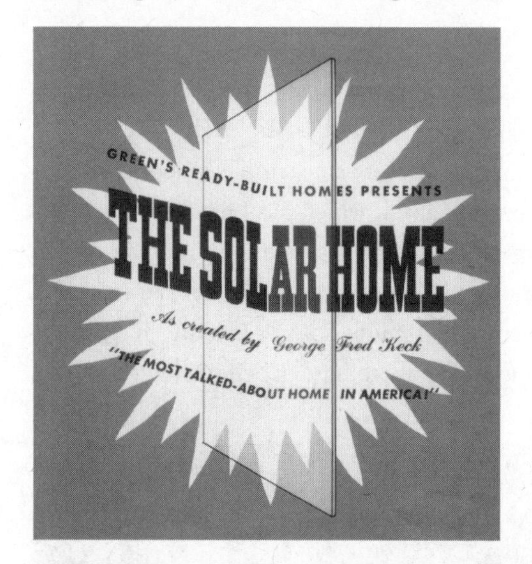

Figure 19.10. Green's Ready-Built Homes touted Keck's solar housing.

Figure 19.11. A recently born baby boomer and parents lounge outside their solar home.

for example, a company like Libbey-Owens-Ford focused on manufacturing coverings for aircraft, tanks, and other types of war craft. And wood that would have been used in peacetime for house building was now assigned mainly to the manufacture of ammunition boxes. This meant Keck had time on his hands to contemplate a book project. But at the end of the war, the Chicago architect noted that "never in American history has there been an opportunity like the present for the development of something new" in architecture, such as the solar house.[26]

When peace finally came, the housing market boomed as never before. Demand for solar houses so dominated his time that Keck did not have a free moment for a pursuit such as book writing. A nationwide construction firm, Green's Ready-Built Homes, hired Keck as its architect and touted the solar home as "the most talked about home in America."[27] Revere Copper and Brass, for example, produced *Homes for Tomorrow's Happy Living*, a booklet featuring the solar house. In its introduction, the company's president hailed Keck as the man who would show the way for the inevitable post–World War II architecture: "Architects, engineers, builders, [and] manufacturers are realizing that we will want homes that adapt to people, that people must no longer adapt themselves to homes."[28]

Many solar-oriented housing developments soon went up, ranging in size from fifteen units to well over a hundred. In the Chicago suburb of Northbrook, developer Howard Sloan, for example, built eighteen two- and three-bedroom houses that sold for fifteen thousand dollars each. The homes on the north side of the street faced south, while those on the south side had large glass windows facing the backyard.[29] At the same time, individual, architect-designed custom solar houses sprang up all over the country.

Controlling the Solar Heat

But even while greater numbers of solar homes went up, Keck readily admitted that "so far only the surface has been scratched in this matter of investigation of solar radiation and its possibilities for future development."[30] One detail that soon became clear was that solar houses built solely with wood would overheat under full sun and immediately cool down when a cloud obstructed the sun. But in those built with concrete floors, the opposite occurred. The floors, acting as a solar battery, soaked in the solar heat when sunlight poured into the house and gave up that heat to the interior as the rooms facing south cooled down.[31]

An innovative Arizona architect, Arthur Brown, developed another type of

Figure 19.12. Architect Arthur Brown.

heat storage system for solar houses. In 1945 Brown built a long, narrow home in Tucson with the broad sides of the building facing north and south. For aesthetic reasons the couple who owned the house wanted one of the outside walls painted black. Knowing that a black wall on the east or west would get too hot in the torrid Arizona summer, he suggested painting the south wall. One winter day while inspecting the house, Brown was startled by the amount of heat radiating from this black wall. "I could feel it five feet away," he remembered, "and I thought that the next time we do a house, we'll paint the wall inside the hall black so that we won't lose that heat."

Brown's next house was also a long, narrow building, but this one had a south wall made entirely of glass, which covered a sun porch and dining room, and a black-painted concrete wall separating them from the bedrooms, living room, and kitchen. During clear winter days, sunlight struck the black central wall, and solar heat penetrated the concrete. Brown had estimated that heat moves through concrete at the rate of 1 inch an hour, so he built the wall 8 inches thick. His estimate turned out to be correct, and by evening solar heat would fully penetrate the wall and begin to radiate from the back side, into the rooms in the northern part of the house — the bedrooms, living room, and kitchen. By morning, the black concrete wall would be cool, ready for another cycle. Brown also recommended that the curtains be drawn soon after sunset to help the rooms retain the solar heat. Although this heat storage system was never tested quantitatively, the house's owners reported that it definitely increased the sun's share in heating the home.[32]

Solar houses became such an item by 1945 that Henry Wright, who had helped early on to promote Keck's ideas in *Architectural Forum*, and his coauthor, George Nelson, devoted a whole chapter to the subject in their 1945 book, *Tomorrow's House*. They suggested that one way to keep heat from passing out a window and being lost was by the use of "a good drapery, lined and interlined with heavy material" to insulate the window at night. The purchase of the item

Figure 19.13. Brown's solar house with the south wall painted black.

Figure 19.14. Sun room with wall inside separating it from the rest of the house.

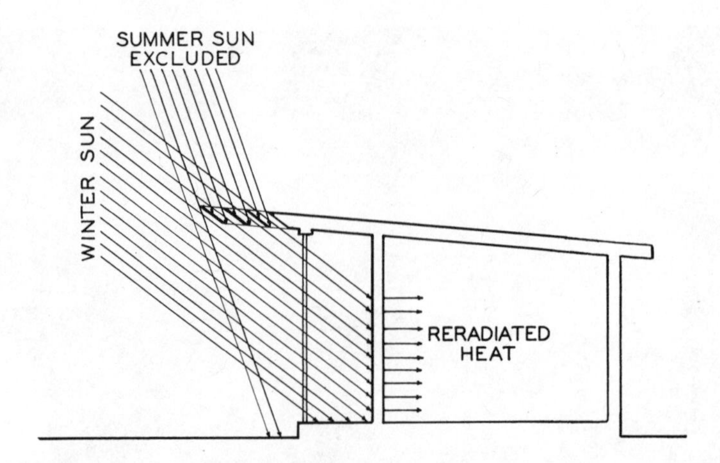

Figure 19.15. Diagram showing how the solar wall heated the interior during winter.

made an excellent investment, the authors concluded, since it "pays for itself in reduced fuel bills long before wearing out."[33]

As time went by, Keck, too, saw additional ways to improve the effectiveness of solar-oriented homes. For example, he suggested making eaves more flexible so they could be narrowed in the spring, when it was still cold, and widened in the autumn, when most people would not desire additional heat.[34]

The renaissance of solar architecture renewed the controversy among engineers about the effectiveness of south-facing glass windows. As one engineer put it, "Substantial reductions in seasonal heating costs are said to have been found from some installations while results of an opposite order are attributed to other installations. Where, in this confusion of claim, opinion and estimate, does the truth lie?" In an attempt to answer this and other questions, Libbey-Owens-Ford sponsored a second study of solar house heating, which was conducted in 1945–1947 by Dr. F. W. Hutchinson, professor of mechanical engineering at Purdue University. Hutchinson found that one of the primary reasons for the variable results was the lack of reliable information on the optimum ratio of glass area to floor area. This ratio should vary with climate, as Keck had hinted in his chapter synopses; if the glass area were too large for a particular climate, the house would overheat and solar heat would be wasted. If the windows were too small, insufficient solar heat would be obtained. In response to this information, Hutchinson developed a set of tables that allowed architects to make more accurate estimates of the proper amount of south-facing glass necessary in their particular region.[35]

The Decline of Solar Architecture

In 1947, Simon and Schuster brought out the book *Your Solar House*, edited by Maron J. Simon, the company's publisher, and sponsored by the manufacturers of double-paned glass Libbey-Owens-Ford.[36] The book's proposed audience were the "millions that seek information" regarding the solar house. Its purpose was to enable the American public "to take advantage of the gifts of nature that lie forever within such easy reach."[37] Its contents included "49 sets of plans and drawings" for solar houses suited to "all parts of the country, by 49 of America's leading architects."[38]

As the publishing plans unfolded, the management of Libbey-Owens-Ford told Keck that the "book will be absolutely 'tops' in its class and should go a long ways in furthering the type of architecture you so ably pioneered."[39] When *Your Solar House* finally hit bookstores in 1948, the glass company found it difficult to market the books as planned. "Most newspapers are holding off publicizing its publication," the management at Libbey-Owens-Ford lamented, because they "feel that spiraling costs have placed the houses in a bracket too high for them to promote editorially[. And] many builders are hesitant, too, some because of materials and labor scarcity, others because they feel a lower cost house is the only practical participation for them."[40]

But the legacy of Keck lived on. The Duncan house, for example, had become the Cohen house by the time the *Chicago Sun-Times* wrote about it in 1979: "The temperature can dip to zero, but if the sun is shining, the Gerald D. Cohens turn *off* their furnace as soon as they get out of bed in the morning. Otherwise the house gets too warm." The *Sun-Times* reporter later visited the house "on one of the most stiflingly hot days of August." Sitting in the living room, she found the interior "pleasant although none of the air conditioners was operating." When comparing heating bills with neighbors living in comparable homes, the Cohens found that they paid 25 to 50 percent less.[41]

Figure 20.1. Southwest view of MIT's fourth solar house in Lexington, Massachesetts. Planned as a prototype for mass construction, it obtained half of its winter heat from the sun.

Solar Collectors for House Heating

During the 1940s, a number of engineers and scientists tried a new approach to solar house heating. They decided to use solar collectors resembling a hot box or the Sun Coil for this task. The low-temperature solar heat from these collectors was well matched to the heating needs of a home. Many of the experimental homes at this time also had a heat storage unit separated from the solar collector. The solar heat collected as hot air or water could be piped to a massive heat storage tank located inside the house and insulated from its surroundings. This way, the heat would be available as needed, whether at night or on cloudy days. The approach resembled that taken by William J. Bailey when he split the functions of solar heat collection and hot water storage in the Day and Night solar water heater. The advocates of solar heat storage for homes thought that an integrated system of collectors and storage would use the available solar energy more efficiently.

Morse's Solar Air Heater

The use of solar collectors for house heating was not a completely new idea. The earliest recorded instance dates back to the 1880s, when Edward Sylvester Morse, a world-renowned botanist and ethnologist and lecturer at the Essex Institute in Salem, Massachusetts, used a hot box for this purpose. In the patent granted in September 1881 for his product, titled "Warming and Ventilating

Apartments by the Sun's Rays," the scientist described the device: "The invention consists...of a casing attached to the outer wall of a building[, as well as] of a blackened surface...protected by glass....The action of the sun's rays upon the blackened surface heats the air in the space" between it and the building, which "as it ascends [is] directed into the room or building, so as to warm the same."[1] The idea came to Morse's attention, he told a scientific audience, while he was "observing [that] some dark curtains hanging before windows upon which the sun shone [had] become quite warm, and induced a very perceptible upward current of hot air."[2]

Figure 20.2. Professor Edward S. Morse.

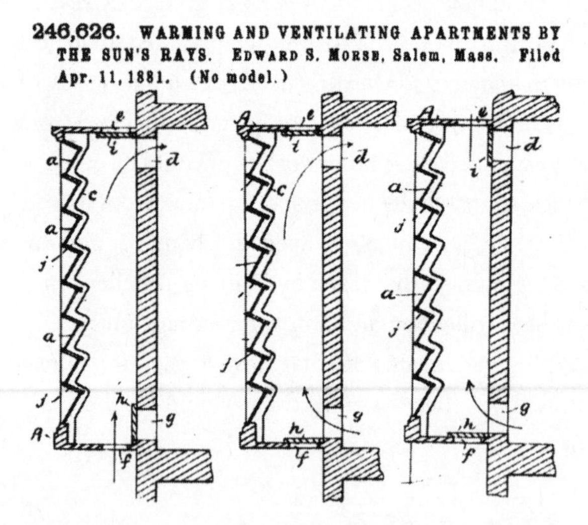

Figure 20.3. Patent drawings of Morse's first Sun-Heater, showing its three different modes of operation.

The first solar air heater built by Morse was installed on the large south wall at the Peabody Museum in Salem, Massachusetts. It consisted of black corrugated iron inside a glass-covered box measuring 13 feet high and 4 feet wide, and was inclined at a 30-degree angle so it would "receive the rays of the sun as nearly direct as practicable," according to the inventor.[3] Since the

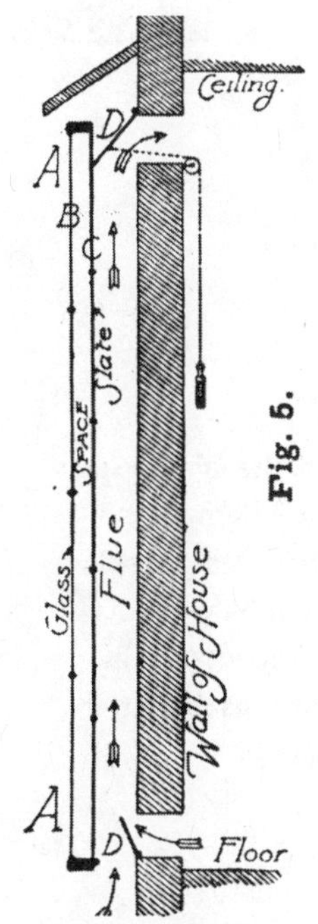

A A, Sun-heater.
B, Glass. *C*, Slate.
D D, Valves, by the operation of which the outer air may be drawn into the room above, or the air of the room may be drawn out below.

Figure 20.4. Cross section of his refined Sun Heater, which replaced the metal absorber with black slate.

hall at the Peabody Museum that the solar air heater was designed to heat was quite large — 100 feet long, 40 feet wide, and 21 feet high — the device had little effect on the interior temperature. What impressed Morse and others, though, was the fact that when the outside air temperature was below freezing, the air warmed more than 26 degrees as it passed through the solar air heater.[4]

Morse told his audience that he had built a second heater, in which he replaced the corrugated iron with black slate and positioned the slab vertically. It heated a much smaller room, one with twenty-four times less air space than the previous venue.[5] On a hazy day, with the floor temperature at 64°F at 10:45 in the morning, the temperature of the air exiting the heater was 31 degrees higher than the outside temperature. The heater's outstanding performance led Morse to conclude that during the usual five-month heating season in the Boston area, a heater such as the one he designed could help heat a home there 80 percent of the time. He based his calculation on weather charts for the period of October 1, 1884, through February 28, 1885, which showed there was sufficient sunshine on 130 days to operate the solar air heater. As for its price, the solar air heater only cost eighteen dollars to build and install.[6]

Scientific American liked Morse's heater. The magazine congratulated him for devising "an ingenious arrangement for utilizing the heat in the sun's rays in warming our houses.... The thing is so simple and apparently self-evident that one only wonders that it has not always been of use."[7]

Newspapers that picked up the story of Morse's solar air heater exaggerated its capability, much to the inventor's dismay. The *Troy Daily Times* called him "a man who thinks he has solved the problem of house heating."[8] *The News of the World* wrote, "It is said that a certain professor down in New England...

warms his home during the daytime...by the heat of the sun alone." The *Salem Gazette* credited Morse with "finding a way to outwit the coal combine."

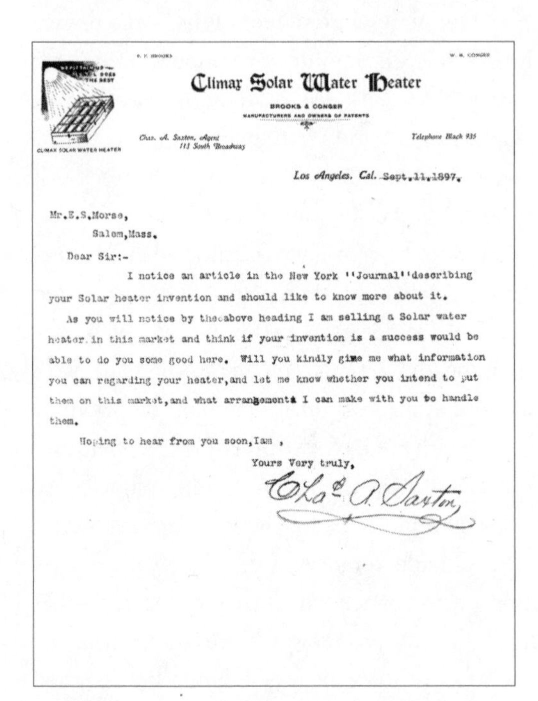

Climax Solar Water Heater

BROOKS & CONGER

MANUFACTURERS AND OWNERS OF PATENTS

Chas. A. Saxton, Agent
113 South Broadway

Telephone Black 935

CLIMAX SOLAR WATER HEATER

Los Angeles, Cal. Sept.11.1897.

Mr.E.S.Morse,
Salem,Mass.

Dear Sir:-

I notice an article in the New York ''Journal''describing your Solar heater invention and should like to know more about it.

As you will notice by the above heading I am selling a Solar water heater in this market and think if your invention is a success would be able to do you some good here. Will you kindly give me what information you can regarding your heater,and let me know whether you intend to put them on this market,and what arrangements I can make with you to handle them.

Hoping to hear from you soon,Iam ,

Yours Very truly,

Chas A. Saxton

Figure 20.5. The Climax Solar Water Heater Company learned about Morse's solar heater through the media and expressed interest in the invention, as this letter demonstrates.

Obviously upset by the public misinterpretation of his work, Morse sarcastically remarked that he expected the next news story to describe his house in the following manner: "A visit to the unique structure in an intensely cold snap with the thermometer at zero revealed the male members of the family roaming round in pajamas, having just gathered a large crop of oranges, the ladies swinging in hammocks and fanning themselves[,] while fronds of gigantic tree ferns formed canopies in every room."[9]

But although Morse parodied the newspapers' inept reports, he sincerely believed that his solar air heater would have useful and important applications. He knew that solar energy could help save fuel, which was especially important in energy-scarce regions: "The results were sufficiently satisfactory to show that in regions such as the great [American] Southwest, where fuel is scarce, temperatures often low, and [there are] many sunny days, the invention could be made of great service." Morse even received a letter from the general manager of the Los Angeles office of the Climax Solar Water Heater Company asking him to let them know "whether [Morse] intends to put them [the solar air heaters] on the market, and what arrangement I [the general manager] can make... to handle them." But despite Morse's prognosis and the extensive publicity he received, his idea fell by the wayside for another half century.

The First MIT Solar House

In 1938, engineers at the Massachusetts Institute of Technology began two decades of research on the use of solar collectors for house heating. A grant of $650,000 bankrolled this extensive study. The money was a gift from Godfrey Lowell Cabot, a wealthy Bostonian described by *Time* magazine as a man of seventy-seven who, in his old age, "broods much about the vast stores of energy in sunlight which man does not utilize."[10]

Hoyt Hottel, a professor of chemical engineering at MIT, became director of this research program, assisted by Byron Woertz, a graduate student, and others at the university. Hottel was aware of the many previous developments in solar energy, and his objective was to test their home-heating potential scientifically. His first object of intensive study was the flat-plate solar collector, the type successfully used to heat domestic hot-water supplies in California and Florida. He favored this device because of its simplicity, its popularity in Florida at the time, and its potential for economical home heating. Hottel also liked the fact that its medium for heat transfer and storage, water, had a very high capacity for absorption and retention of heat.

The research staff set out to conduct a rigorous scientific study of the technical and economic feasibility of heating a building exclusively with solar energy. Their first task was to construct a small building that would serve as the experimental laboratory. Unfortunately, the architect and builder used a compass to determine geographic north — which positioned the building about 7 degrees off a true-south orientation. "Hottel was about ready to bite into a nail" when he learned of this blunder, said Woertz. (Hottel was a perfectionist — 7 degrees would not really have made a difference.)

When the structure was finished, fourteen flat-plate collectors were installed on the south side of the roof, tilted 30 degrees from the horizontal. The collectors occupied a total of 408 square feet — almost the entire south side of the roof. Each collector consisted of a shallow plywood box with six parallel copper tubes inside it soldered to a blackened copper plate. This configuration was very

Figure 20.6. A workman installs flat-plate collectors on the first MIT solar house, 1939.

similar to that used by William J. Bailey in his earliest solar water heaters. On top of the box were three glass covers separated by air spaces, and on the bottom was more than 5 inches of rock-wool insulation (made from minerals). Hot water was pumped through the copper tubes and then down to a 17,400-gallon storage tank that took up the entire basement. The insulation covering this huge tank averaged a little over 2 feet thick. To heat the building, cool air was drawn from the rooms by fans and blown over the hot tank. The warmed air then circulated back into the rooms.

In this "active" solar-heating system, an external source of electrical power was required to activate pumps and fans so that heat could be moved from the collectors to the storage tank and from storage to the rooms. Unlike Keck's solar homes and Morse's solar air heater, this system could not function without power.

The solar-heating system began operating during midsummer in 1940. Hot water was pumped from the rooftop solar collectors to the storage tank continuously. During winter, whenever water in the collectors became cooler than water in the storage tank, a control box shut down the system and allowed the water in the collectors to drain back into the tank. That way heat was not lost

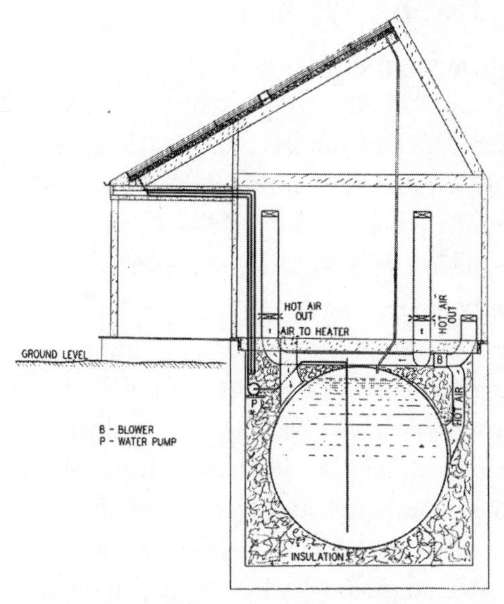

FIG. 1 SECTIONAL VIEW OF SOLAR-ENERGY BUILDING AT THE MASSACHUSETTS INSTITUTE OF TECHNOLOGY

Figure 20.7. Cross section of MIT solar house. Solar heat from the rooftop was stored in a 17,400-gallon water tank that filled the entire basement.

from the storage tank, and there was no danger of water freezing in the collectors and splitting the copper pipes.

From a technical viewpoint, the results of the experiment were generally positive. By means of solar collectors alone, the laboratory remained at a steady 72°F throughout the winter. However, the results of the economic analysis were clearly unfavorable. Relying completely on solar energy meant year-round collection, which required a larger, more costly array of collectors and a massive, heavily insulated storage tank. But aside from demonstrating technical feasibility, the experiment yielded important data. The report on this test, written by Hottel and Woertz, is still regarded as a classic in the field. The engineers isolated the principal factors affecting the performance of a solar collector — the tilt angle of the collector, the light transmission of the glass covers, heat lost through the bottom of the plywood box, the type of absorber plate used, and several other important factors. The MIT study demonstrated that the efficiency of a flat-plate collector drops as the temperature difference between the absorber plate and the outside air gets larger. Efficiency also drops if more than three glass covers on each collector box are used. One surprising discovery they made was that airborne dirt and soot have little effect on collector performance.[11]

A Solar Hot-Air System

Instead of using water to transport and store solar heat from the rooftop collectors, some engineers chose to develop systems relying on solar-heated air. This approach, pioneered by Morse in the late nineteenth century, had long been forgotten. His collectors operated by natural convection of solar-heated air but could not store solar heat for nighttime use. By the middle of the 1940s, two people who had been involved in the first MIT experiment began to work with full-scale solar hot-air systems for house heating.

Dr. George Löf, an associate professor of chemical engineering at the University of Colorado, had done his graduate work under Hoyt Hottel's guidance and was familiar with the MIT experiment. In 1943, he became director of a project to devise an effective, economical solar-heating system. Ironically, while World War II had interrupted the MIT experiments (everyone on the staff had been reassigned to war research), it led to the commencement of this new solar project in Colorado. The War Production Board funded the study because, said Löf, the government was concerned about the possible fuel shortages due to heavy military requirements.

Löf quickly assembled a group of aides and set to work. He decided not to follow the MIT approach, saying, "Probably the main reason... [those at MIT] went into the use of water was that solar water heaters had been used for many years and that was the easiest route — just make them bigger and use them to heat houses. But our rationale was that we shouldn't be constrained or biased by how solar energy had been used before."

A hot-air system was the logical choice, Löf believed, because most homes at the time were heated by hot air, and existing heating systems could easily be combined with solar air-heating systems. And in cold climates, where most Americans then lived, solar collectors using air would have no problem with cracked pipes caused by freezing, and corrosion and leakage would be eliminated. Water does have distinct advantages, however. It carries far more heat than an equal volume of air does, and a water-heating system employs narrower piping and requires less storage volume, reducing initial costs.

The Colorado team sought to cut the system costs by developing an all-glass collector. Two layers of glass covered the collector box, inside of which lay several glass plates overlapping each other — somewhat like roofing shakes. The exposed surface of the overlapping plates was painted black to absorb the sun's rays. Preliminary tests with this collector were encouraging.

But when it was time to fully test the system, the research team found they were unable to build a test site. The war had halted new construction, so they would have to find an existing home in which to test it. Löf volunteered his cozy six-room house in Boulder, which was then heated by a gas-fed hot-air system. More than 460 square feet of collectors went up on the roof of the 1,000-square-foot bungalow. Air from the rooms entered the solar-heating system through a duct in the roof. The temperature of the air passing between the glass plates rose by as much as 110°F before it circulated down to the furnace, where a fan pushed the warm air through the regular heating ducts.

Figure 20.8. Dr. George Löf by the door of his solar-heated bungalow.

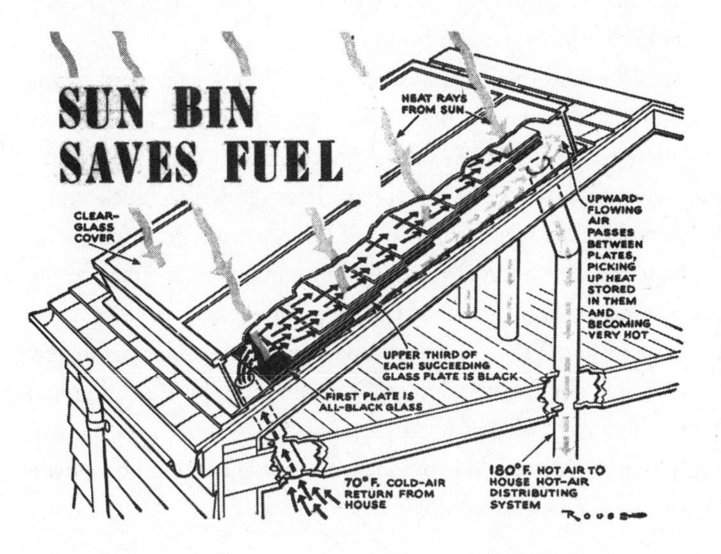

Figure 20.9. Diagram showing how the sun heated air on Dr. Löf's rooftop.

Figure 20.10. Close-up of solar air collectors on the roof of Löf's house.

When solar heat was insufficient, the gas furnace took over. During the first winter, the sun provided about a quarter of the heat needed in the house. To improve this figure, Löf knew, he needed a storage system so that solar heat could be supplied during cloudy days and at night. He considered various materials for absorbing heat from hot air — such as crushed granite, cinder blocks, and hollow tiles. Rock was the cheapest option. Löf calculated that 6 tons in an insulated bed located beneath the floor would hold enough heat to keep the house from getting too cold overnight. The advantage of an even larger storage container, one holding enough heat for two or three days, would have been offset, he felt, by the higher initial cost of building it.

With the rock-bed heat storage, the system provided about a third of the home's heating needs the following winter. This test proved that an ordinary house could be refitted fairly easily with solar equipment. But the all-glass collector did not work out, because the overlapping plates tended to crack under heat stress. And when the Löfs were ready to sell their house after living with the solar-heating system for several years, they felt compelled to remove the collectors. Löf told why: "Nobody at that time was interested in using solar because fuel was so cheap, and if there were any maintenance problems, the new owners of the house would probably feel helpless because a typical heating and air-conditioning installer wouldn't know much about the system." But he had shown convincingly that the sun could provide a substantial heating boost even in regions like Colorado, with its icy cold winters.[12]

Water-Wall Collectors

After World War II interrupted the MIT project, Hoyt Hottel and his team resumed their research in 1947. To develop a more economical system, they sought to combine in a single unit the functions of collecting, storing, and distributing solar heat. They erected a wall of water containers behind a south-facing glass wall — a "water-wall" solar-heating system. In this experiment, 1- and 5-gallon water cans painted black were stacked up just inside a south-facing wall made almost entirely of double-pane glass. Dr. Albert G. Dietz, an engineering professor involved with the project, explained how the system worked: "The sun hit the cans of water through a couple of sheets of glass. The water got warm, and then that energy was transmitted to the interior of the house. It was obviously a much simpler system than the usual one where you have a flat-plate collector and either air or water to take the sun's energy from storage into the house."

The building used to test this water-wall approach was a long, narrow laboratory aligned east-west and facing south; it was divided into seven small cubicles, each sealed off from the next and equipped with a slightly different solar-heating system. Insulating curtains kept the heat from escaping at night. The first cubicle had no water cans behind the large south-facing window. In

Figure 20.11. South wall of the laboratory used by MIT for its experiments with "water-wall" collectors.

the other six cubicles, cans of water stood near a large south-facing window, absorbing the sun's rays and transmitting their heat to the interior. To avoid overheating the rooms, the engineers inserted a curtain between the water cans and the interior in four of the cubicles. The curtain raised automatically whenever the interior temperature fell below 72°F, allowing solar heat into the room. In each of the remaining two cubicles, a fan controlled by a thermostat blew room air past the heated water cans and back into the cubicle, the approach used in the first MIT house.

Figure 20.12. A graduate student monitors a cubicle.

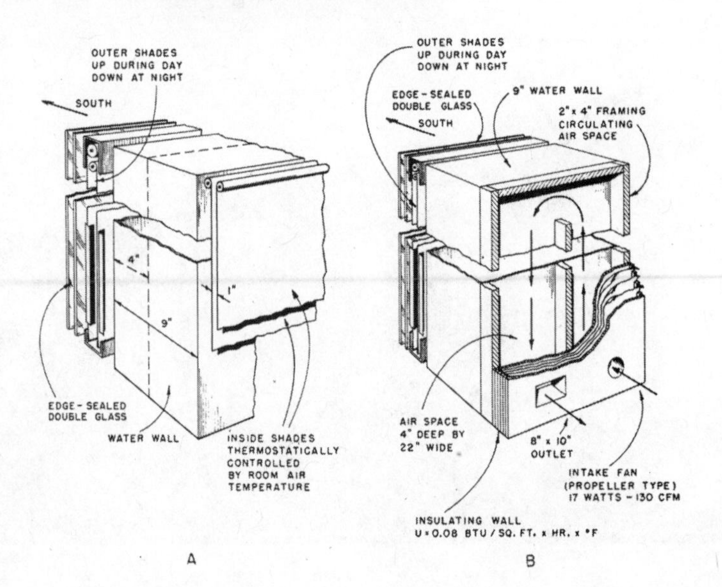

Figure 20.13. Schematic of the MIT solar water collectors.

All of the water-wall systems exhibited one major flaw — a lot of heat escaped through the glass from the warmed water cans sitting right next to it. To reduce this heat loss, the engineers installed aluminum curtains between the cans and the glass; these curtains were drawn as soon as the sun went down. Even so, heat losses ranged from 71 to 84 percent of the solar energy collected. Evidently these curtains failed to block warm air currents from reaching the cold window surfaces.

Despite these heat losses, the water walls were still able to supply between 38 and 48 percent of the heating needs in each cubicle. Better insulation between the water cans and the glass might have made this water-wall approach a successful house-heating option for the cold New England states. But the MIT research team did not pursue the idea any further. Nor did they compare the cost of such a simple heating system with the cost of a more efficient but highly complex and expensive flat-plate collector system. It might have been wise to sacrifice collector efficiency for lower costs and greater reliability.[13]

Combining Collectors with Solar Architecture

Disappointed by the large heat losses from the water wall, the MIT engineers returned to an active system that pumped solar-heated water from rooftop collectors to a storage tank. Two important modifications of their first active system brought down the cost. First of all, the design goal was no longer to heat a home entirely with solar energy. As Dietz observed, "Since the collector is the most expensive part of an installation, it just isn't very economical in most places to have one large enough to take care of all your heating requirements under all conceivable conditions. The concept was that our solar collector should be designed to collect enough heat on a sunny day to carry a house through two or three sunless winter days. Anything beyond that, we would call on the auxiliary heater." This auxiliary consisted of two small electrical immersion heaters placed in the hot-water storage tank. Because of this auxiliary, less hot water needed to be stored for cloudy days and heating at night. So the collectors and tank could be built smaller, which saved money.

The second modification in this third project was the reliance on large south-facing windows, as in Keck's solar homes, to help heat the house. The designers figured that these windows could trap enough heat on a sunny winter day to keep the house warm from morning until nightfall. All the solar heat absorbed by the rooftop collectors during the day could then be stored in the tank for use at night.

This modified solar-heating system was installed in the building that had been used for the second MIT house — the long, narrow laboratory — which was remodeled to become the third MIT house. An A-frame roof was added and sixteen collectors were installed on its south slope, tilting at 57 degrees. These collectors resembled the ones used in the first MIT solar house, except that they had only two glass covers instead of three. Situated in the attic, the storage tank was less than one-tenth the size of the tank used in the first MIT solar experiment. The long, cylindrical tank held 1,200 gallons of solar-heated water; it was insulated with rock wool ranging from 4 to 12 inches thick. Instead of air being blown over the storage tank and distributed to the rooms as in the first house, the hot water itself was circulated to radiant ceiling panels inside the house.

Because the engineers wanted to test the system under normal living conditions, an MIT graduate student and his family moved into the house. Data collected during three years of operation — from 1949 to 1952 — showed that the system was strikingly successful. Solar energy provided almost three-quarters

Figure 20.14. The family that lived in the third MIT house take a stroll along the south side of the house.

of the home's heating requirements. The engineers were particularly surprised by the effectiveness of the large south windows. Hutchinson's tests at Purdue and the second MIT study both seemed to contradict claims by Keck and occupants of his solar homes that south windows contributed substantially to house heating. In this third MIT house, however, the windows supplied on average a whopping 36 percent of the heat required inside — consistent with Keck's claims. If body heat given off by the occupants and waste heat from appliances is added to the heat trapped by windows, the number jumps to 52 percent. The discrepancy between these findings and previous analyses lay in the difference between sterile laboratory tests and actual living conditions. Furniture in the house absorbed some of the solar heat during the day and released it when the temperature cooled off. And the rental contract stipulated that the family had to close its customized insulating curtains by 10 PM every evening to help keep the heat from escaping through the windows.

The solar collectors, which were the more expensive and technically complex part, provided far less warmth to the house than the other, simpler aspects.[14] The solar-heating system operated successfully from 1949 through 1953, when the experiment came to an abrupt end owing to a series of mishaps caused by faulty wiring in the electric heater. When the wiring short-circuited and started to smoke, the fire department arrived. As Hottel recalled, "They got up on the roof, swung their axes, hit metal and insulation, decided this was not the place to open up, and then moved down and hit another. And they went along and opened up enough holes in the roof so the air could move through the attic. From nothing but smoldering, they built a fire for us and ruined the house."[15]

The Dover House

At about the same time that the experiment with the third MIT house was taking place, a novel approach to solar house heating was tested in Dover, Massachusetts, just outside of Boston. This project was organized by Maria Telkes, developer of the solar still that saved so many airmen in World War II. She reasoned that only if a solar-heating system provided all the house's heat would it be economically feasible.[16] But to store enough heat to carry a house through a week of cloudy winter days, a huge volume of water or crushed rock was needed — a volume much too large for practical purposes.

Telkes sought to harness the capacity of certain materials that absorb great

Figure 20.15. Dr. Maria Telkes, inventor of the solar desalinator, and Eleanor Raymond, the architect, check the plans for their Dover, Massachusetts, experiment.

amounts of heat in the process of melting, and that release this "heat of fusion" when they cool and resolidify. For years, she had searched for an inexpensive, readily available material with a low melting point and high heat of fusion. In 1946 Telkes found what seemed like the perfect substance — Glauber's salt, or sodium sulfate decahydrate, commonly used in the manufacture of glass, paper, and detergents. Over a particular temperature range, Glauber's salt could store seven times more heat than the same volume of water, its melting point was about 90°F, and at the time it cost only $8.50 a ton.[17]

Telkes discussed Glauber's salt with Hoyt Hottel, who was also looking for a better way to store the sun's heat. He enthusiastically organized a group of graduate students to test the substance. But here an unexpected difficulty arose. According to Dr. Albert Dietz, "The principal problem was simply this — when you first melt the salts they separate into two phases, the heavier sodium sulfate settling to the bottom and the saturated water solution on top. Then you'd have the problem of remixing them during any subsequent cycle. To get full efficiency, they should be remixed during the freezing cycle. That is the problem — how do you do that?"[18]

Unless the two stratified solutions could be remixed, the solidification would be only partial — and the heat recovery incomplete. At the laboratory they tried to control the stratification by using containers of various sorts, but all to no avail.

Hottel thought it premature to use Glauber's salt in a solar house without further testing. But Telkes was not satisfied with the way the experiments had been run. Disagreeing with Hottel, she found outside support to build a solar-heated house using Glauber's salt for heat storage. She announced that this house would be the first occupied residence to be completely solar heated, an idea that interested Amelia Peabody, the wealthy Bostonian who agreed to finance the project. Telkes asked her architect, Eleanor Raymond, to draw up the plans for a solar-heated house to be built on Peabody's estate in nearby Dover. This house consequently became an "exclusively feminine project," as the *Saturday Evening Post* noted in 1949.

Rather than mounting the collectors on a sloping roof, Telkes stood them upright to form the south wall of the second story. This bank of eighteen collectors ran the full 75-foot length of the house, which was long and narrow — only one room deep — to accommodate the necessary collector area. Fans inside the rooms blew air through the ducts and across the black iron absorber plates in these collectors, and then back down through ducts and past cans of Glauber's salt located between the walls of adjacent rooms. The salt absorbed solar heat from the passing airstream and melted. When the rooms cooled down in the evening, another set of fans circulated room air past the cans of Glauber's salt, which released some of their heat to the airstream, warming the house.

But exactly how well the salt worked is not clear. Did the salt actually melt and solidify? Or did it merely store heat by getting warmer — as would rock, water, and any other substance? This uncertainty occurred because Telkes used an enormous amount of the salt — 470 cubic feet — enough to store plenty of heat without ever melting. And she incorporated this large volume of salt into the plan because she designed the system to have a seven-day storage capacity and did not install an auxiliary heating system.

Unfortunately, the weather refused to cooperate. Right after Dr. Anthony Nemethy, his wife, Esther, and their three-year-old son, Andrew, moved into the house, on Christmas Eve in 1948, they were confronted with eleven cold, sunless days. "The solar heating system was exhausted," recalled Mrs. Nemethy. "As we didn't have any electric or oil heating or anything, we just were practically without heat." But during the remainder of that winter, which was milder than usual, the house stayed comfortable without any need for backup heating.

For two and a half winters, the Dover house functioned fairly well. Only when a stretch of cloudy winter days lasted a week or longer did the solar-heating system fail to keep the house warm. During the third winter, however, the Glauber's salt deteriorated. Some cans began to leak, and the salt separated into two distinct layers — solid and liquid — which meant the salt could not release the solar heat it had absorbed. Esther Nemethy recalled that winter: "Both I and my son had very bad colds and there was a snowstorm and there was no heating in the house. I called up Mrs. Peabody and said, 'I'm sorry, we love Dover, we love you, we love the house, but I'd rather move out from here unless you install electric heaters or do something!' "[19]

After visiting the Nemethys and seeing how cold the house could get, Amelia Peabody installed small electric heaters in the rooms. And in 1953, she decided to install an oil heating system and remove the solar.

Telkes admitted that as the first of its kind her system could not be expected to work perfectly. Perhaps additional, more rigorous testing would have allowed her to dispel the cloud of doubt that surrounded the use of Glauber's salt after the Dover experiment. In retrospect, total reliance on the sun for house heating in a cold, cloudy climate like that of Massachusetts was an unrealistic goal, and a backup heating system should have been installed at the outset.

A Solar-Heated Schoolhouse

Compared with those in Massachusetts and Colorado, the winters in Tucson, Arizona, are like spring; the average daytime temperature drops only to the mid-50s in January. A solar-heating system there could be expected to carry a much higher percentage of the heating load — especially if heating at night were not required. Such was the case in the first solar-heated public building, Rose Elementary School, which was designed by Arthur Brown and built in 1948. When he was first commissioned to work on this project, Brown knew that his main challenge would be to provide a solar-heating system without running up a horrendous bill that would antagonize the taxpayers. His solution was unique and effective.[20]

There was no need for heat storage (a major expenditure in most systems), since classes went from nine in the morning until three in the afternoon — the warmest, sunniest part of the day. Equally important in cutting costs, the roof itself served as the solar collector, and the conventional hot-air system distributed the solar heat. The roof consisted of an understructure overlaid with self-contained aluminum troughs running parallel to each other in a north-south direction, which were covered on top with an aluminum sheet to form a double roof.[21] A fan circulated room air through a duct and into the troughs on the roof. The air warmed up by 10 or 15°F. The warmed air was circulated by a second fan to the classrooms. This solar-heating system provided 86 percent of the school's heat. Even higher temperatures could have been attained if Brown had covered the roof with one or two layers of glass, but his budget was severely limited.

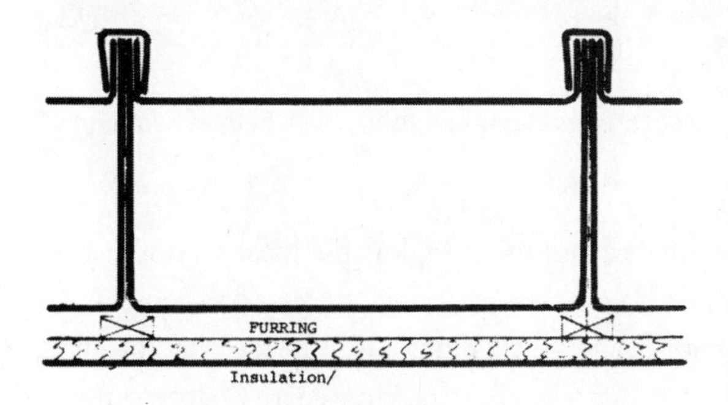

Figure 20.16. View of one of the self-contained aluminum troughs through which air passed and was heated when the sun warmed the aluminum roof.

In the summer, Brown's design helped to keep the school buildings cool with a minimum of artificial air-conditioning. A longtime resident of Arizona, he remembered that many old-time residents put a second roof over their ordinary roof to help keep out the sun's heat in summer. He applied this concept to the school building by relying on the air space between the roof and troughs to help shield the classrooms from the sun's burning rays. The ducts that normally circulated solar-heated air back to the rooms during winter were opened in summer to let the fans blow the warm classroom air outside. Brown also had the roof built so that it extended beyond the south wall of each building, and it shaded the walkway beside the building and the south walls of the classrooms inside.

For ten years Brown's system kept the Rose Elementary School warm in winter and cool in May and September — the two hottest months of the school

Figure 20.17. View of the Rose school and its aluminum roof, which helped heat and cool the school.

year. But when the time came to expand the complex, the local school district decided to replace the system with a gas furnace. Some teachers had complained that the fans made too much noise, and others did not like the fact that the ducts had to be opened by hand. No one thought of saving energy, Brown noted regretfully. "I did these things at a time when gas was so cheap," he remarked, "that people didn't have an interest in solar heating." But his system had demonstrated a workable approach to solar heating for a building located in a mild climate and occupied only during the daytime. Its simplicity kept construction and operating costs to a minimum. Since the collector also served as the roof, and the only extra expense was that of keeping the fans running, the system began paying dividends from the moment it was installed.

The Fourth MIT Solar House

During the 1950s, solar architects and engineers built a second generation of buildings heated by "active" solar-heating systems. Most of these projects were genuine dwellings or office buildings designed to demonstrate that comfort, convenience, and aesthetics need not be sacrificed when using solar collectors for space heating. George Löf used another solar hot-air system, similar to the one he had developed in Boulder, to heat his newly built, ranch-style home in Denver, Colorado. In Albuquerque, New Mexico, the engineering firm of Bridgers and Paxton built the first solar-heated office building in 1956. This system cooled the building in summer.

In 1958, the MIT solar research team erected a fourth solar house — this one a full-scale dwelling built from scratch in Lexington, Massachusetts. Once again they used solar-heated water to warm the house, but this time only about half of the house's heat was supplied by the sun. The MIT staff conducted a sophisticated economic analysis of this system and calculated that, on the basis of the low cost of fossil fuels at the time, the price of the solar-heating system would have to be slashed by 80 percent if it was to pay for itself in ten years. According to Hottel, "With the price of oil as it was then, solar would [have been] economically interesting only if we could have gotten the collectors for a couple of dollars a square foot." This was impossible — given the high cost of aluminum, copper, and glass. George Löf agreed with Hottel's assessment of solar economics in the 1950s.[22]

But despite these pessimistic views, the years of research had shown that

Figure 20.18. Occupants of the fourth MIT house roast hot dogs outside on their solar oven.

solar house heating in a cold climate was technically feasible. The only stumbling block was cost. Perhaps that obstacle could have been overcome by doing away with the major expense — the solar collectors and storage tank — and instead looking at ways to more effectively make use of the solar energy streaming through the south-facing windows tried in the third MIT solar house, or the inexpensive water walls of the second MIT house, or both. The research team failed to realize, for example, that if they had enlarged the cubicles in which they placed the water canisters, or if they had better insulated the windows, less heat would have been lost. But in 1962 the team terminated more than two decades of hopeful research.

PART V

PHOTOVOLTAICS

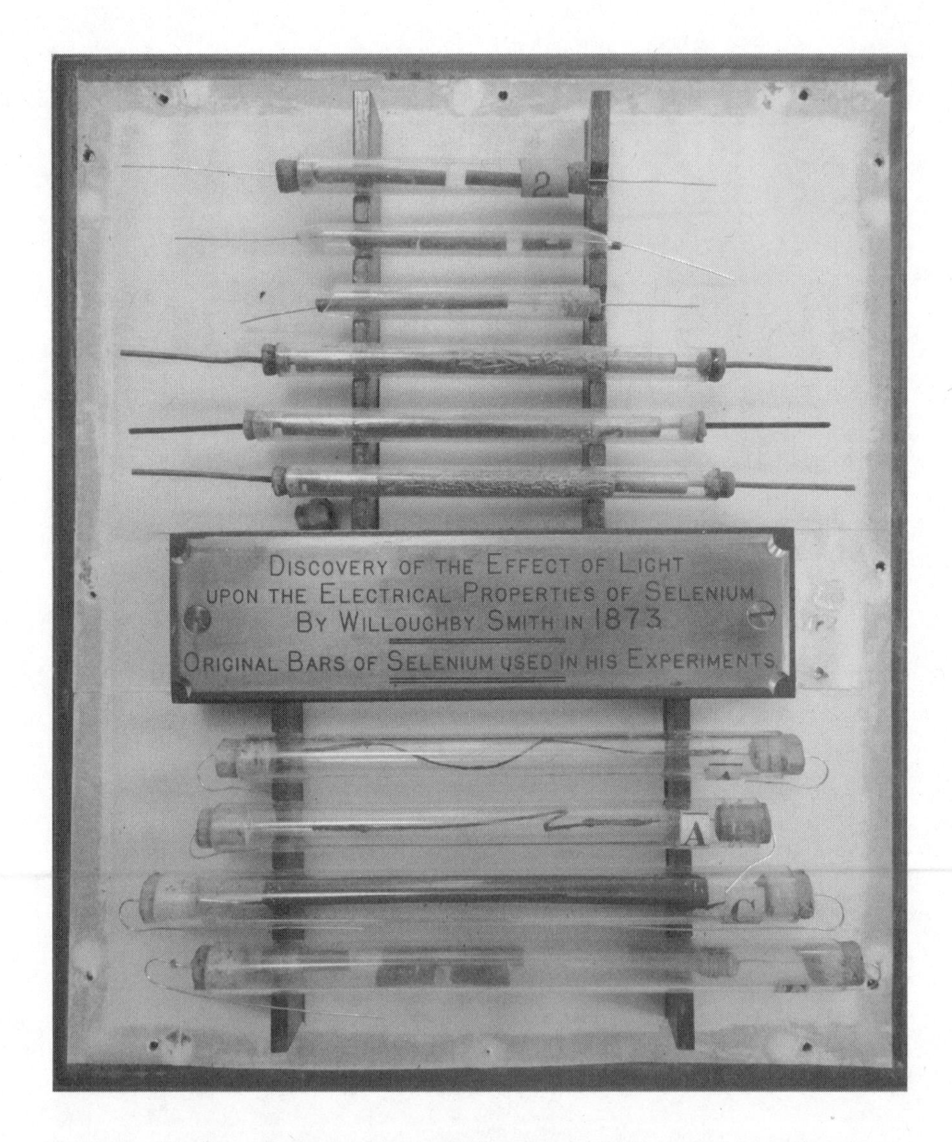

Figure 21.1. The bars of selenium used in Willoughby Smith's research as well as the investigations of William Grylls Adams and Richard Evans Day into behavior of the material when exposed to light.

From Selenium to Silicon

The great Scottish scientist James Clerk Maxwell wrote in 1874 to a colleague: "I saw conductivity of Selenium as affected by light. It is most sudden. Effect of a copper heater insensible. That of the sun great."[1] Maxwell was among many European scientists intrigued by a behavior of selenium that had first been brought to the attention of the scientific community in an article by Willoughby Smith, published in the 1873 *Journal of the Society of Telegraph Engineers*. Smith, the chief electrician (electrical engineer) of the Gutta Percha Company, used selenium bars during the late 1860s in a device for detecting flaws in the transatlantic cable before submersion. Though the selenium bars worked well at night, they performed dismally when the sun came out.[2] Suspecting that selenium's peculiar performance had something to do with the amount of light falling on it, Smith placed the bars in a box with a sliding cover. When the box was closed and light excluded, the bars' resistance — the degree to which they hindered the electrical flow through them — was at its highest and remained constant. But when

Figure 21.2. Willoughby Smith.

the cover of the box was removed, their conductivity — the enhancement of electrical flow — immediately "increased according to the intensity of light."[3]

Discovering the Photovoltaic Effect in a Solid Material

To determine whether it was the sun's heat or its light that affected the selenium, Smith conducted a series of experiments. In one, he placed a bar in a shallow trough of water. The water blocked the sun's heat, but not its light, from reaching the selenium. When he covered and uncovered the trough, the results obtained were similar to those previously observed, leading him to conclude that "the resistance [of the selenium bars] was altered... according to the intensity of light."[4]

Figure 21.3. William Grylls Adams.

Among the researchers examining the effect of light on selenium following Smith's report were two British scientists, Professor William Grylls Adams and his student Richard Evans Day. During the late 1870s they subjected selenium to many experiments, and in one of these trials they lit a candle an inch away from the same bars of selenium Smith had used. The needle on their measuring device reacted immediately. Screening the selenium from light caused the needle to drop to zero instantaneously. These rapid responses ruled out the possibility that the heat of the candle flame had produced the current (a phenomenon known as thermal electricity), because when heat is applied or withdrawn in thermoelectric experiments, the needle always rises or falls slowly. "Hence," the investigators concluded, "it was clear that a current could be started in the selenium by the action of the light alone."[5] They felt confident that they had discovered something completely new: that light caused "a flow of electricity" through a solid material. Adams and Day called current produced by light "photoelectric."[6]

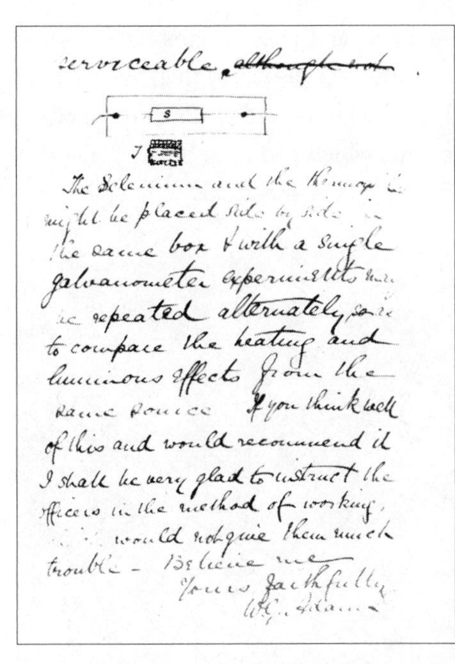

serviceable,

The Selenium and the thermopile might be placed side by side in the same box & with a single galvanometer experiments might be repeated alternately so as to compare the heating and luminous effects from the same source. If you think well of this and would recommend it I shall be very glad to instruct the officers in the method of working, which would not give them much trouble – Believe me
Yours faithfully
W.G. Adams

Figure 21.4. Handwritten page by Adams that includes a diagram of the first selenium solar cell.

The First Module

A few years later, Charles Fritts of New York moved the technology forward by constructing the world's first photoelectric module. He spread a wide, thin layer of selenium onto a metal plate and covered it with a thin, semitransparent gold-leaf film. This selenium module, Fritts reported, produced a current "that is continuous, constant, and of considerable force[,]...not only by exposure

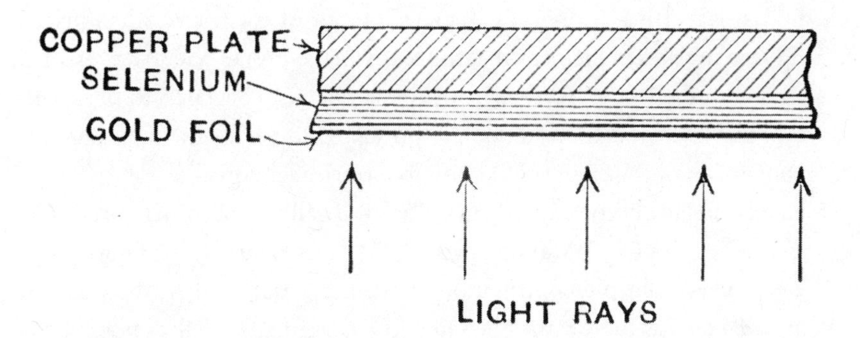

Figure 21.5. Fritts's selenium solar module.

to sunlight, but also to dim diffused daylight, and even to lamplight." As to the usefulness of his invention, Fritts optimistically predicted that "we may ere long see the photoelectric plate competing with [coal-fired electrical-generating plants],"[7] the first fossil-fueled power plants, which had been built by Thomas Edison only three years before Fritts announced his intentions.

Figure 21.6. Charles Fritts put up the first photovoltaic array, in New York City in 1884.

Fritts sent one of his solar panels to Werner von Siemens, whose reputation ranked on a par with Edison's. The panels' output of electricity when placed under light so impressed Siemens that the renowned German scientist presented Fritts's panel to the Royal Academy of Prussia. Siemens declared to the scientific world that the American's modules "presented to us, for the first time, the direct conversion of the energy of light into electrical energy."[8]

Siemens judged photoelectricity to be "scientifically of the most far-reaching importance."[9] James Clerk Maxwell agreed. He praised the study of photoelectricity as "a very valuable contribution to science." But neither Maxwell nor Siemens had a clue as to how the phenomenon worked. Maxwell wondered, "Is the radiation the immediate cause or does it act by producing some change in

Figure 21.7. Werner von Siemens.

the chemical state?"[10] Siemens did not even venture an explanation but urged a "thorough investigation to determine upon what the electromotive light-action of [the] selenium depends."[11]

Few scientists heeded Siemens's call. The discovery seemed to counter all of what science believed at that time. The selenium bars used by Adams and Day, and Fritts's "magic" plate, did not rely on heat to generate energy as did all other known power devices, including solar motors. So most dismissed them from the realm of further scientific inquiry.

One brave scientist, however, George M. Minchin, a professor of applied mathematics at the Royal Indian Engineering College, complained that rejecting photoelectricity as scientifically unsound — an action that originated in the "very limited experience" of contemporary science and in "a 'so far as we know' [perspective —] is nothing short of madness."[12] In fact, Minchin came closest among the handful of nineteenth-century experimentalists to explaining what happens when light strikes a selenium solar cell. Perhaps, Minchin wrote, it "simply act[s] as a transformer of the energy it receives from the sun, while its own materials, being the implements used in the process, may be almost wholly unmodified."[13]

The scientific community during Minchin's time also dismissed photoelectricity's potential as a power source after looking at the results obtained when measuring the sun's thermal energy in a glass-covered, black-surfaced device, the ideal absorber of solar heat. "But clearly the assumption that all forms of energy of the solar beam are caught up by a blackened surface and transformed into heat is one which may possibly be incorrect," Minchin argued.[14] In fact, he believed that "there may be some forms of [solar] energy which take no notice of blackened surfaces[, and] perhaps the proper receptive surfaces" to measure them "remain to be discovered." Minchin intuited that only when science had the ability to quantify "the intensities of light as regards each of [its] individual colours [that is, the different wavelengths] could scientists judge the potential of photoelectricity."[15]

Einstein's Great Discovery

Albert Einstein shared Minchin's suspicions that the science of the age failed to account for all the energy streaming from the sun. In a daring paper published in 1905, Einstein showed that light possesses an attribute that earlier scientists had not recognized. Light, he discovered, contains packets of energy, which he called light quanta (now called photons).[16] He argued that the amount of power that light quanta carry varies, as Minchin suspected, according to the wavelength of light — the shorter the wavelength, the more power. The shortest wavelength, for example, contains photons that are about four times as powerful as those of the longest.[17]

Einstein's bold and novel description of light, combined with the discovery of the electron and the ensuing rash of research into its behavior — all happening at the turn of the nineteenth century — provided photoelectricity with a scientific framework it had previously lacked and that could now explain the phenomenon in terms understandable to science.[18] In materials like selenium, the more powerful photons carry enough energy to knock poorly linked electrons from their atomic orbits. When wires are attached to the selenium bars, the liberated electrons flow through them in the form of electricity. Nineteenth-century experimenters called the process photoelectric, but by the 1920s scientists referred to the phenomenon as the photovoltaic effect.

This new legitimacy stimulated further research into photovoltaics and revived the dream that the world's industries could hum along fuel- and pollution-free, powered by the inexhaustible rays of the sun.[19] Dr. Bruno Lange, a German scientist whose 1931 solar panel resembled Fritts's design, predicted that, "in the not distant future, huge plants will employ thousands of these plates to transform sunlight into electric power... that can compete with hydroelectric and steam-driven generators in running factories and lighting homes."[20] But Lange's solar battery worked no better than Fritts's, converting far less than 1 percent of all incoming sunlight into electricity — hardly enough to justify its use as a power source.

The pioneers in photoelectricity failed to attain the goals they had hoped to

Figure 21.8. Illustration from 1930s *Popular Mechanics* article on solar's contribution to the world's energy mix. The selenium solar cell began the dream that one day photovoltaics would power the world.

reach, but their efforts were not in vain. One contemporary of Minchin's credited them for their "telescopic imagination [that] beheld the blessed vision of the

Sun, no longer pouring unrequited into space, but by means of photo-electric cells... [its] powers gathered into electric storehouses to the total extinction of steam engines and the utter repression of smoke."[21] In his 1919 book on solar cells, Thomas Benson complimented these pioneers' work with selenium as the forerunner of "the inevitable Solar Generator."[22] Maria Telkes, too, felt encouraged by the selenium legacy, writing, "Personally, I believe that photovoltaic cells will be the most efficient converters of solar energy, if a great deal of further research and development work succeeds in improving their characteristics."[23]

With no breakthroughs on the horizon, though, the head of Westinghouse's photoelectricity division could only conclude, "The photovoltaic cells will not prove interesting to the practical engineer until the efficiency has increased at least fifty times."[24] The authors of *Photoelectricity and Its Applications* agreed with the pessimistic prognosis, writing in 1949, "It must be left to the future whether the discovery of materially more efficient cells will reopen the possibility of harnessing solar energy for useful purposes."[25]

The First Practical Solar Cell

Just five years later the beginning of the silicon revolution spawned the world's first practical solar cell and its promise for an enduring solar age. Its birth accidentally occurred along with that of the silicon transistor, the principal component of every electronic device in use today. Two scientists, Calvin Fuller and Gerald Pearson of the famous Bell Laboratories, led the pioneering effort that took the silicon transistor from theory to working device. Pearson was described by an admiring colleague as the "experimentalist's experimentalist." Fuller, a chemist, learned how to control the introduction of the impurities necessary to transform silicon from a poor to the preeminent conductor of electricity. As part of the research program, Fuller gave Pearson a piece of silicon containing a small concentration of gallium. The introduction of gallium had made the silicon positively charged. When Pearson dipped the rod into a hot lithium bath, according to Fuller's formula, the portion of the silicon immersed in the lithium became negatively charged. Where the positive and negative silicon met, a permanent electrical field developed. This is the p-n junction, the heart of the transistor and solar cell, where all electronic activity occurs. Silicon prepared this way needs but a certain amount of outside energy for activation, which lamplight provided in one of Pearson's experiments. The scientist had the

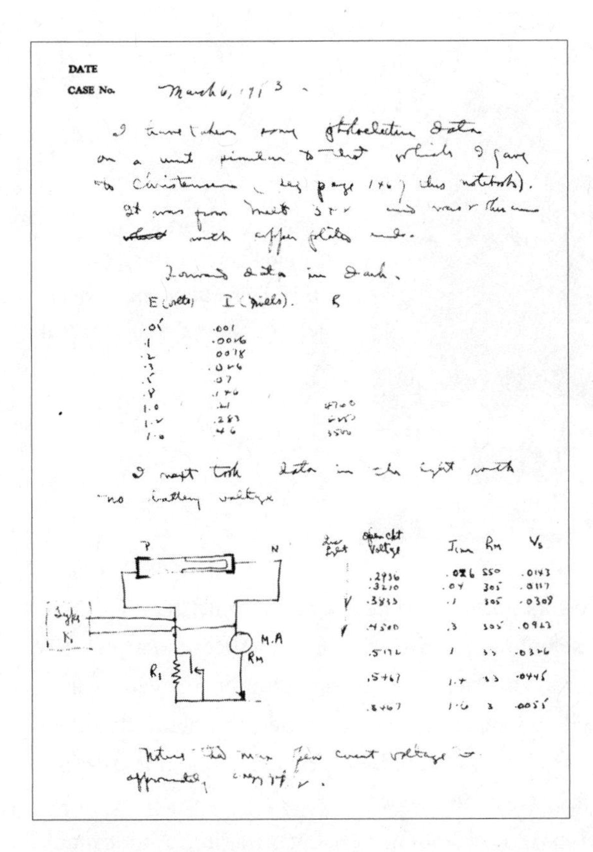

Figure 21.9. A drawing by Gerald Pearson of the first silicon solar cell, serendipitously discovered by him.

specially prepared silicon connected by wires to an ammeter, which, to Pearson's surprise, recorded a significant electrical current.[26]

While Fuller and Pearson worked on improving transistors, another Bell scientist, Daryl Chapin, had begun work on the problem of providing small amounts of intermittent power in remote humid locations. In any other climate, the traditional dry-cell battery would do, but "in the tropics [it] may have too short a life" due to humidity-induced degradation, Chapin explained, "and be gone when fully needed."[27] Bell Laboratories had Chapin investigate the feasibility of employing alternative sources of freestanding power, including wind machines, thermoelectric generators, and small steam engines. Chapin suggested that the investigation include solar cells, and his supervisors approved.

In late February 1953, Chapin commenced his photovoltaic research. Placing a commercial selenium cell in sunlight, he recorded that the cell produced 4.9 watts per square meter. Its efficiency, the percentage of sunlight it could convert into electricity, was a little less than 0.5 percent.[28] Word of Chapin's solar

Figure 21.10. Left to right: Gerald Pearson, Daryl Chapin, and Calvin Fuller, the principal developers of the silicon solar cell.

power studies and dismal results got back to Pearson. He told Chapin, "Don't waste another moment on selenium," and gave him the silicon solar cell that he had made.[29] Chapin's tests, conducted in strong sunlight, proved Pearson right.[30] The silicon solar cell had an efficiency of 2.3 percent, about five times greater than the selenium cell's. Chapin immediately dropped selenium research and dedicated his time to improving the silicon solar cell.

His theoretical calculations of its potential were encouraging. An ideal unit, Chapin figured, could use 23 percent of the incoming solar energy to produce electricity.[31] However, he set a goal of obtaining an efficiency of nearly 6 percent, the threshold that engineers of the time felt it was necessary to reach if photovoltaic cells were to be seriously regarded as electrical power sources.

Chapin, doing most of the engineering, had to try new materials, test different configurations, and face times of despair when nothing seemed to work. At several junctures, seemingly insurmountable obstacles arose. One major breakthrough came directly from knowledge of Einstein's light quanta (photon) work. "It appears necessary to make our p-n [junction] very next to the surface," Chapin realized, so that the more powerful photons belonging to light of shorter wavelengths could effectively move electrons to where they could be harvested as electricity. To build such a cell required collaboration with Fuller.[32] Chapin also observed that silicon's shiny surface reflected a good deal of sunlight that could be absorbed and used, so he coated its surface with a dull transparent plastic. Adding boron to the top of the cell permitted better photon harvesting by allowing for good electrical contact on the silicon strips while keeping the p-n junction close to the surface.[33] Chapin finally triumphed, reaching his 6 percent goal. He could now confidently call the cells he built "power photocells... intended to be primary power sources."[34] Assured of the cells' reproducibility and sufficient efficiency, the trio built a number of arrays and demonstrated them at a press conference and the annual meeting of the National Academy of Sciences.

Figure 21.11. Calvin Fuller places arsenic-doped silicon into a quartz-tube furnace, where he introduced a controlled amount of boron to the material. This resulted in the first solar cell that could generate enough electricity directly from sunlight for practical purposes.

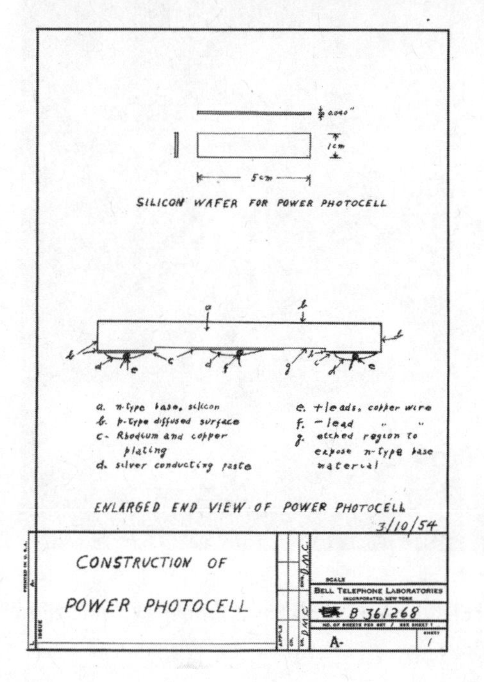

Figure 21.12. A working diagram of the Bell Power Cell.

Proud Bell executives presented the Bell Solar Battery to the press on April 25, 1954, displaying a panel of cells that relied solely on light power to run a 21-inch Ferris wheel. The next day the Bell scientists ran a solar-powered radio transmitter, which broadcast voice and music to America's top scientists gathered at a meeting in Washington, DC. The press took notice. *U.S. News & World Report* speculated excitedly in an article titled "Fuel Unlimited": "The [silicon] strips may provide more power than all the world's coal, oil and uranium.... Engineers are dreaming of silicon-strip powerhouses."[35] The *New York Times* concurred, stating on page one that the work of Chapin, Fuller, and Pearson, which resulted in the first solar cell capable of generating useful amounts of power, "may mark the beginning of a new era, leading eventually to the realization of one of mankind's most cherished dreams — the harnessing of the almost limitless energy of the sun for the uses of civilization."[36]

Figure 21.13. The first batch of Bell's solar cells powered this Ferris wheel, showing the press what the Bell Solar Battery could do.

ADM: ORG: NAS: Meetings:
Annual: Exhibits

news

From BELL TELEPHONE LABORATORIES, 463 West Street, New York 14, CHelsea 3-1000

FOR RELEASE AT 6 P.M.,
SUNDAY, APRIL 25, 1954

A solar battery--the first successful device to convert useful amounts of the sun's energy directly and efficiently into electricity--was demonstrated today at Bell Telephone Laboratories.

Scientists there, with an amazingly simple-looking apparatus made of strips of silicon, showed how the sun's rays could be used to power the transmission of voices over telephone wires. The Bell solar battery also used energy from the sun to power a transistor radio transmitter carrying both speech and music.

Bell Laboratories reported that it was able to achieve a 6 per cent efficiency in converting sunlight directly into electricity. This compares favorably with the efficiency of steam and gasoline engines, in contrast with other photoelectric devices which have never been rated higher than 1 per cent.

With improved techniques the Bell Laboratories scientists said they expected to increase this efficiency substantially. Nothing is consumed or destroyed in the energy conversion process and there are no moving parts, so the Bell solar battery should theoretically last indefinitely.

(more)

Figure 21.14. Press release from Bell Laboratories noting the significance of its new invention — the Bell Solar Battery.

Figure 21.15. Bell scientists ran a solar-powered radio transmitter, which broadcast voice and music to America's top scientists gathered at the National Academy of Sciences' meeting in Washington, DC.

SIMPLE AND EFFICIENT—The ***Bell Solar Battery*** is made of thin, specially treated strips of silicon, an ingredient of common sand. It needs no fuel other than the light from the sun itself. Since it has no moving parts and nothing is consumed or destroyed, the ***Bell Solar Battery*** should theoretically last indefinitely.

New Bell Solar Battery Converts Sun's Rays Into Electricity

Bell Telephone Laboratories demonstrate new device for using power from the sun

Scientists have long reached for the secret of the sun. For they have known that it sends us nearly as much energy daily as is contained in all known reserves of coal, oil and uranium.

If this energy could be put to use there would be enough to turn every wheel and light every lamp that mankind would ever need.

Now the dream of the ages is closer to realization. For out of the Bell Telephone Laboratories has come the ***Bell Solar Battery***—a device to convert energy from the sun directly and efficiently into usable amounts of electricity.

Though much development remains to be done, this new battery gives a glimpse of future progress in many fields. Its use with transistors (also invented at Bell Laboratories) offers many opportunities for improvements and economies in telephone service.

A small ***Bell Solar Battery*** has shown that it can send voices over telephone wires and operate low-power radio transmitters. Made to cover a square yard, it can deliver enough power from the sun to light an ordinary reading lamp.

Great benefits for telephone users and for all mankind will come from this forward step in harnessing the limitless power of the sun.

BELL TELEPHONE SYSTEM

Figure 22.1. The Bell System believed Bell Solar Batteries would one day run every electrical device in the world.

Saved by the Space Race

Few inventions in the history of Bell Laboratories evoked as much media attention and public excitement as the silicon solar cell, known at the time as the Bell Solar Battery. The details about this invention that Bell disclosed at the press conference the company held in April 1954 greatly boosted interest in solar energy. John Yellott, a mechanical engineer who probably knew more about twentieth-century attempts to use the sun's energy than anyone else in America, hailed the silicon solar cell as "the first really important breakthrough in solar energy technology" for that time period.[1] In fact, according to a 1955 *Newsweek* report, many foresaw the solar cell's "development as an eventual competitor to atomic power."[2]

Technical Progress Continues in the Lab

Technical progress continued, and in the next eighteen months cell efficiency doubled. But commercial success eluded solar cells because of their prohibitive cost. With a 1-watt cell costing $286, Chapin calculated that in 1956 a homeowner would have to pay $1,430,000 for an array of sufficient size to power the average house. This led him to the sober assessment that, "however exciting the prospect is of using silicon solar converters for power,... clearly, we have not advanced to where we can compete... commercially."[3]

Hoffman Electronics, owner of the license to commercialize the Bell solar cell, begged to differ. CEO Leslie Hoffman saw its future in powering electrical devices in areas distant from electrical lines. "Consider," he told a gathering of government officials, "a remote telephone repeater station in the middle of the desert.... An array of solar cells and a system of storage batteries is the answer.... Either heavy expendable dry batteries or fuel for a rather unreliable gasoline-engine charging unit has to be packed in to the site at great expense. The solar cell," he emphasized, full of conviction, "is a much better answer."[4]

The Silicon Atomic Battery Temporarily Eclipses the Bell Invention

These officials, like most Americans, however, saw their salvation in a different type of independent power system — the atomic battery. RCA, Bell Laboratories' rival, had come up with a nuclear-run silicon cell, hoping for a perfect fit with America's emerging infatuation with the atom. Instead of using sun-supplied photons, the atomic battery allegedly ran on photons emitted from strontium 90, one of the deadliest radioactive materials.

RCA made a dramatic presentation at its Radio City headquarters in Manhattan: David Sarnoff, founder and president of RCA, whose initial claim to

Figure 22.2. David Sarnoff, founder and president of RCA, telegraphing the message "ATOMS FOR PEACE."

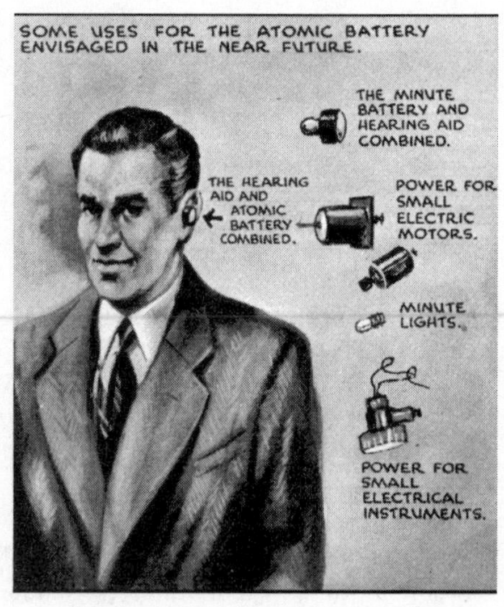

Figure 22.3. Some future uses envisioned for the atomic battery.

fame was being the telegraph operator who had tapped to the world: "THE TITANIC HAS SUNK," hit the keys of a facsimile powered by the atomic battery to send the message "ATOMS FOR PEACE."[5]

What RCA did not tell the public was why the venetian blinds had to be closed during Sarnoff's telegraphy. Years later one of the lead scientists in the atomic-battery project revealed RCA's dirty little secret: had the silicon device been exposed to the sun's rays that afternoon, solar energy would have overpowered the strontium 90. And if the strontium 90 had been removed, the battery would have continued to work on solar power alone! The director of RCA Laboratories didn't mince words when he ordered his scientists to comply with the ruse, telling them, "Who cares about solar energy? Look, what we have is atomic energy at work. That's the big thing that's going to catch the attention of the public, the press, the scientific community," despite the fact that the Bell Solar Battery outperformed its atomic rival by a factor of 1 million.[6]

And right he was. The government persuaded Americans, with the help of the media, to dream of the day when their homes, cars, locomotives, and even airplanes would run on radioactive waste produced by nuclear reactors. The *Illustrated London News* chimed in, calling the atomic battery "a revolutionary invention which in time may have as far-reaching an effect as [Michael] Faraday's" discovery of the means to generate electricity.[7] With this in mind, who needed the sun? Alas, powering novelty items such as toys and transistor radios turned

Figure 22.4. This advertisement from Hoffman shows toys and other novelties powered by the Bell Solar Battery.

out to be the Bell Solar Battery's only commercial success. Martin Wolf, one of the first scientists devoted to photovoltaic work, recalled a company display "where under room lighting... [toy] ships traveled in circles in a children's wading pool. The model of a DC-4 with four electric-motors... turning the propellers was powered exclusively by the solar cell embedded in the wings."[8] With the Bell Solar Battery used solely for playthings, much of the initial enthusiasm generated by the Bell discovery soon waned.

The Bell Solar Battery Finds Its Place in the Sun

How could Chapin, depressed by this turn of events, help but wonder, "What to do with our new baby?"[9] A colleague at Bell Laboratories, Gordon Raisbeck, did not share Chapin's pessimism. He wrote, "Very likely the solar battery will find its greatest usefulness in doing jobs the need for which we have not felt."[10] Dr. Hans Ziegler, a power expert for the U.S. Army Signal Corps who was sent to Bell Laboratories to evaluate the new device, came to a similar conclusion. After months of extensive discussions, he and his staff found one application in particular where the solar cell made a perfect fit for a top secret program — the development of an artificial satellite.[11]

Figure 22.5. Dr. Hans Ziegler.

Freed from terrestrial restraints on solar radiation — namely, inclement weather and nighttime — "operations above the earth's atmosphere [would] provide ideal circumstances for solar energy converters," the Signal Corps believed. Silicon solar cells theoretically would never run out of fuel since they ran on the sun's energy. The other power option — batteries — would lose their charge in two to three weeks. The modular nature of photovoltaics also meant that cells could be tailored for the exact power requirement of a particular satellite. There would be no wasted weight or bulk. A tiny array could provide the small amount of power that the transistorized communication equipment on board a satellite needed, without encumbering the payload. The Signal Corps concluded, "For longer periods of operation and limited allowance for weight…the photovoltaic principle…appears most promising."[12]

The secrecy of America's space plans ended on July 30, 1955, when President Dwight Eisenhower announced America's plans to put a satellite into space. A drawing that accompanied Eisenhower's front-page statement in the *New York Times* showed that it had a solar power source.[13]

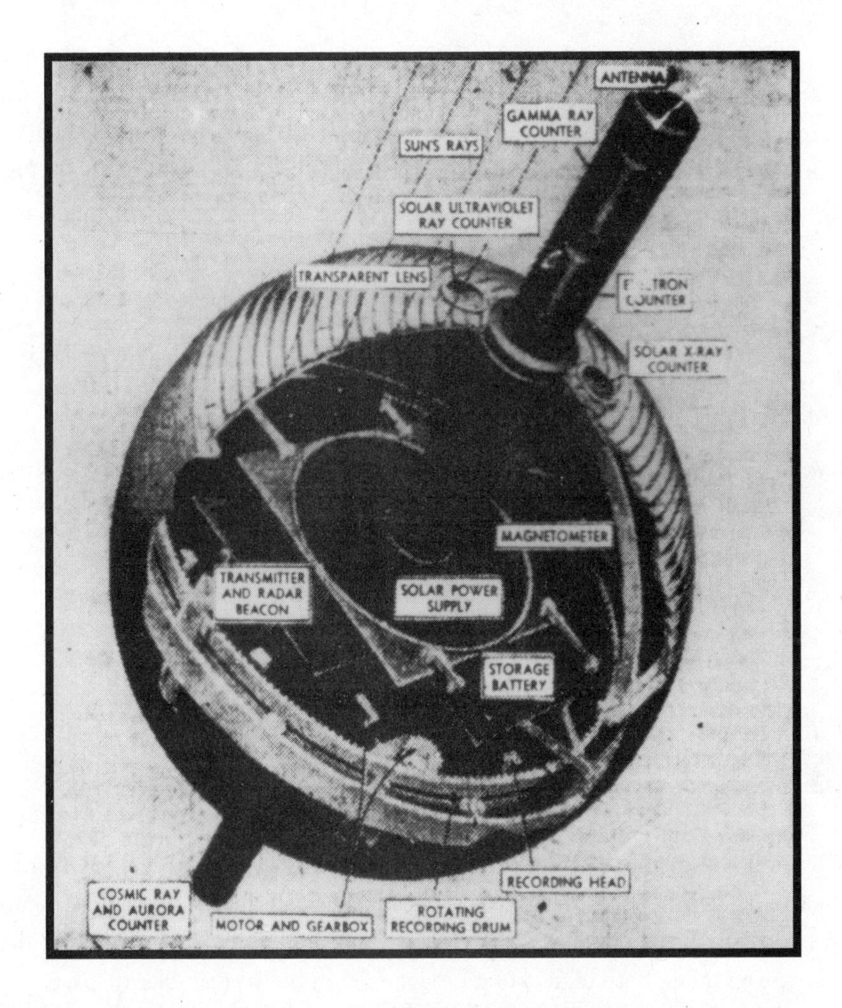

Figure 22.6. The model of the first satellite, appearing with the news on the front page of the *New York Times* and just about every other newspaper in America that the country would soon launch it, shows that the satellite's designer, Dr. S. Fred Singer, planned to use solar photovoltaics as its power source.

The civilian scientific panel overseeing America's space program saw powering satellites with photovoltaics as immensely important, since relying on batteries automatically condemned "most of the on-board apparatus... [to] an active life of only a few weeks." They noted that "nearly all of the experiments will have enormously greater value if they can be kept operating for several months or more" and decided it was "of utmost importance to have a solar battery system" on board.[14]

Based on the civilian panel's recommendations, the Signal Corps was asked to take on the responsibility of designing a solar-cell power system for the program, called Project Vanguard. Its staff readily developed a prototype that clustered individual solar cells on the surface of the satellite's shell. The group designed the modules to provide "mechanical rigidity against shock and vibration and to comply with thermal requirements of space travel."[15] To test their reliability in space, the Signal Corps attached cell clusters to the nose cones of two high-altitude rockets. One rocket reached an altitude of 126 miles, the second 192 miles, both high enough to experience the vicissitudes of space. "In both firings, the solar cells operated perfectly," Ziegler reported to an international conference on space activity held in the fall of 1957.[16] A U.S. Army Signal Corps press release added, "The power was sufficient for satellite instruments... [and they were] not affected by the temperatures of skin friction as the rockets passed through the atmosphere at more than a mile a second."[17]

Figure 22.7. A navy technician attaches a nose cone embedded with solar cells to a rocket that will be launched to test whether the cells can survive the rigors of space.

The Vanguard: The First Practical Use of the Bell Solar Battery

The first satellite with solar cells aboard went into orbit on Saint Patrick's Day in 1958. Nineteen days later, a headline in the *New York Times* revealed: "Radio Fails as Chemical Battery Is Exhausted: Solar-Powered Radio Still Functioning."[18] Celebrating the first anniversary of the Vanguard launch, the Signal Corps let the public know that "the sun-powered Vanguard I... is still faithfully sending its radio message back to earth."[19] By July 16, 1958, of the four satellites in the sky only the solar-powered Vanguard and Sputnik III, which was launched after Vanguard and also had solar cells aboard, were still transmitting. The small Vanguard satellite and Sputnik III proved far more valuable to science than the first two Sputniks, equipped with conventional batteries, which had been silenced after a week or so in space.

The success of the Vanguard and subsequent satellites using transistors in tandem with solar batteries demonstrated the synergy between these two newly discovered semiconductor devices. The transistor allowed for miniaturized and long-lived electronics light enough for launchable payloads and reliable enough for an environment where future maintenance was not an option. Because the transistor had a low power draw, the surface of the satellite was sufficient for enough solar cells to provide the necessary electricity, guaranteeing the longevity required for practical space missions.

Figure 22.8. The Institute of Radio Engineers, now the Institute of Electrical and Electronics Engineers, celebrated at its annual conference in 1959 the first year of flawless service by the solar-powered Vanguard.

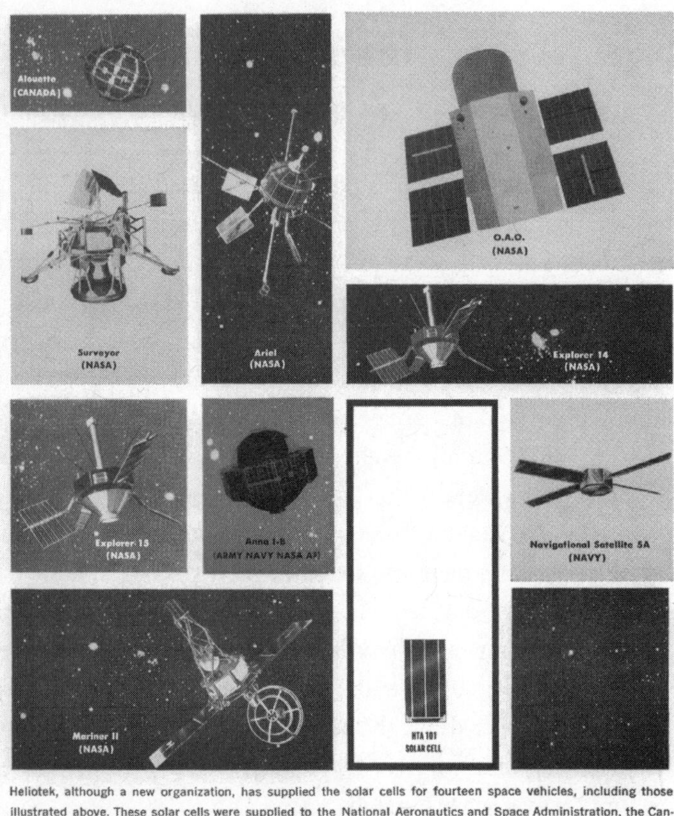

Figure 22.9. Image showing multiple satellites launched prior to 1961 using photovoltaics.

The success of the United States and Russia with solar cells in space led Soviet space scientist Yevgeniy Fedorov to predict in 1958 that "solar batteries will ultimately become the main source of power in space."[20] Events proved Fedorov right. Those working with satellites came to accept the solar cell as "one of the critically important devices in the space program," since it "turned out to provide the only practical power source" for satellites within a certain distance from the sun.[21] As Rear Admiral Rawson Bennett, chief of naval research, succinctly stated, "The importance of [solar] power is that satellites can play a valuable role in warfare."[22] Solar-powered satellites gave America precise knowledge of its enemies' capabilities and intentions both on land and in the sky. These satellites could also guide sea-based ICBMs, which became powerful deterrents. The Soviets would no longer consider a knockout first strike, knowing that the United States could retaliate by sea. Another series of solar-powered satellites, called the Vela Hotel, could detect nuclear explosions on the ground and in space. Their verification capability cemented the trust required for signing the nuclear test ban treaty, vital to the health and safety of the planet.[23]

Figure 22.10. Solar-powered satellite gives soldier instantaneous messaging regarding radar sighting.

The urgent demand for solar cells above the earth by the military opened an unexpected and relatively large business for the companies manufacturing them. "On their own commercially, they wouldn't have gotten anyplace," observed Dr. Joseph Loferski, who spent a lifetime working in photovoltaics.[24] Indeed, as the scientist Martin Wolf contended, "The onset of the Space Age was the salvation of the solar cell industry."[25]

Figure 23.1. Dr. Elliot Berman testing various solar arrays manufactured by his revolutionary photovoltaic company, Solar Power Corporation, that brought solar cells down to earth.

23

The First Large-Scale Photovoltaic Applications on Earth

Between 1953 — when the discovery of the silicon solar cell was announced — and 1961, the solar cell "changed from a scientific curiosity to the only reliable long-term power source for our space program," according to solar pioneer Paul Rappaport.[1] For applications on earth, however, solar cells remained too expensive. Nor did any terrestrial developments seem to be in the cards for the foreseeable future. Dr. Harry Tabor, one of the most respected solar scientists of the time, saw little hope, stating, "Most experts seem to agree that silicon solar cells are not going to get much cheaper."[2]

Bringing the Solar Cell Down to Earth

Then in 1968, Dr. Elliot Berman, an industrial chemist, proposed a completely different type of cell, one that could be produced like photographic film, to bring photovoltaic prices down to earth. Its completely automated processing appealed to executives at Exxon, who were then involved in scenario planning that showed there would be a need for solar electricity sometime during the early twenty-first century. "We found [Berman's ideas] very exciting," one present-day Exxon Mobil executive remarked. "We conceived the end product being very cheap, being used as roofing material."[3]

Not only did Berman convince Exxon to finance his novel solar work, but he also persuaded the company to explore current market possibilities. The search led to his crafting of a silicon solar cell that sold for less than those used for space. Rather than buy expensive semiconductor-grade silicon, he purchased much cheaper wafers that the industry had rejected and would otherwise have been thrown away. He also assembled modules less rigorously than for space — the environment on earth being more forgiving than in the sky, where solar panels had to contend with bombardment by micrometeors and radioactive particles, as well as huge swings in temperature.

With these and other strategies, Berman scaled down the price considerably. Power generated by the new modules still cost around forty times what most Americans paid at the time for electricity from power lines, but Berman and his Exxon-financed entity, the Solar Power Corporation, saw that a demand for electricity at much higher prices existed in remote locations on earth where utility lines did not reach. Such markets — supplied only by expensive batteries or generators — became in 1973 the newly founded company's target.[4]

Solar Cells Power Aids to Navigation

The Solar Power Corporation first set its sights on powering the navigational warning systems — foghorns and lights — attached to Exxon's oil platforms in the Gulf of Mexico. Huge, bulky conventional batteries had run them since the 1940s, when the Coast Guard had ordered Exxon to install the lights and horns on the platforms to warn ships of their presence at night and during inclement weather.

The need for reliability, paramount for safety equipment, made the biweekly servicing and frequent replacement of primary nonrechargeable batteries essential. Moving them on and off the platforms was a chore: the batteries were heavy and highly toxic. And making boat and helicopter trips to the platforms to tend the batteries, bring in new ones, and take old ones back to shore ran up a steep bill, not to mention the initial cost of the batteries themselves. In contrast, when a sun-charged secondary battery went bad, its replacement cost $160, compared to $2,100 for a nonrechargeable battery. Furthermore, the entire photovoltaic-powered system could be transported by a small skiff, but installing a nonrechargeable battery called for a crane boat at $3,500 per day.[5] With such advantages, the oil and gas industry rapidly took to photovoltaics. "[Solar cells] saved us time and money, and that, of course, is better," stated a veteran navigation aids engineer.[6] In the mid- to late 1970s, hundreds of modules were sold for use on the ever-increasing number of oil platforms. All the major oil companies — Amoco, ARCO, Chevron, Exxon, Texaco, and Shell — chose the solar option. Houston became the center of photovoltaic sales in the 1970s. Not long after that, owners of many of the oil rigs around the world followed the Gulf's example.

Figures 23.2 and 23.3. Hoisting and securing a photovoltaic array on an oil rig in the Gulf of Mexico

Figure 23.4. Now a common sight throughout the world, photovoltaics powering buoys and other navigation aids for coast guard services of every nation.

The coast guard soon took up photovoltaics, too. President Ronald Reagan commended the guard for saving "a substantial amount of the taxpayers' money through the conversion of aids to navigation from batteries to photovoltaic power."[7] In many nations, lanterns fueled with acetylene gas lit up buoys and other maritime warning signals. The cost of maintenance on such equipment exceeded what was spent in America. Payback time for the investment in photovoltaics for maritime warning signals was less than a year. The French Lighthouse Service began the replacement of acetylene systems with solar modules in 1981.[8] By 1983 Greece had installed photovoltaics to power the lights on most of its 960 buoys and lighthouses.[9] The technology's reliable service while constantly soaked in the sun and bathed in salt water proved its durability.

The photovoltaics industry also got its first significant opportunity to power land operations with the oil and gas industry during the mid-1970s. Underground aquifers frequently contain salt water, which corrodes well casings and pipelines. Electrical current injected near these casings and pipelines provides protection. But in remote areas like the Hugoton Gas Field, in Kansas, it's cost-prohibitive to bring in utility lines. After reading an article on photovoltaics in a popular science magazine, corrosion specialist Larry Beil, who was assigned to the Hugoton field, bought some solar cells from the Solar Power Corporation to ameliorate the situation.[10] Since their initial installation, tens of thousands of solar cells have provided protection for well heads and pipelines worldwide.

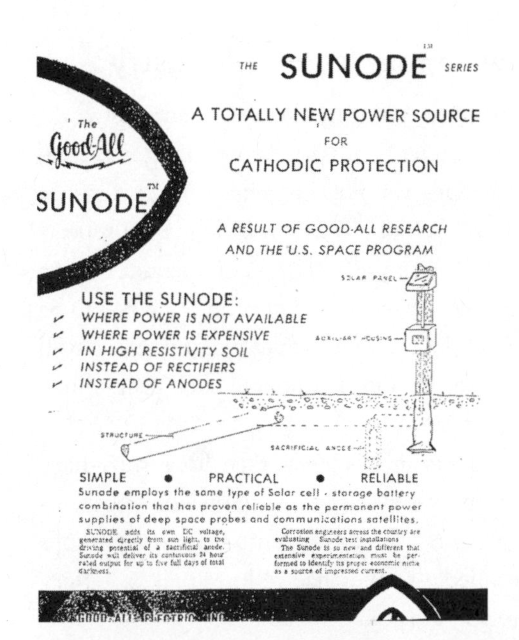

Figure 23.5. This advertisement introduces the use of solar cells as a power source for corrosion (cathodic protection).

Figure 23.6. Solar energy protects this wellhead in Kansas from underground corrosion.

Solar Cells Aid the Telecommunications Industry

Powering repeaters was another early terrestrial market for solar. Networks of microwave repeaters carry radio and telephone signals over long distances to remote locations; the dish-shaped repeaters pick up, amplify, and transmit such signals. These networks usually consist of a series of dish-shaped repeaters mounted on tall towers, which act like relays. In the age of vacuum tubes a relatively large power unit — either huge batteries or a generator — had to be brought in to operate the repeaters and the required power electronics, as well as the temperature-control equipment for the hut that housed such gear, which had to be kept at a suitable operating temperature.

John Oades, a microwave systems engineer who worked for a subsidiary of an American telephone company, was keenly aware of the problem of supplying energy to run microwave repeaters. "Many were up on mountaintops or in the middle of nowhere," he recalled, "so they had to have fuel or battery replacements brought in by helicopter."[11] On a ski trip in the early 1970s, Oades mulled over a potential solution to this seemingly complex problem and had a startling revelation. Perhaps all that gear wasn't necessary. Oades realized that using filters and covering the dish with a metal shield would do away with any interference between incoming and outgoing signals, and that this would eliminate the need for most of the supporting electronic equipment. Bill Hampton, Oades's boss at the time, described the difference between Oades's invention and the old technology as the difference between "a rack of equipment" and "something you could put in your pocket."[12] Or in the words of the inventor: "This was the most minimalist way.... I simplified the technology about as far as you could."[13]

Oades was able to cull whatever gadgetry was actually required by the repeaters from transistor technology developed for satellite-to-earth communications. The operating equipment for Oades's repeater fit into a cabinet about the size of a fuse box, it worked at any temperature encountered on earth, and it required little power and tending. Because the system could operate for years without maintenance and needed only a few watts of electricity, Oades sought a compatible source of energy, one that was reliable and compact. In this case, the choice of solar cells "came right at the beginning. In fact, if it weren't for photovoltaics," Oades noted, "I probably wouldn't have built the repeater."[14]

Navajo Communications Corporation bought the first unit from the

Figure 23.7. John Oades (left) and Bill Hampton check out the solar panels of Oades's sun-powered microwave repeater.

telecommunication company Hampton and Oades worked for. The company wanted to connect the community of Mexican Hat, Utah, with the rest of the world by telephone. The rugged, almost inaccessible terrain of deep canyons and sheer cliffs that surrounded the town stood in the way of the long-distance telephone service that most Americans in the 1970s took for granted. As one telecommunications executive observed, "It was one of those impossible situations of having to plow in miles of cable up and down the mountains and through a valley to a very small community. The investment on such a project would probably never have been returned. Mobile telephones would not have worked either. The high walls of nearby canyons would have effectively blocked out most signals."[15]

So Oades's repeater was installed on Hunts Mesa, overlooking picturesque Monument Valley. The location gave the repeater a clear line of sight from the microwave terminal at Mexican Hat to a microwave link at Kayenta, Arizona, which tapped into the national telephone lines. The repeater and its installation cost Navajo Communications a fraction of what the alternatives would have. "It

Figure 23.8. A view from the Hunts Mesa solar repeater.

took less than two days' time to mount the pole structure, install the equipment, and get it on the air," reported an astounded J. Shepherd, CEO of Navajo Communications.[16] In the fall of 1976, Hunts Mesa became the first solar-powered microwave repeater site in North America and one of the first in the world.

In Shepherd's estimation, the Hunts Mesa installation "represent[ed] a major breakthrough in the economics of serving small population centers with microwave communications." Its success also boosted confidence in photovoltaics as a power source for repeaters and other isolated stand-alone applications. "The problem," Oades pointed out, "was that people were concerned about long-term reliability. Repeaters have to be very reliable. They're out there in the middle of nowhere with ice and snow, just terrible conditions. People didn't know how solar cells would stand up to all this. After a year of successful operation at Hunts Mesa, they had something to point to."

Oades's company sold more than one thousand solar-powered repeaters in six years, resulting in a multimillion-dollar worldwide business. "The big appeal and fanfare," said Bill Hampton, "were the solar panels."[17] Since Oades's novel work intrigued those in the telecommunications business, he received invitations to conferences and meetings throughout the world to discuss his company's work. One excursion took Oades to Australia, where professionals had good reason to be interested: the government decided early in the 1970s to provide every rural citizen, no matter how remotely situated, with telephone and television service comparable to that enjoyed in the larger population centers.

Michael Mack, a power engineer for Telecom Australia, the government body tasked with carrying out the mandate, explained the dilemma his agency had to resolve. "In trying to get telecommunications to outlying properties — and we had a lot of them that are really quite remote — we found that one of the biggest problems was a reliable power supply." Telecom engineers brought in thermoelectric generators, diesels, and wind-driven machines; sometimes they connected their equipment to the customer's own power source. But none of these panned out as well as Telecom Australia had hoped. "They weren't reliable," Mack complained. "The problem was you were dependent on somebody actually maintaining the wind generator or keeping gas bottles replenished. Or if it was the customer's power supply, the quality of that power was very uncertain."[18]

On the lookout for something better, Arnold Holderness, a senior power engineer for Telecom Australia, kept his eyes open while on trips to America, Europe, and Japan. At the headquarters of the Sharp Corporation, the Japanese electronics equipment manufacturer, Holderness saw some literature about the photovoltaic modules that the company had just started to produce, and he ended up taking some Sharp panels home. "We were alive to the need for remote power supplies, and with this government mandate to hook everyone up, we had to think innovatively," Holderness recalled.[19]

The high price of the Sharp modules, about one hundred dollars per watt, severely limited their usefulness to Australia Telecom's mission. But a year or so later, the appearance of the Solar Power Corporation's modules, which were five times cheaper than the Japanese product, gave Telecom Australia the green light to put photovoltaics to work. The program started off with twenty systems, each consisting of a telephone, a transmitter, and a receiver powered by low-wattage Solar Power modules. Calls were beamed to and from a telephone exchange that was connected to the national network. The solar-powered phones worked so well that, after several years of service, the staff at Telecom Australia could state with great confidence, "Using direct photovoltaic conversion has become a viable and preferred power source for [remote telephones]."[20]

After demonstrating through these microinstallations "a cost benefit over

Figure 23.9. Solar-powered telephone in the Australian Outback.

any other primary power source suitable for the job," the initial solar program of Telecom Australia was deemed successfully completed by the summer of 1976. Telecom engineers had gained confidence in and experience with the new power source and were ready to design and develop large solar power systems that would link distant towns with the rest of Australia's phone and television networks.[21]

The dramatic drop in the amount of electricity needed to run large repeaters prompted Telecom Australia to consider powering entire telecommunications networks by the sun. A large dish repeater that had required a kilowatt to operate now took only 100 or 200 hundred watts. Improved electronic power equipment — a result of silicon transistors — brought about the breakthrough, as did housing the circuitry equipment underground, where temperatures remained mild year-round, eliminating the need for air-conditioning or heating.[22] These repeaters consumed much more power than did Oades's invention,

Figure 23.10. One of Telecom Australia's many solar-powered microwave repeaters, which have provided rural Australians with telephone and television service.

though, because the Australians needed to relay television and telephone signals over hundreds of miles, something Oades's design could not do. By 1976, photovoltaics could economically handle the power necessary to run a large-scale repeater — one that could carry hundreds, if not thousands, of calls at a time. As Telecom Australia reported that year, "To supply 100 watts in a remote area, without [utility] power, solar means are cheaper than any other source."[23]

Telecom Australia began its first solar-powered telecommunications system with thirteen solar-powered repeaters that connected the former diesel-powered network at Tennant Creek to the hub at Alice Springs. In the process, it linked such colorfully named intermediate points as Devil's Marbles and Tea Tree to Australia's national telephone and television service. People in these and neighboring towns, such as Sixteen Mile Creek and Warraby Knob, no longer had to dial the operator for long distance and shout into the phone to be heard. Nor did they have to wait for news tapes to be flown to their local stations, viewing them hours or sometimes days after the rest of Australia had.

The Tennant Creek–Alice Springs solar-powered system worked well. "The concept proved so successful," Michael Mack and colleague George Lee reported ten years later, "that Telecom Australia went on to install seventy similar solar power packages throughout its network."[24] All the installations had "gratifying results," because "there had been no system failures" after ten years of operation.[25] The Kimberley Project, which linked distant, but booming, northwestern Australian towns, stood as the world's largest sun-driven microwave system. Forty-three solar-powered repeaters, spaced about 35 miles apart, spanned 1,500 miles.[26] Not only did the repeaters bring long-distance telephone service to the towns along their line, but also people at sheep stations and isolated homesteads and communities within 30 miles of their range could connect to them via solar-powered phone systems.[27]

At the 1984 International Telecommunications Energy Conference, Mack told the attendees, "We have advanced from the stage where solar power was considered an exotic source to where it now plays an important role alongside conventional power supplies."[28] Or in the words of power engineer Arnold Holderness, "We were showing the world how solar power could be used in a big way out in the field."[29] And indeed, the world caught on. One telecommunications expert declared in 1985 that photovoltaic systems had "become the power system of choice [for] remote communications."[30]

PART VI

THE POST–OIL EMBARGO ERA

FROM THIS PIPELINE WILL COME
MORE INDUSTRIES, MORE BUSINESS, HAPPIER HOMES!

The arrival of Natural Gas in the Pacific Northwest and British Columbia completes the basic economy of the entire Pacific Coast. What Natural Gas has done for the well-being and growth of Southern California will undoubtedly be duplicated in the Pacific Northwest. For—from the new pipelines now serving the Pacific Northwest will come not only the miracle of Natural Gas, but also more industries, more business and happier homes.

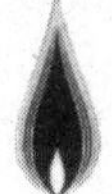

SOUTHERN CALIFORNIA GAS COMPANY
SOUTHERN COUNTIES GAS COMPANY

Ad No. M-77-56
Western Gas
September, 1956
Advertisement prepared by
McCANN-ERICKSON, INC.
Los Angeles

Figure 24.1. A gas company ad to encourage more gas use.

[1945–]

24

Prelude to the Embargo

For almost three decades after the end of World War II, the United States had few problems with its energy supply. Its industry, commerce, and homes all had ready access to oil and gas from both domestic and foreign sources. Most of the oil was close to the surface, easy to tap, and economical to extract. Foreign governments sold their oil to American companies at extremely low prices, and the U.S. government also helped to keep oil prices low and profits high. Depletion allowances permitted oil companies to write off a portion of their income taxes against the cost of discovering and drilling for oil in the United States. Royalty payments on foreign oil companies could also be deducted. These and other government subsidies helped to keep the price of oil below three dollars a barrel during the late 1950s and throughout the 1960s. Natural-gas prices during this same period were low, too — below one dollar per 1,000 cubic feet.[1] Often obtained as a by-product of oil exploration, gas enjoyed the same tax advantages as oil. The cost of natural gas was also kept down by government regulation of gas prices and by improved pipeline networks that made supplies much more accessible.

The falling prices of fossil fuels during this period also reflected the growing availability of supply. Crude-oil reserves rose steadily between 1952 and 1964, as did natural-gas reserves.[2] Corporate spokespeople assured the public that this rosy situation would continue almost indefinitely. As one gas company executive stated, "The industry discovers more gas every day than is consumed every day in the United States. I don't think in our lifetime we will see the depletion of our product."[3]

With fuel apparently so abundant and cheap, electric companies began to expand to meet growing postwar demands. Liberal government policies made it easy to procure the needed capital to build larger and more efficient power plants. Soon after World War II the electric rates plunged as consumption grew and power-plant efficiency increased. The electric utilities encouraged greater consumption because the costs of building new plants and installing electric lines could be recovered more quickly if their customers used more electricity. "Once you had the lines in, you hoped that people would use as much electricity as possible," an executive for one electric company remarked. "You wanted to get as much return on your investment as you could."[4] The gas companies had a similar objective. As one employee explained, "If you sell more you make more."[5]

Figure 24.2. An electric company refrigerator magnet to encourage consumption.

Both gas and electric utilities promoted consumption through advertising campaigns and preferential rate structures. The Gold Medallion Program, a national promotional campaign conceived by General Electric and later taken over by regional electric utilities, urged Americans to buy more electric appliances — and consequently, use more electricity. Lower rates for increased use of electricity also stimulated consumption. The gas companies had similar rate structures that made it cheaper to use more gas. They also launched their own proconsumption campaign, which used a blue flame as its slogan and symbol.

The energy companies' publicity and the enticement of lower prices worked. A desire to "live better electrically" led families to opt for homes with electric house-heating, water heating, ovens, and many other appliances. A growing affluence that allowed people to indulge their appetite for new electric appliances, combined with the postwar baby boom, helped increase electricity generation by over 500 percent between 1945 and 1968. Natural-gas consumption, too, zoomed upward as gas heating and conveniences such as clothes dryers became more popular. Natural-gas production nearly tripled, from 6 to 16 trillion cubic feet between 1950 and 1965. U.S. fuel consumption as a whole more than doubled between 1945 and 1970.[6]

A Note of Caution

The frenetic pace at which America was gobbling up its energy resources alarmed only a few farsighted individuals. Eric Hodgins, editor of *Fortune*, called the careless burning of coal, oil, and gas a terrible state of affairs, enough to "horrify even the most complaisant in the world of finance." Writing in 1953, he warned that "we live on a capital dissipation basis. We can keep this up perhaps for another 25 years before we begin to find ourselves in deepening trouble."[7] But such warnings were generally treated with derision or merely ignored. Those predicting energy shortages were labeled pessimists. "Not many in industry wanted to hear such talk," commented Charles A. Scarlott, then editor of the Westinghouse Company's technical publications. "They were making too much money on energy sales."[8]

A few scientists and engineers took the same dim view as Hodgins and sought an alternative to the fuel crisis that they saw was inevitable. In 1955, they founded the Association for Applied Solar Energy and held the World Symposium on Applied Solar Energy in Phoenix, Arizona. Delegates from all over the world attended, presenting research papers and exhibiting solar devices. Israel displayed its commercial solar water heaters, and representatives from Australia and Japan discussed their nations' increasing use of the sun. To many, the symposium represented the dawn of a new solar age. But the careless confidence of energy-rich America squelched that hope in this country. Solar energy received virtually no support in the ensuing years, and by 1963 the association found itself bankrupt. "They couldn't even pay my final salary," noted Scarlott, then editor of its publications.[9]

The governments of Israel, Australia, and Japan deliberately aided the solar industry, but the U.S. Congress and the White House sat on the sidelines while the hopes of a prescient few foundered. True, as early as 1952 the President's Materials Commission, appointed by Harry S. Truman, came out with a report, *Resources for Freedom*, predicting that America and its allies would be short of fossil fuels by 1975. This report urged that solar energy be developed as a replacement. "Efforts made to date to harness solar energy are infinitesimal," the commission chided, despite the fact that "the United States could make an immense contribution to the welfare of the free world" by exploiting this inexhaustible supply. The commission predicted that, given the will to go solar, there could be 13 million solar-heated homes in the nation by the mid-1970s.[10]

Atoms for Peace

Although the President's Materials Commission advocated a fifty-fifty split for nuclear and solar contributions to America's energy future, the U.S. government lavished billions on atomic power research while spending a pittance on solar.[11] International cold war politics more than technological advantages accounted for the nuclear preference. President Eisenhower's adviser on psychological warfare suggested in 1953 in a top-secret report called Operation Candor that the Eisenhower administration level with the American people in several fifteen-minute television presentations outlining the perilous times the nation faced. The Soviet's growing military might and possibility of nuclear warfare would dominate the proposed talks. Rather than scare Americans, the president chose a more positive approach. He decided to give nuclear weapons a happy face by introducing the peaceful atom.[12]

Eisenhower found the occasion to do just that on December 8, 1953, at the United Nations. He announced to the world that America "would be proud to take up... the development of plans whereby [the] peaceful use of atomic energy would be expedited.... The United States," Eisenhower assured the world body, "pledges before you — and therefore before the world — its determination to help solve the fearful atomic dilemma — to devote its entire heart and mind to find the way by which the miraculous inventiveness of man shall not be dedicated to his death, but consecrated to his life." The crowd listening to Eisenhower numbered over three thousand and represented almost every nation in the world. When the president proposed the peaceful use of the atom — "to

Figure 24.3. Image of President Eisenhower on a State Department brochure pushing the peaceful atom.

apply atomic energy to the needs of agriculture [and] medicine...[and] to provide abundant electrical energy in the power-starved areas of the world" — everyone sprang up and applauded and kept on cheering.[13]

Eisenhower's speech earned rave reviews. The *New York Times* lauded the president for sweeping "away the last vestiges of the Soviet propaganda theme that the United States rather than Soviet Russia obstructs progress toward easing international tensions. And [this plan] has made President Eisenhower the unchallenged leader of the 'camp of peace.'" The premier of France concurred with the assessment of the influential newspaper, stating, "Systematic propaganda of Soviet inspiration tried to give the great powers a guilt complex about their large stocks of atomic materials. President Eisenhower's speech has radically changed this situation."[14]

Eisenhower's UN presentation made Theodore Streibert's job a lot easier. Streibert was in charge of spreading America's propaganda overseas. He no longer had to waste his time answering Russian accusations. The president's suggestion of turning the bomb into a weapon for peace had the Soviet Union squirming. Abbott Washburn, Streibert's assistant, triumphantly told the *Times* that it was "unquestionably the best idea in the cold war so far."[15]

Someone called Eisenhower's plan "Atoms for Peace," and the phrase stuck. Selling the peaceful atom as the world's future energy source suddenly became America's number one priority. The nation now had to follow through, and follow through it did. "More than 60 chemical, power, engineering and equipment making companies" were ready to jump on board, according to *Life* magazine, asking the government "for a change in the Atomic Energy Act that would allow them to own atomic fuels."[16] Eisenhower supported the idea. Congress did, too, passing the Atomic Energy Act of 1954, making available at no cost to industry "the knowledge already acquired by 14 years and 10 billion dollars worth of [government] research." In this act, the government pledged to undertake for the private sector "a program of conducting, assisting, and fostering research and development to encourage maximum scientific and industrial progress."[17] As Nobel laureate Hannes Alfvén observed, these corporations, many previously involved in building the bomb, committed to promoting the peaceful atom "because the government paid [all the expenses] and took all the risks."[18] There was no parallel "Solar Energy Act."

As an indication of world support for nuclear power compared to that for solar power, the United Nations sponsored an international conference on the peaceful uses of the atom, which the world's top statesmen and scientists would attend. So, in August 1955 the political and scientific leaders of the world converged in Geneva to discuss the peaceful uses of atomic energy. The United States even flew in a reactor for the conference. In the ultimate public relations gesture, the meeting opened on the tenth anniversary of the atomic bombing of Hiroshima. "It was on Aug. 6, 1945, a Monday, that the first atomic bomb transformed the city of Hiroshima into a gigantic mushroom-topped cloud of life-destroying radioactive dust," wrote the *New York Times*' science editor on page one. "It is on the second Monday of August, 1955 [the opening of the Geneva conference], that the peoples of the world may look upon in the future as the day heralding the ultimate deliverance from fear of atomic annihilation, the day that marked the beginning of the atomic realization of the biblical prophecy of beating swords into plowshares."[19]

(In contrast to the worldwide governmental and scientific sponsorship and news coverage of the Atoms for Peace meeting in Geneva, the First World Symposium on Applied Solar Energy held just three months later had to settle for

10 E THE NEW YORK TIMES, SUNDAY, JULY 24, 1955.

The glow of a giant 75,000-watt lamp signaled the start-up of a turbine-generator using atomic energy at West Milton, near Schenectady, N. Y., on July 18, 1955. Housed in a 225-foot steel sphere, the reactor prototype of the power plant of the Navy's submarine Seawolf supplied steam to power a generating plant provided by General Electric at its own expense, which is now in operation. Income from the sale of electricity goes to the U. S. Government.

General Electric switches on first commercially distributed

Atomic Electric Power for Homes, Farms, Factories

How the Atomic Power Plant Works

Inside the giant steel sphere is an atomic reactor. Heat generated by atomic fission is transferred to liquid sodium which is piped to a heat exchanger where steam is generated. Steam flows through pipes outside the sphere to drive a turbine-generator which produces electricity. For 16-page illustrated booklet, "Putting The Atom to Work," write to General Electric Company, Section 192-1, Schenectady 5, New York.

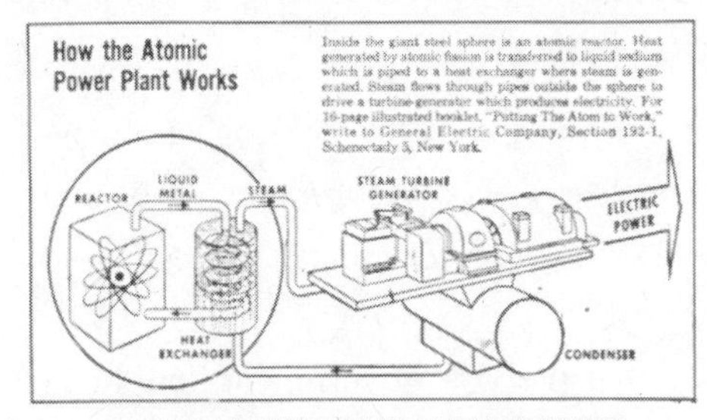

OTHER G-E MILESTONES IN ATOMIC PROGRESS

1940 Drs. Kenneth H. Kingdon and Herbert C. Pollock, scientists of the General Electric Research Laboratory, made up one of two independent groups which succeeded in separating U-235 from natural uranium and so helped to lay the foundations for atomic power.

1941 Scientists from General Electric contributed towards the achievements of Manhattan District of the Army Corps of Engineers, then charged with the development of the first atomic installations in this country.

1946 G.E. appointed by the War Department to operate the Hanford Atomic Products Operation at Richland, Wash., for the U. S. Government.

1947 General Electric, under contract with the Atomic Energy Commission, started operation of the Knolls Atomic Power Laboratory.

1950 General Electric began development work for the AEC on a submarine intermediate reactor.

1951 G. E., at the request of AEC and Air Force, joined in aircraft nuclear propulsion program.

1955 General Electric established Atomic Power Equipment Department to build and market a complete line of components and equipment for research and electric power production.

1955 General Electric contracts to build world's largest all-nuclear power plant for Commonwealth Edison Company of Chicago and Nuclear Power Group, Inc.

1955 General Electric launches first sales campaign for nuclear research reactors.

1955 General Electric starts up prototype power plant for SSN575 *Seawolf*.

Developmental submarine power at West Milton, N.Y., demonstrates practical value of atomic power stations to come

The atom was harnessed for peace-time electric power use on Monday, July 18, near Schenectady at West Milton, N. Y.

In co-operation with the Atomic Energy Commission, General Electric placed in operation America's first atomic electric power generator capable of supplying 10,000 kilowatts of power for public use, or enough to meet the needs of a community of 20,000 people, if operated continuously.

Electricity from this plant went out on the power lines of the Niagara-Mohawk Power Corporation, helping to serve many customers over a wide area.

This atomic generating plant will demonstrate the practical possibilities of generating electric power from the atom and will stand as a milestone in the use of atomic energy for peaceful purposes.

The production of electric energy from the atom is important not only to America but to the entire world. For while America has always had an abundance of low cost electricity, a large portion of the world's population has never had enough conventional fuels and other resources with which to improve their standards of living.

Where Communism alone has failed, Communism *plus* atomic power might succeed. America's course is therefore clear. Our progress with the atom is encouraging to *all* the world's people.

And the co-operation of our Government with private industry is an important reason why so much has been accomplished in so short a time.

Figure 24.4. GE celebrated nuclear power as part of the free world's arsenal against communism.

sponsorship by a few private industrialists and foundations and received scant coverage by the media.)

People from every national and political inclination agreed that the world had arrived at the threshold of the atomic age, "the third great epoch in human history," which was preceded by the agrarian revolution in the Euphrates, Indus, and Nile river valleys and the Industrial Revolution of the nineteenth century.[20] A few, though, had second thoughts about the wisdom of the crusade for the peaceful atom. The Nobel Prize–winning chemist Dr. Glenn T. Seaborg, who later became the head of the Atomic Energy Commission, argued that the difficulty of finding sites for disposal of the dangerous radioactive wastes created by atomic power plants would severely limit their development.[21] Worse, experts agreed that the owners of an atomic power plant could quickly convert their fissionable material to build bombs. Dr. Robert Oppenheimer, one of the lead developers of the nuclear bomb, testified to the U.S. Senate that a large atomic power plant could be changed over to atomic arms production in a matter of months.[22] Even members of the Eisenhower administration admitted having "some unhappy second thoughts... about the President's 'atoms for peace' program" turning into "atom bombs for all." They confessed that the specter of nations of the underdeveloped world arming themselves atomically was "terrifying."[23]

What about Solar?

Dr. James Conant, the American scientist who oversaw the making of America's first nuclear weapons, agreed that nuclear power was too dangerous and expensive. He urged the nation to instead create a program like the Manhattan Project for the development of solar energy.[24]

The *New York Times* also suggested that the U.S. government "ought to transfer some of [its] interest in atomic power to solar."[25] But the attitude of Washington and the private sector mirrored that of a nation hypnotized by seemingly limitless supplies of cheap fossil fuel and by the almost magical aura surrounding nuclear energy. As late as 1972, the National Science Foundation declared, "There has been little Federal support of solar energy utilization.... Commercial and philanthropic support of solar power generation has been almost negligible."[26] This, despite the fact that the developer of the modern nuclear reactor, Farrington Daniels, argued, "There is no gamble in solar energy use; it is sure to work."[27]

But the new nuclear age had little or no need for solar technologies. *Newsweek* judged "the sun's diffuse radiation" as "paltry" when compared with what nuclear could do. A Marxist scientist, parroting the new Soviet line, gave an even bleaker assessment. We do not "need...to seek for a way of trapping...solar energy," John Desmond Bernal wrote, because of "the discovery of how to produce and usefully direct the enormously concentrated energy" of the atom.[28] The best that solar enthusiasts could hope for, according to the prevailing wisdom of the middle and late 1950s, was to plan for far-off energy needs. The *New York Times* articulated this point of view well, predicting, "Electricity from the atom will keep industry turning and homes lighted for centuries into the future. And the energy of the sun...will be available after the last atomic fuel is gone." Or as *Life* magazine put it in an article aptly titled "The Sun: Prophets Study Rays for Far-Off Needs," "A few farsighted scientists are dreaming of ways to save the U.S. when coal, oil, gas and uranium run out. That may be from 200 years to 1,000 years away."[29]

George W. Russler, chief staff engineer at the Minneapolis-Honeywell Research Center, suggested that solar energy could better tackle the growing need for the replacement of oil by providing heat for houses and office buildings. He pointed out that the low-temperature heat required in homes and office "ideally matches the low-grade heat derived from the simplest and most efficient solar energy collectors." This was the perfect way to start putting solar energy to widespread use and ameliorating the ominous circumstance that the number of new oil discoveries in the United States had fallen every year after 1953, while the reliance on imported oil kept growing. In fact, in 1967, for the first time in the nation's history, crude-oil reserves declined. And the renowned oil engineer Marion King Hubbert predicted in 1956 that American petroleum production would peak between the late 1960s and the early 1970s. Most in the oil industry ridiculed his work. But in 1970 the laughing ceased. His prediction had come to pass.[30]

Figure 25.1. Solar air collector feeding into a hippie dome at the aptly named Drop City, Colorado.

25

Solar in the 1970s and 1980s

Thirty million people in America celebrated Sun Day in 1978 to launch, in the words of one of the organizers, "a new age — the solar age."[1] The huge turnout demonstrated how popular solar energy had become in the 1970s. Two writers for *Science* magazine and its book division, William Metz and Allen Hammond, observed that "large numbers of consumer and reformist groups" embraced solar for several reasons:

- It offers a new way of doing things at a time when many people are disenchanted with the old.
- It promises renewable energy when importing depletable energy is seen as a national liability.
- It promises clean energy when the degradation of the environment is becoming readily apparent.
- It offers individuals a degree of energy independence unprecedented in the industrial age, and it has sufficient potential for changing the institutional structures of the country so that some of the powers find it threatening.

The support for solar, the authors concluded, "has some of the attributes of a political [and social] movement" joined wholeheartedly by those who had

Figure 25.2. Celebrating Sun Day in 1978.

opposed the Vietnam War and had been involved in the counterculture begun in the 1960s.[2]

Solar Pool Heating

It is not without irony that solar pool heating got the American solar industry started. And that it was the only solar application "making significant inroads into the U.S. consumer market in the 1970s," according to the most extensive study of all solar technologies between 1973 and 1984, conducted by the Solar Energy Industries Association.[3] To pile irony upon irony, the affordable solar pool heater, the first major breakthrough in solar technology since the discovery of the silicon solar cell, had its beginnings in Atherton, one of the wealthiest suburbs of San Francisco.

Freeman Ford, a former navy aviator who had flown missions over Vietnam, noticed one day that his gas-fired pool heater had stopped working. Ford could have simply replaced it with a newer model but did not want to spend another penny for fuel. He recognized that the sun and pool heating was a marriage made in heaven. No extra money would have to be spent for storage of the solar-heated water, because the swimming pool itself would hold the heated water. He would not have to invest in a pump to move the heated water from solar collectors to the pool, since his current pump still worked well. And because he didn't want to boil himself or his friends alive in the pool, a water temperature between 80 and 90°F would suffice. Since he needed to maintain only a relatively low temperature, Ford could rule out, much to his relief, the typical solar water-heating system made of glass and metal, which would have made his business model financially untenable. Suddenly, a plan came to mind: make the solar collectors out of plastic and dispense with the glass.

He sent his idea for a solar swimming-pool heater to renowned experts in the field. "You know, plastics," everyone he talked with responded, "do not conduct heat very well. Besides, they degrade in sunlight. Stick with flat-plate collectors, the kind used in California and Florida to heat water." He ignored the advice, figuring that by increasing the collector area he could compensate for the material's poor conductivity and yet still spend a lot less than he would if using the traditional method of solar water heating. To deal with the issue of plastic degradation, Ford put into the polymer mix ingredients that would protect plastics from damages caused by ultraviolet radiation.

It took multiple tries to get the right combination. He also ran into problems when configuring the tubing on the solar panels. In the first models, water pressure would perforate the narrow channel extrusions through which the water flowed, and the whole system would spray like a fountain. With better chemistry and design, Ford brought to market an affordable and durable product in the early 1970s, right around the time the first oil embargo hit America.[4] Concern over impending energy shortages led many levels of government to ban the use of fossil fuels to heat swimming pools in their jurisdictions. They felt that such a

Figure 25.3. Freeman Ford testing one of many configuration of his revolutionary plastic solar pool-heating panels.

frivolous use of scarce resources had to be stopped, and so in many cases having a heated pool required solar. By 1977, fully 60 percent of California's 250,000 pools were solar heated.[5] Ford's company, Fafco, sold half a million systems throughout the world in the company's first decade of operation.

Figure 25.4. Freeman Ford, founder of Fafco, next to one of his plastic solar collectors for swimming-pool heaters.

Solar for Hippies

Drop City, Colorado, one of America's first countercultural communes, was a far cry from wealthy Atherton, where solar pool heating was developed. But pool owners and hippies shared a common interest — getting free energy from the sun. Peter Rabbit, the city's scribe, had this to say about one prominent feature of his community: "The people [nearby] call us Dump City. [But] the most beautiful place in Drop City is the dump.... It's our greatest resource."[6] From that dump the people of Drop City scavenged Buick windshields to serve as south-facing coverings for their Grow Holes — pits dug out of the hillside in which makeshift greenhouses kept the commune's crops growing during cold but sunny Colorado winter days.

It was Steve Baer who had taught the Droppers all about the value of junked windshields for solar applications such as greenhouses and smart domes that heated a building's interior with free energy from the sun. And like the other Droppers, Steve Baer was one of those "dirty hippies — as they were known — who dropped out of consumerist America's mainstream and began settling in the arid reaches of this country's Southwest," according to *Mother Earth News*. "[While,] undoubtedly, some of those dropouts really were the ne'er-do-wells their parents thought they were," the magazine continued, "others in the crowd, like Steve Baer, were genuine visionaries, Renaissance men and women."[7]

Baer had started his work motivated by the pure joy of feeling the warm air flow from a space exposed to sun and covered only with glass or clear plastic. "You just get hooked and can't stop," he said. There was nothing more satisfying, according to Baer, than "when you take these dead materials — this glass and metal and insulation — and place them together in very simple, easy-to-build forms and — suddenly! In the middle of winter! — there's *warmth*."[8]

Baer got his start in the field by reading Farrington Daniels's book *The Direct Use of the Sun's Energy*, which Baer described as "the best book on solar energy that I know of." He also credits solar inventor Harold Hay for turning "my whole head around on how to best attack these problems" of heating and cooling with natural forces.[9] He counted "Harold Hay's experiment with a styrofoam ice chest as among the most exciting and illuminating ever done."[10] In this experiment, Hay bought a simple ice chest at a supermarket in Phoenix, Arizona, filled it with water, closed the top with its insulated cover during the broiling days of summer and opened it to the cold, radiant sky at night. The water stayed cool day and night. In winter he reversed the process and the water kept warm 24/7.

With his concept proved, Hay successfully applied the ice-chest experiment to heating and cooling a house in Phoenix, where the average high temperature in July is 106°F and the average January low is 46°F. At a major solar conference he described his work: "A moveable insulated roof permits containers of water in thermal contact with a metallic ceiling to absorb solar energy during the daytime in winter and radiate that heat to the interior when needed[,] and in summer [to] absorb the solar heat from the building and dissipate it to the night sky, ready the next morning for another cycle."[11]

The scientific crowd just laughed. "You mean that's all you're doing. You're moving some insulation back and forth?" one of them said to Hay.[12] Despite the lack of enthusiasm and outright derision for Hay's work, Baer praised him for inspiring "simple solutions to heating and cooling [buildings] in most climates."[13]

Others, too, saw the value of Hay's work. Renowned solar expert John Yellott, who monitored the Phoenix house, reported, "Data taken every 12 minutes ... during an 18-month period... showed that the building could be kept within the comfort zone during an Arizona winter, with temperatures as low as 25F (–4 C), and during two Arizona summers, when the temperature rose to 115F (46 C)."[14] As for the economics involved, General Electric's program manager for solar heating and cooling of buildings, A. D. Cohen, judged the cooling component as competitive with conventional air-conditioning in Los Angeles and Phoenix.[15]

One of Baer's first inventions, the drum-wall heating system, was derived directly from Hay's work — relying on movable insulation to aid the heating of his domes in Corrales, New Mexico. The drum wall consists of stacks of 55-gallon black-painted metal drums that formed the south wall of the Baers' house. "Just

Figure 25.5. Drum-wall installation.

outside the wall," his wife, Holly, explained, "is a pane of glass[,] and on the outside of that is a big door. On winter days we open the door[,] and the sun shines through the glass and heats the water in the drums. When we close the door at night, the water [inside the drums] radiates warmth into the house.[16]

The Baers were part of a group that converged on Santa Fe, New Mexico, and shared a common interest in tinkering with solar applications. Another member of the group was David Wright, a licensed architect who had left a good job in Santa Cruz, California, to apprentice himself to Bill Lumpkin, a master adobe builder in Santa Fe. The solar collectors that Lumpkin, Baer, and some scientists from Los Alamos National Laboratory were experimenting with seemed too complex to Wright. The collectors circulated their heat via the forces of nature, but their construction required a substantial amount of additional materials, such as sheet metal, pipes, and water. Why not just make the house its own solar collector?

Rediscovering Solar Architecture

An earlier trip in the 1960s to Betatakin in the Navajo National Monument gave birth to Wright's vision. High above a canyon that Wright and others hiked through stood an entire village nestled in a cave that had been carved out of a cliff by the erosive forces of nature. The low winter sun, their interpretative guide explained, poured into the south-facing cave and heated up the heavy-walled structures inside. The heat moved through the walls and into the interior by nighttime, keeping the village's inhabitants sufficiently warm without recourse to any other fuel. Such architecture, the ranger explained, allowed them to survive where there was little wood to burn — owing to what he called their ancestors' shortsightedness.

Wright saw Betatakin with an architect's eyes and appraised the deserted village as "a very sophisticated use of low-energy materials and natural elements. It is a maximal use of obvious resources with an absolute minimum of technology, and it worked, as the ancient cave dwellers had only the cave and rock with which to make their buildings energetically self-sufficient." That fact he would never forget. "That impressed me," Wright recalled.[17] Further research revealed to the architect that other ancient peoples throughout history had done "variations on [the] same theme." "Look[ing] back and draw[ing] from history, it becomes increasingly apparent that the really hard work has already been

Figures 25.6 and 25.7. Distant and close-up views of cliff dwellings at Betatakin.

done," he concluded. Modern building technology, Wright recognized, could enhance what the ancients had accomplished with climate-sensitive architecture. "Just think how much more comfortable those Greek or Indian dwellings would have been," the young architect mused, "had they had double- and triple-paned glass on their south side to collect the solar heat or modern insulation to keep the heat in."[18] When Wright considered that he had at his disposal a far greater range of materials to work with, he knew that he could succeed in his mission.

With such insight, Wright realized that instead of using Steve Baer's water-wall idea he could substitute floors and walls that, if built properly, would retain the same amount of heat. He envisioned his building as a thermos bottle. He would keep the heated part — in this case the interior — warm by wrapping insulation around them. And he would aim the openings of his "thermos" house — the windows — to the south and "cork" it at night by closing them with movable insulation so the heat remained inside.[19]

In 1972 Wright designed and built a house according to this plan. A two-story double-glass wall formed the entire south side. The depth of the house was calculated so that the sun would penetrate the entire interior in winter. The house's thick adobe walls were insulated on the outside with polyurethane foam to protect those living inside from the winter cold and the summer heat. The roof was also heavily insulated, and the floor consisted of a relatively massive

Figure 25.8. Wright's pioneering passive solar house, built in 1972. The auxillary heating system in the foreground were made unnecessary by the passive design and materials.

layer of adobe brick or sand with rigid insulation beneath. Tight-fitting, folding shutters opened and closed according to need.

Before Wright had finished the house, though, his scientific friends from Los Alamos dynamically modeled its future performance and became worried. "It might not work as well as you think!" one of them warned, and they advised him to install an auxiliary heating system. In response, he placed high-temperature tubing in the sand beneath the floor and planned to eventually connect it to solar collectors outside. The collectors never had to be connected, though. His Franklin stove generated the minimal amount of extra heat needed for comfort. Wright had to burn only a fraction of what others usually burned during the cold, snowy Santa Fe winters, and still the interior temperature held steady at a comfortable level. He was especially pleased with the loft where he and his wife slept, which, he discovered, "is great at bedtime as it is four to five degrees warmer upstairs."[20]

Sitting inside the house he called his sun ship, not long after the house was completed, Wright and some of his solar-energy-minded friends, including Steve Baer and the two scientists from Los Alamos — Herman Barkman and "Buck" Rogers — were chatting together. Wright asked the group: "So what type of system have I built?" Rogers replied, "Well, in mechanical engineering, we call systems that use pumps or fans 'active,' so I guess this must be a passive solar design." Wright later recalled that it was the first time he had ever heard the term *passive solar*, adding that "today *passive solar* design is known and practiced around the world."[21]

The 1973 Oil Embargo

Events far away in another arid region of the world should have spotlighted the work of these solar tinkers, or so it seems. In October 1973, the Arab oil-producing states cut back their exports to America in reaction to the nation's support of Israel in the Yom Kippur War. The ensuing shortage, *National Geographic* reported in 1974, brought about the greatest disruption in peacetime in the United States and much of the rest of the developed world since the Great Depression of the 1930s: "Factories shut down, workers were laid off, lights dimmed, buildings chilled, gasoline stations closed, Sunday driving was banned, fuel prices soared, stock markets fell." Petroleum geologist Marion King Hubbert, whose work had suggested such a crisis would one day occur because we were approaching the point of "peak oil," told the magazine, "We've had an oil shortage in this country for more than twenty years and didn't know it." As far back as 1947, "our domestic production slipped below our consumption, and we became a net importer of oil."[22] "When political events cut back oil imports from the Middle East in 1973," the *National Geographic* reporter commented, "King Hubbert's 26-year-old oil shortage came alive."[23]

When asked about possible solutions, Hubbert suggested solar energy. "We have it already," he asserted, and to prove his point he took from his pocket a propeller driven by a small motor powered by solar cells. When Hubbert faced the array toward the sun pouring in from the window, it "began to whirl," fueled solely by the "clean, pure energy from a source at least as long lasting as man's occupation of the planet."[24]

Figure 25.9. When the oil embargo struck, many people began, for the first time, to consider solar energy as a major alternative fuel.

Solar Collectors in Space, Power Towers, and Parabolic Mirrors

The question left unanswered by Hubbert was how to turn the power of the sun into usable energy. One scientist, Peter Glaser, came up with what he believed was the ultimate solution, a solar power station in space. The challenge of how to transport it into outer space — with its multiple, butterfly-shaped solar wings, each of which would measure 6 by 7 miles — kept the proposal in the realm of science fiction.[25]

Honeywell, a large aerospace concern, had its own particular solar solution, one far more complicated than the work done by Baer, Wright, and their cohorts in New Mexico. The company proposed placing somewhere in the American desert seventy-four thousand flat mirrors, each measuring 10 by 20 feet. Each mirror would focus sunlight throughout the day onto a single water tank located atop a tower. The water would become hot enough to make steam, which would then be piped to nearby turbines to produce enough electricity for forty thousand homes.

Figure 25.10. In the power tower, flat mirrors called heliostats concentrate solar heat onto a tower that contains a boiler, producing steam for running a conventional generator.

Called a power tower, its development dominated solar research budgets in the United States and elsewhere throughout the late 1970s and the 1980s.

Figure 25.11. A view of multiple parabolic troughs used in the Mojave desert.

Nonetheless, the power tower concept was never a commercial success. The only solar steam-powered plant that succeeded looked remarkably similar to the parabolic troughs Frank Shuman had built in Egypt seventy years before. And its story eerily paralleled Shuman's Egyptian experience. Many welcomed the new solar-concentrating installation in California's Mojave Desert with great enthusiasm. Newton Becker, the head of the company that constructed the plant, called it "an oil well that will never run dry."[26] A solar expert judged that it represented "the high point of the 1980s" for solar thermal electricity.[27] Another lauded the project as "cost competitive with conventional fuel power

Figure 25.12. A close-up of one parabolic trough.

stations" and believed "its potential [was] virtually unlimited." With oil costing close to forty dollars a barrel in the early 1980s, and additional escalations in price predicted, it is no wonder investors placed their money in a solar power plant. But just like Shuman, the investors found they were in for a surprise. The price of fossil fuels hit rock bottom in the late 1980s, and once again it appeared there was no end to the supply. As a consequence, the company responsible for building the plant went bankrupt. The plant, though, continues to operate. It generates enough electricity during daytime hours to power about 230,000 homes.

Solar Collectors for House Heating

Other solar engineers and scientists ran with house-heating ideas copied from earlier work done by Hoyt Hottel at MIT or by George Löf at the University of Colorado. According to Martin McPhillips, a government employee who was privy to such plans, they crammed onto rooftops "large banks of solar collectors," while "storage tanks and bins were crowded into basements. Connecting them were pipes, valves, thermostats, and a multitude of other gadgets to move the sun-heated fluid from collector to tank or bin. With all these moving parts and mechanical stuff," McPhillips observed, "Murphy's Law reigned supreme[, because] they were too complex, too expensive and didn't work quite right anyway."[28]

According to a study published by the American Association for the Advancement of Science, the problem with such solutions, mostly sponsored by the federal government in the mid-1970s, was that they were modeled "from the mainstays of today's energy economy"; they did not represent "novel ways" in which solar energy could contribute. For this reason, the report continued, the government "has been painfully slow" to recognize solar's "potential for decentralized applications," and in particular "the benefits of passive solar heating."[29]

Solar Pool Heating and Solar Architecture Rule the Day

While technocrats in Washington and experts in academia scoffed at simpler approaches like those taken by Baer, Hay, and Wright, who were known to most as the "lunatic fringe" of solar-energy experimentation, word of the latter group's success started to travel beyond Phoenix and Santa Fe. Wright's two Los Alamos scientist buddies started to tell other scientists at the lab about how well Wright's building had performed. As it happened, a team headed by Dr. Douglas Balcomb had recently overseen the installation of a collector system for heating the lab's library. "It was an engineering tour de force," Balcomb recalled. Though technically successful, his group soon learned that such a system — with its pumps and valves — "wasn't going any place fast. Complexity was one reason. Expense was another."[30]

Rather than sour on solar, Balcomb's group started to do calculations and modeling of passive designs, using Cray supercomputers. "From those very first analyses," Balcomb remembered, "it was apparent that well-designed passive [solar] architecture seemed to work just fine."[31] These initial positive results led to further simulations of performance under different weather conditions. The results made Balcomb "tremendously enthusiastic about the prospects for these new concepts. What intrigued me most," he continued, "is that they work so well."[32]

The results obtained by his group so intrigued Balcomb that in 1976 he and his family moved into the first solar suburb built since the Howard Sloan solar heating developments in the early 1940s. Not only the architect, David Wright's mentor, Bill Lumpkin, but also the developers, Susan and Wayne Nichols, belonged to the "lunatic fringe." Lumpkin had attached a solar greenhouse to the home's south side in the same fashion that nineteenth-century architect-builders had joined conservatories to the south faces of Georgian or Victorian houses. The Balcombs loved their solar house. The sun provided almost 90 percent of their heating in winter in a location where the temperature drops below freezing a good part of the time, and they also had the added advantage of growing flowers and even cherry tomatoes in their interior year-round. Operating the heating system demonstrated the simplicity of passive architecture. When the greenhouse, better known as the sunspace, became cooler than the house, they just closed the door that separated it from the living area. "My eleven-year-old daughter learned to do that without any instructions," quipped Balcomb. "If it's warmer you leave the door opened. You didn't need a user's manual for this house."[33]

That place became known as the Balcomb house, but not because Doug had built it. In publicizing its virtues, he brought it to the public's attention. *Life* magazine featured it, and the media in general turned it into an icon of the emerging new solar architecture. Thirty years later, *Houses*, the 2006 special

Figure 25.13. View of the south side of the Balcomb house in winter.

Figure 25.14. Sunlight entering the greenhouse of the Balcomb house.

issue of the journal *Fine Homebuilding*, featured the Balcomb house as "one of the 25 most important houses of America."[34]

The data on the performance of passive housing that Balcomb and his group collected, analyzed, and published legitimized the new approach in the eyes of engineers, scientists, and government officials. Computer programs they generated allowed builders to simulate scenarios for best practices in solar building. Balcomb also used his clout with the federal government to organize in 1976 the first solar conference devoted solely to passive solar design. A retrospective study of solar-energy activity in the seventies described it as "a landmark event."[35] Solar energy's "lunatic fringe" and assorted hippies, architects, engineers, national laboratory scientists, and bureaucrats finally came together to exchange information and learn more about the ascendant solar technology for heating and cooling buildings. Books soon appeared that taught architects, builders, engineers, and anyone else interested in the subject how to build in harmony with the sun and climate. Titles such as *The Solar Home Book* and *The Passive Solar Energy Book* sold hundreds of thousands of copies.

Popular magazines, too, promoted passive solar design. *Sunset* magazine led the way with articles like "Solar — Here Is a Serious Look at Tomorrow" and "Passive Solar: Yesterday Is Tomorrow." These articles advised readers that "the passive approach [is] usually your best [building] bet," and that "as we approach the 1980s passive-solar systems are making a dramatic return to the Western home."[36] A 1980 article in *Life*, "Sun-Conscious Houses: Primer on Passive Solar Heating," demonstrated the extent to which solar design in general, and passive solar design in particular, had entered the mainstream. Just twenty-six years earlier the magazine had given solar technologies the brush-off, labeling them as a solution to energy problems that would not emerge for another 250 to 1,000 years. Now it rhapsodized that "during the energy-short years ahead, the only fuel that is free — the sun — will become the favorite way to heat our homes, and the sun is one furnace that doesn't break down."[37] *Sunset* went even further, suggesting that passive solar design "offers a key approach to a solution of our national energy problems."[38]

As passive solar architecture spread beyond the borders of New Mexico, new ideas emerged to adapt it to different environments. Balcomb suggested, for example, that "for cold, cloudy places super insulation wins the game[,] and you should put most of your energy in that and a little bit of solar."[39] The Saskatchewan Conservation House, built in the late 1970s, did just that. "It differs from many passive buildings," according to those in charge of its operation, "in that it has levels of air-tightness and insulation which greatly exceed" most others on record. Its main windows faced south but were equal to only 6 percent of the floor area. Still, they accounted for 44 percent of the indoor heating in a location where temperatures could drop as low as –13°F.[40]

Figure 25.15. Mike McCulley, who became dean of architecture at the University of Illinois, stands at the south side of his properly oriented, well-insulated, low-energy houses fit for the cold Illinois weather.

Testimonials by those using newly built passive solar structures demonstrated their effectiveness and ubiquity. Frank Hendrick, a priest at Saint Mary's Parish in Alexandria, Virginia, appreciated that the passive solar design for the gym at the parish's high school consumed two-thirds less energy than the one it replaced. He also liked the "pleasant, comfortable atmosphere" it provided. Judy Borie, executive director of the Girl Scouts of Greater Philadelphia, praised the organization's Shelley Ridge Center, with its passive solar design, as "wonderful." It offered, she said, "an excellent opportunity to teach scouts about architecture and energy conservation."[41]

Changing the focus of a troubled housing program developed by the U.S. Department of Housing and Urban Development — from the more complex solar-heating systems to the simpler passive option — calmed those in charge. Previously, according to Martin McPhillips, the agency had had to contend with all sorts of unexpected problems, such as "corroding pipes and outrageous costs that refused to come down." Relieved HUD officials experienced just the opposite with the houses they monitored that incorporated passive solar-energy systems, happily reporting: "Performance has been excellent and occupants have been very satisfied."[42] The number of passive solar homes in the United States exploded after 1976, from a few hundred to hundreds of thousands. "They sprouted everywhere," Balcomb recounts.[43]

Pool solar heaters and passive solar houses have a lot in common. Both have heat storage built in, so the issue of solar intermittency becomes unimportant. The temperature demand in each case also matches well with the heat provided by the sun. As Freeman Ford observed, "The lower the temperature [demand] of the application, the more cost-effective the conversion from solar to useful heat becomes."[44] In fact, the heat that such solar-energy systems harvest is far more compatible with house and pool heating than the energy that fossil fuels and nuclear power plants supply. Very little energy is wasted, since the collection occurs on-site, doing away with the huge infrastructure required for supplying fossil-fuel and nuclear energy. And when solar heating takes the place of electrical heating, it does away with the need to initially raise temperatures hundreds of degrees to run turbines, and the need to transport the resulting electricity hundreds, if not thousands, of miles in order to deliver the power to homes — which require a temperature increase of only 30 or 40 degrees, if not less, for household comfort.

Solar pool heating and solar architecture also demonstrate the power of aggregation. Each solar pool-heating system is small, but the combined heat produced by all such systems as of 2013 is equivalent to that produced by approximately five nuclear power plants.[45] Writing in the same vein, physicist Raymond Bliss demonstrated in the *Bulletin of the Atomic Scientists* in 1976 the energy savings that could be realized by building a massive number of passive solar homes rather than heating and cooling the same number of homes with fossil fuels or electricity derived from oil. Houses heated by fossil fuels or electricity require three times the number of Btu's (British thermal units, a measurement of heat) required by well-insulated passive solar houses, as a large amount of energy also goes into extracting the fuel, refining it, transporting it, and converting it into a form useful for warming the house. When contrasted with a house built with a conventional heating and cooling system, each solar house would save 100 million Btu's over its lifetime. Bliss projected that if passive solar architecture were to dominate construction from 1976 (when he wrote the article) through 1988, America would save more than three times the amount of oil drilled on Alaska's North Slope. The title of Bliss's article nicely articulates the whole energy problem to this day — "Why Not Just Build It Right?"![46]

Figure 26.1. Marshall Hunt (right), one of the principal figures in helping Davis, California, become the first solar city in America, discusses work on a passive-solar house that uses water columns (in background) to store solar heat.

America's First Solar City

In March 1984, French president François Mitterrand arrived by helicopter when he visited Village Homes, a solar subdivision in Davis, California. Six years earlier Rosalynn Carter, America's First Lady, had ridden her bike through this cluster of solar homes, the largest planned aggregation of passive sun–oriented living spaces since the construction of Olynthus. But these and other visiting dignitaries no doubt did not realize the long solar history of Davis, of which Village Homes was only the latest manifestation.

Early Interest at Davis in Solar Energy

Farmers around Davis had cultivated rice well into the nineteenth century. They kept the water in the fields shallow enough for the sun to effectively heat the paddies and maintain temperature stability in places where cold water entered from irrigation ditches. This upheld production levels in the affected areas.[1] In addition, according to research done by the University of California in the 1920s, "many people have found solar water heaters practical and satisfactory for supplying hot water in the warmer parts of the interior valleys of California," including Davis and its surroundings.[2] To answer the "numerous requests" for information that farmers sent to the University of California's Agricultural Experiment Station in Davis "regarding the construction and performance of solar heaters," the university decided to begin a detailed study of the technology at its Davis facility in its "desire to assist the California farmer in obtaining an economical means of warming water for his household and dairy operations." The results were published in a 1929 pamphlet authored by Arthur Farrall, assistant professor of agricultural engineering at Davis, titled *The Solar Heater*. It provided interested farmers with instructions, diagrams, and a list of materials they could use to build their solar water heaters.

Farrall also measured the temperature of water flowing into a sink from a Day and Night solar water heater installed near Davis, recording temperatures ranging from 118 to 130°F in the late afternoon on three different October days. The high performance intrigued him enough to wonder how a better-built sun absorber might perform. At the sprawling Agricultural Experiment Station that would one day become the Davis campus, Farrall constructed, in the late 1920s, a hot box similar to those built by his predecessors. To his amazement, when he added extra insulation and a second pane of glass the temperature inside the device rose to almost 300°F.[3]

Professor Frederick Brooks continued the study of solar water heaters in California after Farrall's departure from the University of California. Deciding to take a comparative approach, Brooks tested various types of solar water heaters over the next five years, including the three commercial types — the

Figure 26.2. Davis visitors viewed these three types of solar water heaters outside of Frederick Brooks's lab.

Climax, the Improved Climax, and the Day and Night. At the 1934 Picnic Day, the annual open house at the Davis campus attended by students, their parents, and many members of the Central Valley farming community, Brooks let visitors see the commercial models at work.

His follow-up study, *Solar Energy and Its Use in Heating Water in California*, came out two years later. Like Farrall's booklet, it gave farmers sufficient information to build their own solar water heaters. It also gave them a choice of design by providing test results for the various models. The report verified Day and Night's claim that using remote storage allowed for sufficient hot water after sunset and the next morning. It showed that the Climax could supply water above 130°F on a very hot day, and the water would remain at that temperature until evening, when the outside air began to cool. He also had a section on how these heaters could be connected to auxiliary equipment to ensure a continual supply of hot water regardless of the weather. "With such a combination system," Brooks wrote, consumers had the best of both worlds — "never to be bothered by lukewarm water, yet... [they could] save heating expense when the sun shines."[4]

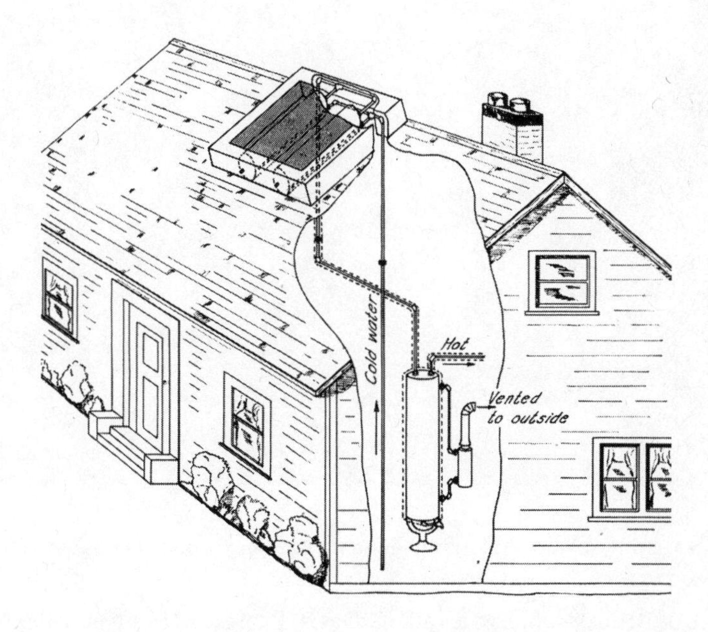

Figure 26.3. A combination system illustrated in Brooks's solar-water-heater booklet, in which two Climax-like solar water heaters sit on the roof and are connected to a conventional solar water heater that will step up the heat should the day be overcast or hot water is needed early in the day.

Unlike Farrall, Brooks broadened his solar investigations by also looking at other examples of heating with sunlight. He discussed, for example, the practice of Russian farmers who spread coal dust over the snow to absorb the heat of the sun and advance spring melting, expanding their growing season.[5] Another discovery by Brooks was that some California farmers spread kraft paper — strong wrapping paper — over recently planted seeds to collect solar heat while also minimizing soil evaporation.[6] He also told of farmers in the Coachella Valley, near Palm Springs, California, who had developed a system reminiscent of the fruit wall. They planted tomatoes, squash, and peppers in east-west rows on south-facing slopes, nestling the seedlings against kraft paper they had attached to an angled fence.[7] One side of the paper radiated solar heat onto the growing plants in one row while the other side focused sunlight onto the neighboring row. The paper also slowed down heat loss from the planted area to the night sky.[8]

Figure 26.4. Using solar techniques to enhance seedling growth in Coachella Valley.

One of Brooks's colleagues at UC Davis, Professor Tod Neubauer, worked on heating barns and other animal housing by orienting these buildings toward the midday sun. Chicken coops, Neubauer observed, "have generally faced south [with] large wall openings."[9] Solar energy pouring into a hen house, aided by good insulation and the heat produced by the birds, provided livable winter temperatures inside poultry houses in areas that regularly received frost but no snow or other severe winter weather, Neubauer observed.[10]

Figures 26.5 and 26.6. A manual on passive farm structures, published by the Libbey-Owens-Ford Glass Company.

Cooling the hen houses in summertime came up in Neubauer's studies as well. The indoor temperature of a hen house should never rise above the average outdoor temperature in hot regions during the summer, which in the Davis area reached more than 100°F. Neubauer advocated shading the east and west walls with trees and adding an overhang extending about 3 feet from the coop opening, which was located on the south side of the building.[11] Hog pens, Neubauer noted, should be "similar in design as poultry houses."[12]

Solar Heat Control

When it came to living comfort, solar specialists in Davis, where hot, dry weather predominates in the summer, focused their interest on keeping the sun out of buildings. For instance, in his 1936 solar water-heating pamphlet, which touched on other aspects of solar design as well, Brooks explained that "whitewashing...is used in Egypt and Arabia for keeping buildings as cool as possible," because it reflects sunlight and, in doing so, "further reduces the solar absorption" capability of the finished plaster.[13] In another article, he complained, "Too often architects specify large glass areas facing west or east, then hand the design to air conditioning engineers to figure the heating and cooling loads."[14] Such actions demonstrated, he told students at the University of California, Los Angeles, in 1956, that the current generation of architects had "a lack of feel" for solar energy. To prove his point, he noted that a poorly sited structure "was given an Award of Merit by the American Institute of Architects for excellent orientation." The building in question "has a wide, two-story lobby with the entire east wall [constructed] of glass," continued Brooks. "Even though the temperature outside remained in the 70s, the offices facing east are virtually unlivable — 92 degrees F at 8 AM."[15] People working in the building resorted to keeping the solar heat out by covering the offending windows with wrapping paper. "Shade for intentional control" of heat appeared to Brooks to be almost completely neglected by builders in North America and Europe.[16]

Another of Brooks's colleagues at UC Davis, Dr. Robert Deering, observed that materials such as wood, stone, and asphalt in the human-built environment take in great quantities of solar heat and, as the temperature around them cools, reradiate that heat into the atmosphere, keeping built areas hotter at night than natural environments. Deering suggested that, to create built environments that are more comfortable in hot weather, "these materials need some sort of protection against the sun." As for pavement, he recommended "shad[ing] as much of the paving as possible with trees" planted along the sides of roads. He also made a more radical suggestion in his unpublished report "A Study for the Moderation of Extremes of Heat, Cold and Undesirable Winds": "Keep paving to a minimum in hot climates."[17]

West-facing walls, more than any other orientation, overheat a building's interior and exterior. Deering recommended adding a shaded terrace along a building's western wall to cool down the building. This could be accomplished by extending the western portion of the roof and attaching a trellis to the far edge of the overhang. Deciduous vines climbing on the trellis would entirely block the afternoon sun, creating a delightfully shaded terrace. Plants, the UC Davis professor pointed out, not only passively obstruct but also actively remove solar heat from the vicinity through photosynthesis and transpiration.[18]

Solar Houses Revisited

The first academic at Davis to closely study the year-round natural heating and cooling of buildings was Tod Neubauer. In his upper-division agricultural engineering course on farm buildings, Neubauer introduced his students to an article in *Architectural Forum* featuring the Duncan house in his lecture, "What Will the Saleable Post-War House Be Like?" The prototypical postwar house, he told them, might be the residence designed by George Fred Keck.[19] He also pointed to the current folly of disregarding the "best sun exposure" for the ubiquitous rural sun porch by mechanically placing it "at the front of the house." And he urged those attending his class to read the Libbey-Owens-Ford booklet *Solar Houses with Thermopane*, which was based on George Fred Keck's work, in preparation for a class discussion on "heating the farm house."[20]

The third MIT house, with its large, south-facing windows that helped to heat it, also caught Neubauer's attention. In the notes he prepared for a lecture to a similar class five years later, he pointed out that "every square mile of California ground receives, during each summer, about as much energy as can be produced by all the power plants of one of the largest public utility systems in the state." With this in mind, and using the MIT house as a teaching tool, he described the essential features of a solar house: orientation, large windows, and sun control. To collect as much solar heat as possible, Neubauer told them, the walls and floor should consist of materials that "retain the trapped sunlight and slowly let the heat out as the house cools." Other features of the MIT experiment he discussed were the house's well-insulated walls and the heavy curtains that the family drew across the windows at night to reduce the loss of solar heat after dark. He also presented two facts that make a solar house a compelling concept: "The highest average heat value of sunlight occurs about [the time of day] when it is most needed — at mid-day on the winter solstice," and "a south wall receives almost *five* times as much heat from the sun in winter as it does in summer" (Neubauer's emphasis).[21]

Anticipating in 1951 that "solar houses in ten years will be available at no greater cost than ordinary homes" in most parts of the country, Neubauer began to devote the majority of his time to integrating solar heating with shade control.[22] To the press and his colleagues, he demonstrated, in the words of one journalist, "a great new gadget" that would allow architects and builders to easily model "the best way to face your house weather-wise."[23] In Neubauer's modest account, the portable device could be mounted on a drafting board and would simulate the "positions and movement of the sun about a [small model] house[,]...varying with the time in hours, months, seasons and latitudes."

In doing so, it "illustrate[d] the approximate angles and extent of shades and shadows."[24]

By 1959, Neubauer had developed a blueprint for a "sun-exposed house," which he presented to the July 1959 meeting of the Western Region Farmhouse Plan Exchange Committee. The aim, he told them, was not to attempt "complete heating or air conditioning by solar methods," since "large heat collectors or heat storage facilities...are obviously too expensive to be practical or economical." He recommended creating a house oriented so that it would "receive the delightful comfort of direct solar radiation on cold days." He also suggested designing homes with "some heat (or cold) storage by use of heavy masonry floors, walls, or storage partitions" and recommended "the use of shades, blinds, shutters, or low cost drapes[,]...white roofs[,]...[and orienting homes away from the] cold north sky."[25] In a paper published a decade later, he further simplified his design for a solar house. After much experimentation, Neubauer had decided that the design of the old-fashioned hen house or hog house with a shed roof facing south may work for humans as well.[26]

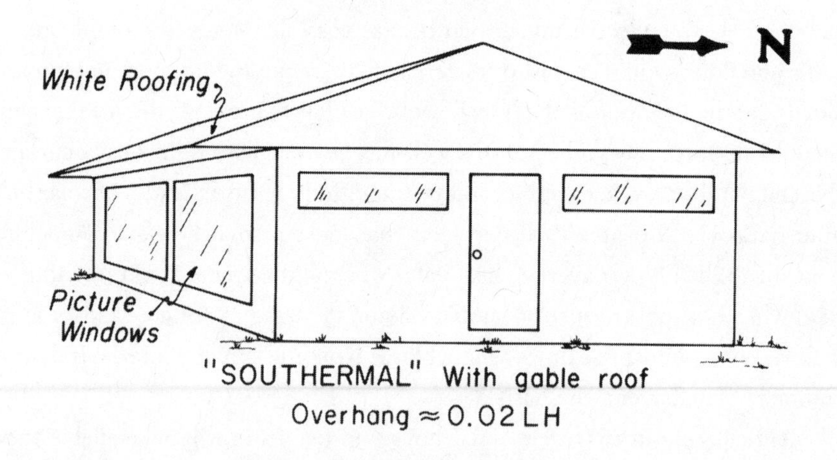

Figure 26.7. Neubauer's design of a simplified solar house.

Despite Neubauer's efforts, no house or building based on his or his colleagues' ideas was built during their academic careers at UC Davis. They left behind only test structures proving the importance of building with the microclimate in mind. Neubauer plainly lamented this fact, writing in *Farm Building Design*, his textbook coauthored with Harry B. Walker: "The possibility of taking advantage of environmental conditions, orientation, and solar heating is neglected."[27]

From Academic Learning to Social Action

Brooks, Deering, and Neubauer did not labor in vain, however. The social movements of the 1960s and 1970s stirred up great interest in the solar ideas these men had devised or inherited. Davis itself went through great changes during these years, growing exponentially as it evolved from an agricultural school to a university. By 1970 it had more than six thousand students of voting age, after passage of a constitutional amendment lowering the voting age to eighteen. The number of university students equaled the number of other residents in the city eligible to participate in elections. Steeped in the social movements of the time, including the civil rights marches, antiwar demonstrations, and Earth Day, UC Davis students were part of a generation eager to translate academic learning into social action.

A group of ecology graduate students — members of the first graduate school in ecology formed in America — urged Bob Black, the most popular student body president ever to serve at the campus and a fervent antiwar activist, to form a coalition of candidates for city council opposed to building an overpass to the city center. The coalition won, wresting political control from the Davis establishment.

Black's roommate, Jon Hammond, who had recently graduated from the ecology department, suggested to the newly elected council member that the

DEAR VOTER,
DAVIS IS AT A TURNING POINT.
THE CHALLENGE OF THE NEXT FEW YEARS IS WHETHER WE CAN REALIZE OUR OWN UNIQUE POTENTIAL AS A COMMUNITY.
WE ARE ON THE BRINK OF BECOMING ANOTHER OF THE THOUSANDS OF AMORPHOUS SUBURBS THAT CHARACTERIZE THIS STATE.
WE MUST DECIDE WHETHER OUR DISTINCTION WILL BE OUR GROWTH RATE (163% IN THE 1960'S AND RISING) OR THE MANNER IN WHICH WE HAVE USED OUR RESOURCES TO FASHION A TRULY FINE COMMUNITY.
WE MUST DECIDE WHETHER THE CITY COUNCIL WILL CONTINUE TO RESPOND TO THE INSISTENT PRESSURES

OF A FEW OR TURN FOR GUIDANCE TO THE WILL AND WISDOM OF ALL THE PEOPLE.

THESE ARE MY PRIMARY CONCERNS; THEY ALL RELATE TO THE QUALITY OF OUR LIVES:

1. That we find ways to build together. This means greater participation by students, young people, East Davis residents, Seniors, women- all those not now represented.

2. That the community should be in control of its growth in terms of quantity, location and type. This means aggressive planning.

3. That the City Council should really serve the people and provide leadership. The Council should be interested in needs like:
a. Day-care
b. Consumer protection and education
c. Lobbying for property tax reform
d. Housing for the elderly and low-income people
e. A recreational facility for Senior citizens

4. That the Council should be able to think creatively and encourage innovation. I'm talking about ideas like:
a. Hiring young people to develop city parks
b. A farmers' and craftsmen's market
c. Using city parks for Sunday afternoon concerts
d. A small-grant fund for cultural and beautification pro jects
e. A research and information coordinator

Figure 26.8. Bob Black's campaign literature.

city should require new building developments to adopt environmental principles in construction. "I set up a student group focused on ecological planning to make the city more environmentally sound," Jon recalled. "I especially wanted to do solar."[28]

Hammond heard that Neubauer had done a great deal of solar experimentation during his tenure at Davis, so he invited him to a forum to speak about what he had done. Neubauer began his presentation by discussing the local weather: hot and dry from July through August, with temperatures dropping off considerably at night; winters were cool and wet but with sunlight between storms; spring and fall featured moderate weather, requiring little heating or cooling. The challenge, then, came in summer and winter. Using his "great new gadget," what Neubauer called the Solaranger, he demonstrated the changes in the sun's apparent movement throughout the year and its effect on housing oriented to the various cardinal points of the compass. The exhibition pointed to the necessity of giving a home a long east-west axis, as illustrated in the floor plan he presented in his next slide, which, he told the audience, "limits the east and west exposures[,] and [the amount of] roof exposed to the hot summer sun, and increases the size of the south walls and windows, for heating by the low winter sun."[29]

Neubauer's reasoned analysis put the more conservative people in the audience at ease regarding the plans of the new city council. As Bob Black recalled, "Neubauer gave credibility to what the younger wild-and-crazies" advocated. "It was so important to have someone so respected in academia and the community to present these things."[30] With the city's blessing a team consisting of Jon Hammond, Marshall Hunt (a former student of Brooks, whose microclimatology teachings had brought Hunt to UC Davis), Neubauer, and his colleague Richard Cramer went into the field to gather hard numbers to support Neubauer's claims.

First, they measured the temperatures and energy consumption in various household settings during the different seasons. They found that gas consumption peaked in winter, and that electrical use was highest during the summer. They decided to concentrate their temperature studies on apartments for the simple reason that many identical apartments, facing the four cardinal directions, were vacant and available for study.[31] The group found that, during winter, the apartments whose windows faced north, west, or east "warmed up only a little, [while] the south[-facing] apartments averaged 72 degrees F [and were]

Figure 26.9. Jon Hammond working on a model of a passive house he would later build.

often warmer." The opposite occurred in summer. A second-story apartment facing south averaged about 85° over the summer and fall, while its east- and west-facing counterparts stayed at almost 100°.[32] The numbers spoke volumes and led Neubauer to conclude that "the effect of window size and orientation may indeed be the major factor in house comfort and energy consumption."[33] He came up with one simple equation for success, a simple formula that Neubauer called the two-percent rule for finding the length of the required overhang on the south face: multiply by .02 the height of the south-facing window(s) by the latitude.[34] Hammond praised Neubauer's work for being basic, easy to explain, and, as a result, more likely to be accepted.

As the team began writing up their findings and recommendations, Hammond recalled, "the oil embargo happened. It was perfect timing. There was a huge interest in the results of our study."[35] The data collected and interpreted were translated into America's first solar-energy ordinance, a reflection of Neubauer's design ideas. Journalist James Ridgeway succinctly explained the new ordinance: "The basic idea... is that new housing built in Davis shall not experience an excessive heat gain in summer nor excessive heat loss in winter."[36]

It allowed builders two choices. The first was a prescriptive path that stipulated a south orientation; the majority of windows would be on the south side, and a minimum of windows would be on the east and west sides and shaded either by eaves, drapery, or vegetation. The second path permitted more leeway as long as the building conformed to the designated heating and cooling loads set by the city.

Figure 26.10. Neubauer's "semi-solar" house realized at Davis, California, under its solar ordinance.

Some builders remained skeptical, such as veteran contractor Ron Broward. In fact, he followed the first pathway when constructing the first several houses he built after the ordinance passed, just so he could prove that what "the consultants... proposed wasn't going to work." Broward put thermostats in these houses, and to his surprise every reading showed that, although the temperature outside always stayed between the high 90s and low 100s, the temperature inside remained under 75°F. Those thermostat readings turned a doubting builder — who had sixteen years in the business and about five hundred houses under his belt — into one of the ordinance's staunchest supporters. "I would like to stress the importance of early adoption of the standards [such as mandatory lot orientation]," the new convert stated to the city council. "Planners should be encouraged to design as many north and south facing lots as possible."[37] In fact, many developers did just that voluntarily "when it became apparent," according to the city's building consultants, "that it would be much easier to comply with the new building code on properly oriented lots."[38]

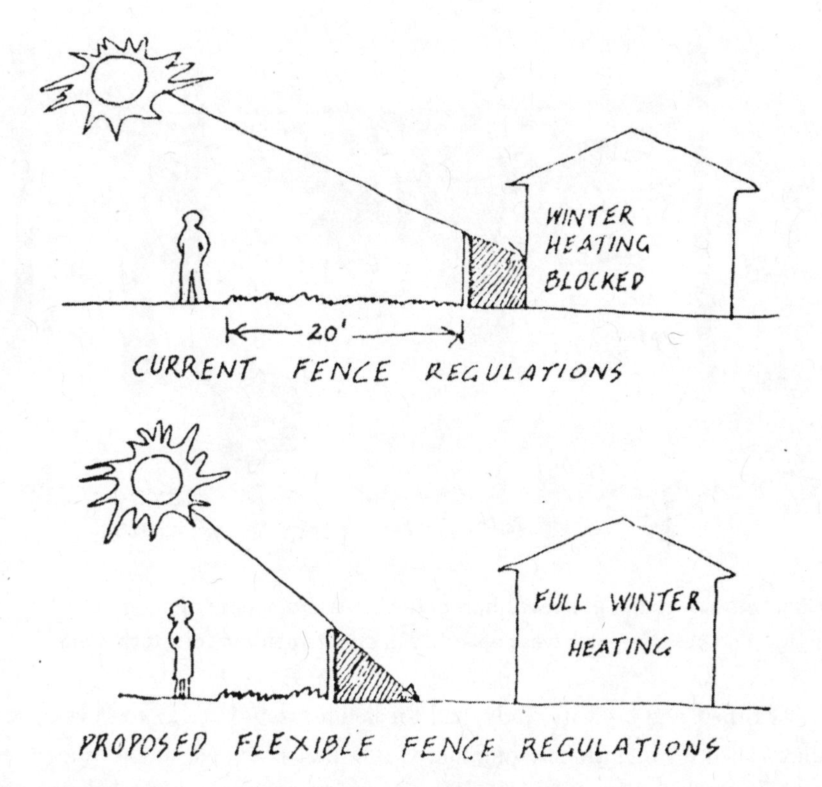

Figure 26.11. Davis's solar ordinance requiring fencing setbacks on the south side of a lot guaranteed full winter solar heating for a properly oriented house.

Village Homes

One builder, Mike Corbett, stood out from the rest for his enthusiasm for the solar aspects of the new ordinance, as shown by his subdivision Village Homes. The Davis consultants praised Corbett for being "particularly interested in the application of techniques proposed in this study to his development."[39] Corbett started out as an architectural design student at California Polytechnic State University in San Luis Obispo, but the school asked him to leave because of his strong opposition to the Vietnam War. He developed an interest in solar energy after reading *Perils of the Peaceful Atom: The Myth of Safe Nuclear Power Plants*, by Richard Curtis and Elizabeth Hogan, and his ecological interest blossomed after finishing Rachel Carson's classic *Silent Spring*.

Figure 26.12. With this image, authors of the Davis Ordinance scoff at those suggesting more research was needed before implementing solar techniques.

Corbett and his wife, Judy, had formed a group in 1972 to develop what they called an intentional community, one that included neighborhoods that were unfenced, bike paths, edible landscapes, and street gutters that drained onto open fields. Village Homes was a result of these efforts. Most of the development's streets ran east-west, but not in the usual checkerboard fashion that created a uniform look. Rather, they gently meandered in that direction, giving a more rural feel to the community, although the houses faced directly south, nonetheless. The streets had the look of pathways, and native deciduous trees completely shaded them, reducing surrounding summer temperatures by about 10 degrees and diminishing the need for home air-conditioning.[40] Nor were the houses slavishly placed close to the pathlike streets. Instead, Corbett set them deep on the lot to create a courtlike effect, so each house had unobstructed access to sunlight in winter. The absence of fences reduced the number of obstructions that would have prevented houses from receiving their solar fuel. Solar rights written into the subdivision's covenants guaranteed solar fuel for all. As resident David Bainbridge wrote, "The planning of the neighborhood was designed to facilitate solar heating and natural cooling, and the very simple steps necessary for solar utilization were among the first made and most vigorously pursued.

Figure 26.13. The pathlike quality of the streets became apparent early in the construction of Village Homes.

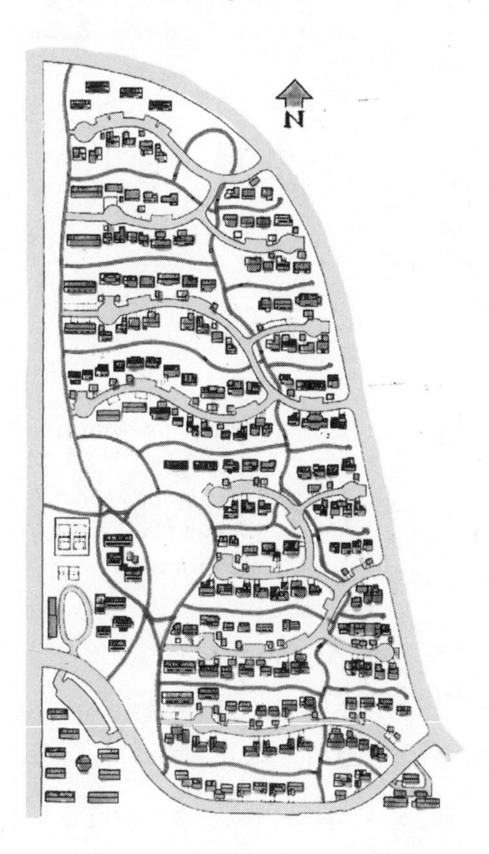

Figure 26.14. Overview of Village Homes.

These included: street orientation, lot orientation, setbacks, solar access, and landscaping."[41]

Almost every conceivable type of solar house was built in the development, "from small solar cottages with little more than good south-facing glass and simple insulation...to...what is undoubtedly the most energy efficient house in Village Homes, the city of Davis, or [the] entire county," according to Rob Thayer, a professor of landscape design at UC Davis and resident of the community.[42] Village Homes and the City of Davis demonstrated to the world, in the opinion of David Bainbridge, coauthor of the book *Village Homes' Solar House Designs*, "that heating and cooling using solar energy and climate control is not complex but a simple, easily applied, and well-tested approach."[43]

That Village Homes has not been replicated may be a result of timing. As that neighborhood really started to take off, Ronald Reagan took office, and his administration's dim view of solar energy still haunts us today. An example of that administration's antisolar bias is its reception of the document *Review*

- Some data from controlled field tests are prerequisite to planning a demonstration. Such data can indicate when the technology is ready to demonstrate, suggest what system selection criteria to use, and support development of consensus standards.
- Goals and objectives for a demonstration of a dispersed renewable technology are often diverse and conflicting. The relative importance of different goals and objectives should be decided early, so that an appropriate implementation strategy can be planned.
- Planning should allow for evaluation throughout the course of the program. Program managers must build in mechanisms which allow feedback and redirection, such as industry review boards.
- To leverage program resources, existing institutions should be used as much as possible in implementing training and information programs. The mass media is an efficient means of reaching the general public, for example.
- The federal government has an important role to play in promoting renewable technologies. A comprehensive strategy must be developed, including traditional areas like funding R&D and providing incentives as well as raising public awareness and conducting special information and training programs. Solar industry sources strongly emphasize the need for education and information efforts.

Figure 26.15. Conclusions of the report by Arthur D. Little that enraged the incoming Reagan administration.

of the Demonstration Program of Solar Heating and Cooling Technologies, which arrived at the White House during Reagan's inauguration. The Department of Energy had paid the highly reputable consulting firm Arthur D. Little a quarter of a million dollars to complete the study. The lead author did not consider the study controversial. It outlined high expectations for what solar energy could accomplish if properly funded. "The following day," one of the members of the staff that produced the report recalled, "word came from the Reagan team: 'Do not release this report...copies are to be destroyed...no secret printings ...no discussions.'" And this was accompanied by a threat: "If any word gets out, Arthur D. Little will not be compensated." The staff member added, "I had never witnessed anything so brutal. There were no pretensions of free speech. It was swift and ruthless. One of the chilliest moments of my life."[44] Under the Reagan administration, "solar budgets got decimated," recalled Edgar DeMeo, director of photovoltaic research at the Electric Power Research Institute in the 1980s. "Reagan dealt the renewable movement a crippling blow,"[45] he added. Doug Balcomb summed up the destruction brought about by Reagan: "The president said in the 1980s, 'The energy crisis, it's been solved; there weren't any problems left.' So people weren't concerned about it anymore, [since] people tend to follow that kind of a lead. The few of us left working in the solar field in the 1980s were pretty lonely. The momentum had evaporated."[46]

Figure 27.1. Archbishop Makarios III, the first president of Cyprus, views solar water heaters imported from Israel at the island's first national trade show.

[1973–]

27

Solar Water Heating Worldwide, Part 2

Almost everyone in academia and government had bet on flat-plate collectors for solar house heating. But in doing so they lost out to the solar underground, who had put their money on passive solar design. The manufacturers of flat-plate collectors, expecting a multibillion-dollar housing industry to purchase their products, had to settle for the much smaller water-heating market. And the economics of that market didn't look very good. The success of the California and Florida solar markets had rested on the fact that there existed at the time no good alternatives. During the 1970s that was not the case. Most homeowners could choose inexpensive natural gas. Additionally, the price for solar units far exceeded the cost of conventional water heaters by a factor of ten to twenty.[1]

Only subsidies in the form of tax credits, and the big jumps in oil prices in 1973 and 1979, kept the solar water-heater industry growing.[2] It grew from only twenty thousand solar water heaters installed during 1978 to nearly a million total by the end of 1983. In 1985 the Solar Energy Industries Association lamented, "Now that the alarm over the energy crisis has faded, the rush of enthusiasm has ended."[3] Another solar enthusiast similarly expressed his pessimism: "The man on the street will tell you that solar is too expensive and — given today's plentiful oil supply — isn't needed.[4] When government support for solar energy in the way of tax credits ended on January 1, 1986, and simultaneously oil prices dropped precipitously, the industry went into a tailspin, with sales falling 80 to 90 percent.[5]

Japan and Australia

The solar water-heater markets in Japan and Australia followed a similar trend. Japanese sales rose dramatically following the two oil price shocks in the 1970s, peaking in 1979–1980 and then collapsing in sync with the fall of oil prices throughout the 1980s. Discoveries of gas in western Australia, along with the burning of the country's vast coal deposits to generate cheap electricity, had a similar negative effect on the growth of its solar water-heater market during the 1980s.

Israel

In other countries, the solar water industry saw better times. Israeli success in the Yom Kippur War brought on the infamous oil boycott of 1973. Israelis responded with mass purchases of solar water heaters. The government also intervened, requesting the Standards Institution of Israel to test and rate solar water heaters produced in Israel and then publicized the results. The Ministry of Energy and Infrastructure itself set up information centers in major Israeli cities. Itzhak Shomron, representing the ministry, noted that his agency "places a great deal of its resources and efforts into information dissemination [in order] to provide accurate answers to questions."[6] Such proactive efforts, combined with solar water heaters costing three to six times less than those produced in the United States, had a positive effect. By 1983, 60 percent of the population heated their water with the sun. When the price of oil dropped in the mid-1980s, the Israeli government did not want people backsliding, as had happened in other parts of the world. And so it required citizens to continue heating their water with the sun, by mandating the use of solar water heaters in buildings with more than four stories — in which the majority of Israelis live.[7] At the time of this writing, more than 90 percent of Israeli households own solar water heaters, making Israel the second-largest per capita user of such heaters.[8]

Cyprus

Cyprus, like Israel, has only one natural energy resource — the sun. The Cypriot solar water-heater industry began with a visit in 1956 by Kypros Psimolophitis, one of the island's leading industrialists, to the Miromit solar water-heater factory in Israel. The technology so impressed him that he had four Miromit solar water heaters shipped to Cyprus, where his company, Metalco, began manufacturing solar panels under license from Miromit. After the country won its independence, Psimolophitis had a working model of a solar water heater installed at the first national trade show, in 1961. At the trade show, the charismatic first president of the island, Archbishop Makarios III, tested the water coming out of the system. It almost burnt his hand. Surprised and impressed, he became a powerful advocate of solar water heaters, especially since the island was otherwise totally dependent on imported oil for producing electricity to heat water.

Under Makarios's leadership the national government committed to installing solar water heaters on all state buildings to serve as an example for its citizens. Unexpectedly, the government's role in promoting the heaters led to installation of a huge number of solar water heaters on the island when, in 1974, Turkish forces invaded and occupied the northern half of the island. Tens of thousands of Greek Cypriots fled to the south, uprooted and homeless. The onus of housing them fell on the government, and since the law obligated it to put solar water heaters on all buildings it constructed, such heaters became ubiquitous. When the remaining population saw how well the heaters worked, they soon installed them on their own homes, too.[9] Cyprus leads the world in per capita installations of solar water heaters, followed by Israel, according to the most recent worldwide solar water-heater census.[10]

Greece

The Turkish invasion of Cyprus caused many Greek Cypriots to leave the island for Greece. One refugee, Panos Lamaris, a mechanical engineer, founded the first solar water-heater company on the mainland. As in Israel and Cyprus, the Greeks had been heating their water with electricity generated by oil-fed generators. But with oil at record prices, Lamaris's company, SOLE, flourished. Others soon followed. The Greek government pitched in with soft loans, tax credits, and national advertising campaigns.

The solar water-heater industry in Greece grew exponentially from 1976 until 1987, when the government began solar-phobic policies that included lowering electrical rates, levying exorbitant taxes on solar water heaters, and withdrawing all incentives for their purchase. As if to make up for its past sins, however, the Greek government has recently mandated, as did the Israeli government more than thirty years earlier, the installation of solar water heaters on all new buildings. Many believe that this new law will generate a renaissance of the Greek solar water-heating industry, as happened in the Jewish state.

The experience gained by building solar water heaters has led to a very successful export business as well. As an example, a field of Greek-produced solar water heaters sits on the rooftop of the world's highest building, the Burj Khalifa tower in Dubai.[11]

Barbados

Barbados is the third-largest per capita consumer of solar hot water.[12] In the 1930s, Florida solar water-heater companies exported their products to the Caribbean. Some of these were sold in Barbados, but the Barbados solar water-heater story didn't really get started until 1964. That year a churchman, the Reverend Andrew Hatch, then rector of Saint Mary's Church in Bridgetown, the capital of Barbados, envisioned solar energy as the appropriate way of heating water in the Caribbean and as a way of getting the unemployed youth of the inner city of Bridgetown working. Hatch sought technological advice from McGill University's prestigious Brace Centre for Water Resources Management, in Montreal, a multidisciplinary and advanced research and training program in water resources management. Supplies for building solar water-heater systems, such as piping, came from local businessman James Husbands.[13]

The first heater that Hatch's fledgling company produced did not, however, meet the expectations of homeowners. Only a dozen went up on rooftops. Husbands, seeing the potential for solar water heating on an island totally dependent on imported fuel, eventually formed his own company, Solar Dynamics. In the early 1970s the company refined the design to ameliorate the older product's shortcomings. It began using glass with greater heat tolerance, and it replaced the former wooden collector housing with aluminum cabinetry and the galvanized piping with a copper flat plate and copper tubing. Solar Dynamics' new entry became the first solar water heater in the world to guarantee temperature performance adequate for all domestic-hot-water needs.

In 1974 the sterling example of Solar Dynamics' solar water heater came to the attention of Barbados's prime minister, Tom Adams. Looking for ways to reduce the stranglehold of fuel imports on his country, he and his plumber built a cheap copy of the Solar Dynamics solar water heater. As Husbands tells it, "Adams was not very happy with the resulting performance, and invited our company to install a unit at his St. George residence." Its stellar performance favorably impressed the prime minister. In fact, he saw its potential for playing an important role in his plans to make his country less dependent on imported oil and gas.

Adams's government ended all duties on imported equipment and materials used to build solar water heaters. It also imposed a 50 percent tax on all electricity and gas consumed by conventional water heaters. A consumer incentive for the purchase of solar water heaters soon followed, allowing homeowners to fully deduct the purchase price of a solar water heater from their income tax. The commercial sector chipped in by providing easy credit for the solar option.[14] The strong support by the government and private sector for solar energy prompted approximately 50 percent of the island's people to switch to solar water heaters.

Both the government and its citizens have greatly benefited from these proactive policies. In return for the approximately 22 million Barbadian dollars that the government allocated for tax incentives from 1976 through 2002, the island

The Barbados Advocate

Established October 1895

2012 Estimates debate begins today

Page 5

Monday March 12, 2012

$1

SOLAR SAVINGS

Almost $58 million saved annually in foreign exchange

By Janelle Riley-Thornhill

THE solar water heater industry is saving the country almost $58 million annually in foreign exchange and can save additional millions, if the stakeholders can make a greater dent in the market.

This sector is often considered to be one of the success stories in Barbados' manufacturing sector, but industry estimates have shown that it has only seen growth of 1.1 per cent per annum over the more than three decades in operation; and a director of one of the leading market players suggested that this rate of development is unsatisfactory and he wants to see the thermal water heater industry given its just deserts.

Sales Director of Sunpower (1998) Ltd., Henry Jordan, maintained during an exclusive interview with *The Barbados Advocate*, that there is no doubt that the thermal sector plays a significant role in this economy, but with the penetration rate still less than 50 per cent among domestic households, the country is spending much more on fuel to operate the electric heating alternatives than it should be.

The ultimate goal for the industry, he said, is to push solar water heater penetration to 75 per cent over the next eight to nine years.

SAVINGS on Page 9

MAJOR DAMAGE: Fontabelle Main Road was the scene of a major accident yesterday afternoon, when eight utility poles were pulled down by a passing truck, leaving the road impassable and knocking out electricity for many homes and businesses in the area. Luckily, no one was injured in what could have been a much more serious accident, as this street is one of the main routes into and out of the City. Here, personnel from the Barbados Light & Power assess the extensive damage. INSET: This skip truck initiated the incident when it hooked onto a power line while turning onto the main road. (See Page 3)

Great potential for Eagle Hall, Cheapside markets

By Patricia Thangaraj

ONE non-governmental organisation is taking the initiative to increase the income-generating potential of fruit and vegetable sellers, develop more activities for tourists coming to Barbados to be involved in by linking agriculture with tourism, provide more opportunities for larger companies to advertise their products to the general public and provide exposure to artistes, entertainers, school choirs and other persons.

Speaking with *The Barbados Advocate* recently, President of Change for Agriculture Must Come Incorporated, Melvina Jones, said that she has passed through the markets at Eagle Hall and Cheapside and noticed that on any given day, there is a scarcity of persons passing through the area and she felt that something needs to be done to increase traffic so that the persons selling their fruits and vegetables there can get customers. She also noted that a similar drive was necessary for the markets at Oistins, Fairchild Street and Palmetto Square.

One of the ways that she plans on doing this is by having

MARKETS on Page 6

Figure 27.2. Article extolling the virtues of solar water heaters on Barbados.

nation saved almost 400 million Barbadian dollars — the amount that would have otherwise been spent on fossil fuels. In addition, owners of solar water heaters experience an annual 30 percent return on their investment, which translates into free hot water after about three years.[15]

It is interesting to note that, unlike Americans in the 1970s and 1980s, whose solar water heaters usually relied on pumps and complex controls, most Israelis, Cypriots, Greeks, and Barbadians have chosen simple systems that naturally circulate sun-heated water from collector to tank, as in the Day and Night.

Austria

It might appear that solar success stories depend on sunny climates. Austrians, who rank just below Barbadians in per capita use of the sun for heating their water, have proven the fallacy of such a perception.[16] As in the United States, Japan, and Australia, the use of solar water heaters in Austria rose and fell with the price of oil — until the early 1980s, when two rural Austrians and some of their friends formed a do-it-yourself group near Graz. According to the pair, "Our primary aim was to build a collector that was inexpensive and easy to build for every one of us. Having become aware of the finiteness of natural resources…[, we focused on] saving energy, environmental protection, and community building."[17]

Thirty-two people participated in the first workshop in 1983. A year later more than two hundred had joined the group. The solar collectors that the do-it-yourselfers installed on rooftops cut costs significantly, because the group had done away with the metal housing of previous models — which also improved the appearance of installations and, as a result, boosted public acceptance of solar water heaters. Not all the self-constructed systems were properly sized, though. The water heaters that were too large produced more hot water than needed, and people began piping the excess hot water into their household heating systems. In response, the self-build group came up with an economically viable solar-energy system that fulfills a household's hot-water needs and supplements its conventional hot-water house-heating system, a design that became known as the combi. Half of all solar installations in Austria today are combi systems.[18]

From the self-build groups emerged a national grassroots movement called

Figure 27.3. Two Austrian members of a self-build group put together a solar thermal collector.

the Renewable Energy Working Group (Arbeitsgemeinschaft Erneuerbare Energie). The working group set up information centers and workshops to inform the public about, and to teach them to build, solar water heaters. Classes began with an introductory lecture, and then instructors took the students to visit installations. They also taught students how to make calculations to ensure that systems would be adequately sized. Finally, they divided the classes into construction groups, which then worked together to build a solar water-heating system for each member of the group. From these self-build groups emerged the infrastructure necessary for the orderly development of a vibrant solar water-heater industry.[19]

Eventually the subsidies that Austria commonly gives to people purchasing a home included assistance for also purchasing solar equipment for the home. With this kind of encouragement to go solar, it's no surprise that one out of eight Austrian households currently makes use of the sun to heat water not only for washing purposes but also to help heat their homes.

Figure 27.4. Ad selling solar water heaters in Austria.

Ærø Island, Denmark

The largest solar water-heating plant in the world is situated on Denmark's Ærø Island, at roughly the same latitude as Juneau, Alaska, and it heats water for the island's largest town, Marstal. Interest in the solar alternative began here in the 1970s, with the country's "No to Nuclear" movement. The two oil shocks of that decade, when petroleum prices leapt and availability plummeted, added to solar energy's attraction.[20]

Private citizens — a blacksmith, some teachers, a farmer, and a bank manager — established the Æerø Energy Office, which became the focal point for information on renewable-energy resources. Those clever with their hands built homemade solar hot-water collectors. Then came the Danish government's Energy Plan '81. The policy designated Ærø Island and some other remote spots to be recipients of renewable-energy plants, because planned natural-gas pipelines would bypass them.

To kick off the renewable-energy program the Danish government began an educational initiative to gain public support. Renewables needed community backing to succeed, because each town on the island would run its own energy plant. The islanders had learned, for example, that their Swedish neighbors had great success in the late 1970s with large fields of solar hot-water collectors, which provided most of the summer heating necessary — mostly hot water

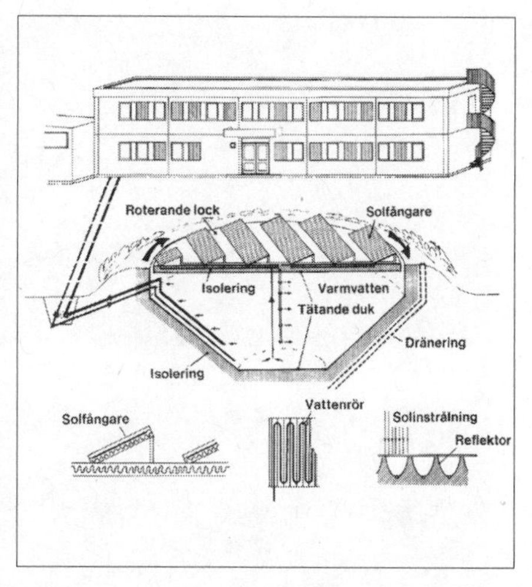

Figures 27.5 and 27.6. A brochure describing the first district solar heating plant in Sweden.

— for several housing complexes. The flats in Sweden got their heat from hot water flowing through pipes connected to a shared boiler, the same plan the Danish would use in the towns on Ærø.

The Swedish experience motivated those living in Marstal. The bank of collectors installed in Marstal covers almost 200,000 square feet. The collectors generate the thermal equivalent of 8.2 billion watt-hours per year for the Marstal District Heating agency, supplying all the hot-water needs of the town's fourteen hundred households. Previously the town had relied solely on highly polluting heavy oil for fueling its boilers, which made the solar choice an easy one.

Other district heating complexes on the island have followed Marstal's example. In 1998 Ærøskobing District Heating installed 52,743 square feet of solar collectors; in 2001 the Rise District Heating built a 43,000-square-foot solar plant; and most recently, the Søby District Heating went solar with a "solar farm" — an array of solar collectors — taking up 24,000 square feet. All told, the island boasts 46 square feet of collector area per person, the highest per capita in the world, far exceeding that in places like Southern California or Arizona, which receive at least twice as much solar radiation as Ærø Island.

The success on Ærø Island so impressed all of Europe that the European Union decided to fund the doubling of the collector area of the Marstal solar farm, add to it a boiler that would be heated by locally grown willow chips,

Figures 27.7 and 27.8. Two views of solar thermal panels for district heating on Ærø Island.

and a reservoir to hold the excess solar heat collected in summer for winter use. It would demonstrate to the world the efficacy of district heating solely with renewable energy.

China

No country has experienced more explosive growth in solar water heating than the People's Republic of China. The first solar water heaters to arrive in China came in the mid-1970s from a Japanese company that distributed Australian solar water heaters to the Asian market.[21] By 1982, the country had about thirty thousand solar water heaters, which primarily provided hot water for barbershops, bathhouses, hotels, restaurants, and laundries.[22] Thirty years later, their number has increased a thousandfold, to more than 30 million, and they can be seen on rooftops throughout China.[23]

At first the Chinese used the tried-and-true natural-circulation flat-plate collector system, but the very low temperatures of Chinese winters caused the water heaters to underperform during winter and spring. "Flat plate collectors can only operate six to eight months" out of the year, a Chinese solar expert complained in the mid-1980s. To provide solar-heated water year-round a new type of collector was built. Each installation has multiple pipes. Every pipe consists of two glass tubes, one inside the other; an evacuated air space separates them and creates a vacuum that insulates the inner tube. Water flowing through the black-coated interior tube heats up and naturally rises into an elevated storage tank. The same expert remarked, "Preliminary experiments indicate it can work in windy, cold weather during the winter and spring months owing to reduced heat losses."[24]

This new solar water-heating system, called an evacuated tube collector, is now sold on the Chinese market. The Chinese, with the help of Australian scientists, took an American invention and made it more efficient and robust. They now manufacture the tubes with a durable, heat-resistant alloy that maximizes performance at high temperatures, and they developed a special coating for the collector's inner tube that has superior antireflective properties effective for absorbing a wider spectrum of wavelengths than in earlier designs. Chinese factories producing the improved solar water heaters are able to manufacture them at a cost equal to that of electric water heaters.[25] The big difference to consumers is that heating water with electricity costs them around two hundred dollars per year, while heating water with the sun is free.[26]

The nearly 54 million solar water heaters installed worldwide by 2009 saved energy equivalent to the amount of energy produced by forty nuclear power plants and prevented 53 million tons of carbon dioxide from entering the atmosphere.[27]

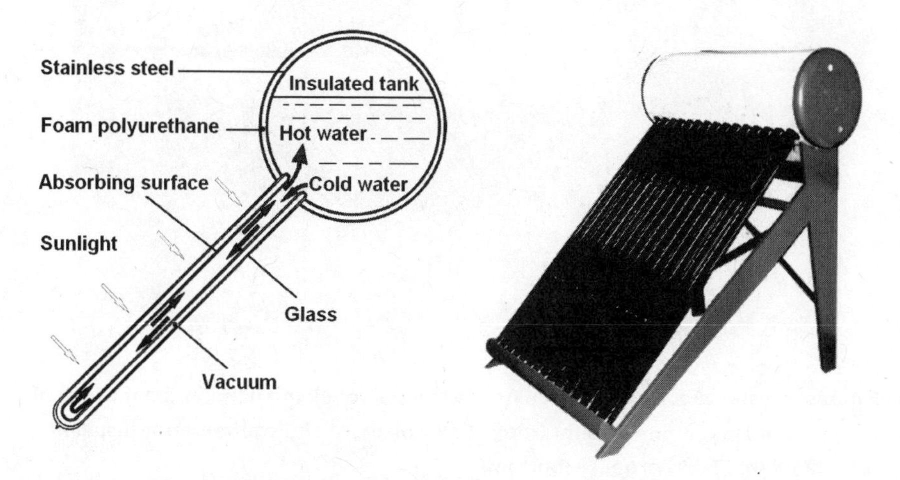

Figure 27.9. Typical Chinese evacuated-tube system.

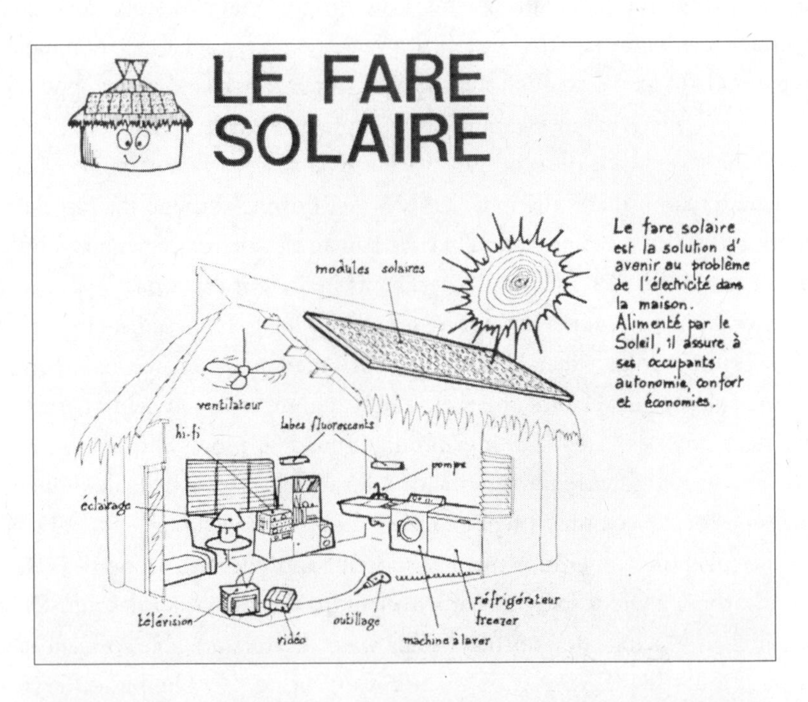

Figures 28.1 and 28.2. An illustration from a manual to sell the Tahitians on the idea of solar photovoltaic home systems (top); "Fare Solaire," a photovoltaic array installed on a traditional Tahitian house (bottom).

[1978–]

Photovoltaics for the World

For decades, rural electrification experts planned to power the developing world in the image of the developed world — by building centralized generating plants and running wires to consumers. But because installing poles and wires has proven too costly, one-third of humanity, over 2 billion people, still live without electricity. Stand-alone power systems have become their only means of generating necessary electricity. Diesel generators were once considered the answer, but the high price of diesel fuel, owing to the expense of transportation and sharp rises in petroleum costs since the early 1970s, has forced most owners to drastically curtail the number of hours they run their generators or give them up. And when breakdowns occur, as they always do, it may take weeks or months before repairs can be made, because parts and skilled mechanics are few and far between. Such problems have led many energy experts to conclude that the diesel engine is a less elegant solution for rural electrification than one would think.[1]

The hundreds of millions of people in the developing world who do not own diesel generators spend tens of billions of dollars a year on ad hoc solutions such as kerosene lamps, candles, or even open fires for lighting, and on batteries for radios, TVs, and cassette players. But none of these options can compare to uninterrupted electricity. A candle, for example, gives off 1 lumen of light, and a simple oil lamp provides 10 lumens, while a 10-watt fluorescent tube provides 500 lumens.[2]

For reliable power, people in unelectrified rural areas must find an indigenous energy source. For the few who live where a river flows year-round or the wind blows steadily through every season, water or wind power might serve them best. The sun, however, shines almost everywhere. A residential photovoltaic system takes less than one day to get up and running, costs a few hundred dollars, and, when connected to a battery to store electricity for sunless periods, operates as a self-sufficient unit day and night on a readily available fuel — the sun.

Tahiti Gets the First Solar Rooftops in the Developing World

The use of photovoltaics for individual remote homes in the developing world was pioneered by the French in Tahiti. Ironically, it was the French Atomic Energy Commission that initiated the program in 1978. The agency's nuclear testing in Polynesia had not endeared it or the French government to the Polynesian people. Public opinion had to be shored up. "We wanted to be popular," admitted Patrick Jourde, who worked for the commission in Polynesia at the time. "We wanted to be known not only for nuclear tests but also for helping out the people."[3] Bernard McNelis, a British colleague of Jourde's, described the agency's intentions more bluntly: "The original motivation for the program was a bit doubtful because it had to do with justifying the French nuclear presence. The attitude was, give the people electricity to keep them happy."[4]

Jourde was in charge of finding a way to bring electricity to those who lived on the isolated atolls and islands scattered over the South Pacific Ocean. The distance between each atoll and island, and between the small populations that lived on them, forced Jourde to look for new ways to provide electricity to them. He experimented with various renewable technologies that could use locally available fuel. Jourde first considered gasification — burning coconut shells and the like to produce a gaseous fuel — but the difficulty of collecting sufficient quantities ruled that out. Wind was also considered, but high maintenance and the significant variability in supply doomed this choice. Photovoltaics alone seemed to fit the bill.

Having made the decision to electrify the outer islands with photovoltaics, Jourde first helped establish a for-profit business, GIE Soler, to assemble the components necessary for solar electric houses to function smoothly, and to design and manufacture energy-efficient appliances specifically for photovoltaic-powered homes. These appliances minimized the number of solar panels — the most expensive item in the system — that each house would need, without compromising comfort. Jourde also organized a photovoltaics information, research, and training center to give customers excellent service.

When the infrastructure was in place, Jourde targeted sales to the most well-respected people on the islands — one of his first customers was Marlon Brando. "Everyone else followed their example," said Jourde. The affluent purchased 400-watt panels that would run lights, a TV, hi-fi, refrigerator, washing machine, fans, and power tools. Together these appliances and the photovoltaic system cost around ten thousand dollars. Those with less money could buy a smaller package for approximately two thousand dollars. A 20 percent subsidy and a low-interest loan payable over five years made such purchases more appealing. And given "the other choices — for example, a diesel generator — the photovoltaics option was cheaper than any other form of energy," Jourde attested.[5]

Kenya

Halfway around the world were millions of middle-class rural Kenyans, such as the Mugambi family, whose parents taught school in the countryside and had wanted electricity for the longest time but had given up hope of ever connecting to the power lines. "If we have to wait for the government [the sole supplier of utility-produced electricity in Kenya]," Mrs. Mugambi complained, "we will be old and grey."[6]

People like the Mugambis want the amenities that run on electricity, such as TV and good lighting — and they have the means to purchase them. Some move to the cities in order to have these conveniences. Others, however, have stayed put but still manage to enjoy some electricity-powered amenities. Until the mid-1980s, this described Joseph Omokambo, a Kenyan civil servant. Though not connected to a power line, Omokambo found that he could get the electricity he needed by hooking a car battery to his television set. Once a week or more often, depending on how much TV he watched, the battery would run out. Recharging meant lugging the battery 3 or so miles to town and leaving it overnight with an enterprising townsman who had electricity and who had set up a makeshift charging station in his garage. Omokambo would return the next morning, pay the man about two dollars, and carry the battery home. As time went on and the battery wore down, the trips became more frequent. Although he considered it burdensome, Omokambo continued to cart the battery back and forth. Without it, he believed, he would have to give up television. Then he learned from friends that there was a way he could keep his battery charged at home. A nearby schoolhouse, he found out, had a gadget that when placed on the roof produced electricity.

Mark Hankins, an American Peace Corps volunteer, had convinced the local school board to install a photovoltaic system purchased in Nairobi, the capital of Kenya. Once the system was up and running, the students' eyes no longer teared from the smoke of the old kerosene lanterns or strained under poor illumination. They no longer had to crowd around the lamp like so many moths to find enough light to read by. The trial run went so well that the school administrators lit the laboratory administration office, the remaining classrooms, and four

Figure 28.3. Solar electricity allows this Kenyan family to read without the fumes, fuss, and eyestrain of kerosene lighting.

Figure 28.4. The module providing the indoor lighting installed on a tin roof in Kenya. It is extremely small compared to those typically used in the developed world, which demonstrates how little power people of the developing world consume.

teachers' homes with solar modules.[7] Visiting educators, once they saw how well illuminated the school was, requested solar electricity for their institutions. Nighttime passersby could not help but notice the new technology, and the more curious stopped to inquire about it.

For his own home, Omokambo bought a typical system consisting of a module of only 12 watts, as did the Mugambis. Thousands of such units have been purchased each year since 1994. These minimodule kits are now sold alongside televisions and radios in rural general stores, because "shopkeepers have realized that if people buy a TV, they need something to power it," observed the Peace Corps volunteer.[8]

With 2 percent of its rural populace relying on solar power for their electricity, Kenya became the first country where more people plug into the sun than into the national rural electrification program. What is more amazing is that photovoltaics' ascendancy occurred without government help. The Kenyan experience suggests that photovoltaics will play a significant role in the generation of electricity in Africa and the rest of the developing world, especially in rural areas lacking grid power.

Figures 28.5, 28.6, and 28.7. Whether in the Dominican Republic, India, Vietnam, or anywhere else, the developing world is discovering the advantages of photovoltaics over candles, fires, generators, or kerosene lamps for lighting.

Rooftop Solar in the Developed World

In the mid-1970s and early 1980s, when the governments of developed countries began to fund solar-energy programs, they tended to favor large-scale, centralized photovoltaic plants over small, autonomous, individual rooftop units, because that is how electricity had been produced with other power sources. "The vision was huge solar farms — gigawatts of solar cells," attested a participant in the design of these large-scale photovoltaic installations.[9]

Figure 28.8. The first large photovoltaic farm went up in the early 1980s.

The American Association for the Advancement of Science took issue with the megaplant approach. "Despite the diffuse nature of the [solar] resource, the research program has emphasized large central stations to produce solar electricity in some distant future," the scientific organization complained, "and has ignored small solar devices for producing on-site power — an approach one critic describes as 'creating solar technologies in the image of nuclear power.'" The government took that approach, according to America's most prestigious science body, because of "twin assumptions that traditional" methods of producing power "should determine the shape of new energy systems and that 'big is better.'"[10] In contrast, solar cells, owing to their modular nature, can be placed on-site where the electricity is needed, and tailored to meet the exact needs of the consumer.

As early as the 1870s, solar pioneer John Ericsson articulated this special aspect of solar power by observing that the sun's energy "falling on the roofs of houses of Philadelphia" could operate "more than 5000 steam engines each of 20 horsepower," leading him to conclude that "one precious virtue of this new energy source is that it can be gathered without occupying useful space."[11] Photovoltaic pioneer Charles Fritts came to a similar conclusion, that photovoltaic systems are better suited for point-of-service placement. When Fritts boldly predicted in 1885 that his selenium solar panels might soon compete with Thomas Edison's coal-fired power plants, he had no intention of constructing large-scale generating stations. Rather, he said, solar arrays were "intended principally for what is known as 'isolated' working, i.e., for each building to have its own plant."[12] One hundred and twenty-eight years later, the Shell Oil Company came to the same conclusion that Fritts had: "In our opinion, the dispersed generation of [photovoltaic] energy — in shopping centers, small manufacturing plants, homes, and apartment complexes — affords the earliest opportunity for photovoltaics to contribute to our [America's] growing energy needs."[13]

Figure 28.9. An illustration from a 1929 French encyclopedia depicts solar modules on a residential rooftop and ground mounted in the front yard. Notice that the illustrator has the modules oriented in various directions. Usually, they should all face south.

The debate over how to situate photovoltaics intensified after a California utility built, with government funding, a megawatt photovoltaic plant at the end of 1982. That led Markus Real, a Swiss engineer, to demonstrate that dispersed photovoltaic units on already-built structures could be an alternative to centralized photovoltaic stations. Real formed Alpha Real, a small company that installed photovoltaic systems. The company became well-known in Real's native Switzerland after it won the world's first solar-car race, staged in Europe in June 1985. Almost every Swiss had seen the "Mercedes Benz powered by Alpha Real" devastate the competition.

Capitalizing on its name recognition at home, Alpha Real continued making milestones in photovoltaic applications. For example, the company built the world's first photovoltaic-powered tunnel-lighting system high in the Alps. As Real and his company gained experience, they made an important discovery about the centralization of solar power production. In a traditional power plant, each incremental enlargement of its turbo generator results in a threefold increase in power. Therefore, size is a major factor in the cost of generating electricity, and it encourages the building of bigger power plants, which centralizes power production for a large area. The same rule, however, does not apply to electricity generated by solar cells. The price of energy produced by photovoltaics decreases only as the cost of the production drops, which happens as the number of modules manufactured rises.

To prove to skeptics the value of placing photovoltaic units on rooftops instead of installing them in large, faraway generating plants, Alpha Real initiated its revolutionary Project Megawatt. Markus Real called the program "the answer to [the] large multi-megawatt installations" that had gained favor throughout the world. Alpha Real advertised over Switzerland's airwaves and through the nation's newspapers in 1987 for "333 power-station owners.... Having a rooftop exposed to the sun is the only prerequisite."[14] People responded enthusiastically. Soon 3 kilowatts of photovoltaics went up on each of the 333 rooftops for a total of 1 megawatt of dispersed power. A special electronic device called an inverter changed the direct current from the panels to the alternating current used on power lines. The device used by Alpha Real allowed ratepayers and a utility company to interact in the production of electricity: not only did the electricity that derived from the sun power the houses, but also when people had more than they could use they sent it to a utility company through the utility lines and received payment for it. At night or during bad weather, when the panels did not generate sufficient power, the homeowners bought electricity from the utility. Electrical generation became a two-way street.

Figure 28.10. Publicizing its drive to solar electrify 333 Zurich homes, Alpha Real announced in this advertisement that the firm "is looking for 333 power-station owners."

Nor did the consumer have the added expense of requiring batteries for the times when the sun did not shine.

Alpha Real's 333 installations constituted at that time the largest demonstration of photovoltaic systems on buildings. People saw the logic: rooftops in Switzerland had the slope optimal for solar-energy harvesting, and these power plants were located right next to the buildings' electrical connections.

Project Megawatt also helped to revolutionize the buying and selling of electricity between miniproducers and utilities. When participants in Project Megawatt produced more electricity than they needed, they sold it to the local utility for an amount equivalent to two cents per kilowatt-hour. But when they needed electricity — for example, at night — the local utility charged them six times as much. Real attributed the imbalance to a flaw in the utility company's thinking: "The idea that electricity doesn't flow solely from the central power station to the consumer, but had become a two-way street with each consumer also becoming a producer, was, of course, something new for utilities to chew on." The unfairness of such pricing infuriated Alpha Real's clients, whom Real described as people of influence: "They were not 'green' or left, but doctors and lawyers." And when these professionals protested, Real recalled, "there was so much pressure from the public sector" that the electric company relented.[15] It agreed to buy from and sell to the independent electricity producers at the same price. Net metering, the establishment of equitable rates when large and small electrical producers interact, has also become the accepted way of doing business in many U.S. states. It makes photovoltaic household systems more economical by significantly reducing the payback time. It also gives a psychological boost to homeowners who install solar cells on their rooftops. "The idea of being able to spin a utility meter backward [that is, sell electricity back to the utility] really appeals to people," observed one photovoltaics engineer.[16]

Since then, more people have come to realize that if each building could act as its own electrical producer, it would eliminate much of the capital cost of building a centralized power plant. Steve Strong, an architect and longtime advocate of residential photovoltaic systems, described the added expenses incurred in putting up a field of photovoltaic modules for large-scale electrical generation. "You've got to buy land," he pointed out, "do the site work on it, dig lots of holes, pour lots of concrete, dig trenches, bury conduits, build foundations and support structures, buy a huge inverter to change the photovoltaic-generated DC current into AC, construct a building in which the inverter is placed, [and] purchase switch gear and a switchyard and transformers. And because your station is usually far away from where people live, you have to spend money on transmission lines to get the electricity where the need is. You have spent a great deal of money, and you have yet to buy a single solar cell!"[17] None of the outlay in time, money, or effort is necessary if the solar modules are placed on the buildings where the electricity will be used. "It makes sense, absolute sense," argued Real. "The roof is there. The roof is free. The electrical connections are there."[18]

Donald Osborn, formerly director of alternative-energy programs at the Sacramento Municipal Utility District, in California, outlined other advantages of on-site photovoltaic electrical generation, for both the consumer and the utilities. "You reduce the electricity lost through long-distance transmission," Osborn stated, which runs about 30 percent on the best-maintained lines. Structures with their own photovoltaic plants decrease the flow of electricity through distribution lines at substation transformers, "thereby extending the transformers' lives." "And for a summer-daytime-peaking utility," Osborn added, "you can offset the load on these systems when the demand for electricity would be greatest," helping to eliminate "brownouts in the summer and early fall." On-site photovoltaic-generated electricity also makes renewable energy economically more attractive than power generated by a large solar electric plant, because it "competes at the retail level rather than at the wholesale level" with other producers of electricity.[19]

Alpha Real's Project Megawatt sparked a revolution in the use of photovoltaics. Photovoltaic experts around the globe learned of Real's work when, in 1990 and 1991, he presented the results of Project Megawatt at the two most important international photovoltaic conferences and published those findings in the conference proceedings.

Sacramento's municipal utility replicated Real's work in 1993 with its innovative Photovoltaic Pioneer Program, the first public rooftop initiative. The utility installed 4-kilowatt power plants on the roofs of 440 volunteering ratepayers, for a total of almost 2 megawatts. Since then, the utility has helped put up 1,504 grid-connected photovoltaic systems, with a total capacity of 14 megawatts.[20] The utility calls the installations "low-cost power plant sites." The program in Sacramento substantiates what Real had been saying all along. Skip Fralick of San Diego Gas and Electric, for example, called rooftops " 'free land'... [since] it needs no development, environmental impact statements, or extensions of transmission lines."[21] Roof installations, Osborn concurred, have allowed his utility to site "photovoltaic power plants with little trouble or expense."[22]

The mantle of leadership in photovoltaics passed to the Japanese, though, despite American scientists' high praise for the technology. Farrington Daniels, for example, had stated in 1955 that the solar-cell discovery at Bell Laboratories represented "an invention of real promise."[23] *Science* magazine was even more effusive. The world's most prestigious scientific journal proclaimed in 1977: "If there is a dream solar technology it is probably photovoltaics. These devices convert sunlight directly, bypassing thermodynamic cycles and mechanical generation altogether. They have no moving parts and are consequently extremely reliable, and easy to operate. Photovoltaic cells are a space-age electronic marvel, at once the most sophisticated solar technology and the simplest, the most environmentally benign source of electricity yet conceived."[24] For these reasons, even MIT solar house heating innovator Hoyt Hottel — who became disillusioned and remarked in 1976 that "the present 'unreserved enthusiasm' about solar energy may be remembered as a period of midsummer madness brought on by the sun" — regarded "the success of solar cells [to] be of such value that research and development should continue."[25] Likewise in 1976, the newly formed research arm of America's investor-owned utilities, the Electric Power Research Institute, concurred that photovoltaics could become a powerful technology for its members, even though at the time it was still just a notion in the minds of a few visionaries.

The American government, though, has consistently held a contrary view, ever since the Bell Laboratories announced its breakthrough. In 1973 the Nixon administration had tasked the Atomic Energy Commission to publish a report, *The Nation's Energy Future*, allegedly based on the National Science Foundation's analysis of the best energy choices for America. *The Nation's Energy Future* proposed spending over $4 billion for the nuclear option — almost $3 billion for the breeder reactor and nearly $1.5 billion for fusion — out of a $10 billion energy budget over a five-year period. The same report recommended a mere $36 million for solar cells, suggesting that, according to the National Science Foundation, photovoltaics' contribution to America's electrical supply would be almost nil through the year 2000.[26]

Budget:

	Dollars in Millions					
	1975	1976	1977	1978	1979	Total
Heating and Cooling of Buildings ..	12.8	13.6	10.7	6.5	6.4	50.0
Solar Thermal	5.0	7.0	7.5	8.5	7.5	35.5
Wind Energy	6.2	6.7	7.2	7.5	4.1	31.7
Ocean Thermal	1.9	3.5	4.5	7.2	9.5	26.6
Photovoltaic	4.2	5.6	7.0	8.0	11.0	35.8
Bioconversion	2.4	3.5	4.5	4.5	5.5	20.4
TOTAL	32.5	39.9	41.4	42.2	44.0	200.0

Figure 28.11. Table from *The Nation's Energy Future* shows the budget allocation for photovoltaics.

Dr. Barry Commoner, a distinguished scientist and strong solar advocate, was "surprised and troubled by the smallness of both the proposed solar-research budget and expected results."[27] He wanted to see the data from the National Science Foundation that supported the Atomic Energy Commission's dismal view of the future of solar power, especially since Solar Subpanel IX, the scientific panel that appraised photovoltaics' contribution, was made up of, in Commoner's judgment, "a distinguished group of experts." A report by Solar Subpanel IX contained their findings, the scientist learned. When asked by Commoner to see a copy of the report, the Nixon administration denied that such a report existed. Not believing the response credible, Commoner enlisted the support of Senator James Abourezk of South Dakota, a strong supporter of solar energy. He, too, received the same runaround. Finally, a solar-energy-friendly "Deep Throat" told the senator that a copy existed and could be found at the Atomic Energy Commission's document reading room. According to Commoner, "This turned out to be a dim photocopy of a hazy carbon; but it has brilliantly illuminated" the discrepancies between the science and politics of energy.[28]

2. The budgets for the accelerated and minimum programs, shown in Fig. 24 and 27, are seen to be about a factor of 5 different. The major milestones are given in Figs. 22, 23, 25, and 26.

C. Implementation

1. As indicated in Figs. 28, 29 of this writeup, the achievement of the cost goals of this program will result in the production of economically competitive electrical power (cost of 10 mils per KWH) by the year 1990. The projected rate of implementation of this solar energy concersion technology will produce more than 7% of the required U.S. electrical generating capacity by the year 2000.

2. The time schedule for implementation of the two primary appli-

Figure 28.12. Paragraph from the National Science Foundation's Solar Subpanel IX report that discusses the contribution that photovoltaics, with proper funding, would provide to the U.S. electrical grid in 2000.

Unlike the author of *The Nation's Energy Future*, the subpanel recommended an outlay of almost six times more money than the Atomic Energy Commission had requested for research and development of solar cells. Furthermore, the National Science Foundation had great expectations for solar electricity, predicting that with its suggested outlay of funds for photovoltaics, solar cells would supply "more than 7% of the required U.S. electrical generation capacity by the year 2000," even though the expenditure for the solar option would be sixteen times less than for the nuclear choice. The subpanel also found the solar option more appealing because "in contrast to problems incurred by nuclear plants, photovoltaic systems would find wide public acceptance because of their minimal impact on the environment." However, the report warned, if underfunded, "photovoltaics will not impact the energy [situation]" in future times.[29] In perverse fashion, Dr. Dixy Lee Ray, the chair of the Atomic Energy Commission and author of *The Nation's Energy Future*, took the subpanel seriously and made sure photovoltaics received but a pittance, to fulfill her prophecy that solar would always remain, in her words, "like a flea on the behind of an elephant" in America's energy mix.[30]

While Jimmy Carter was in office, the Federal Energy Administration's Task Force on Solar Energy Commercialization came up with a novel way of increasing the utilization of photovoltaics. Rather than look for breakthroughs, it sought out already existing government markets for solar cells at their current, 1977, price. This strategy would speedily expand production, which in turn would significantly lower the price of photovoltaics because of enhanced economies of scale. As a consequence, producers of photovoltaics could gain access to even larger markets. Greater production would also raise the learning curve of the industry, which, translated into layman's terms, means, "the more you make, the better your product becomes." "The primary benefit would be," the task force stated, "to stimulate the accelerated establishment of a viable, highly competitive photovoltaic industry capable of supplying the private sector with a major source of clean, non-depletable electrical energy."[31] The plan differed from the National Science Foundation approach in that it proposed providing a real market, which would substantially lower the federal investment necessary for the same expected outcome.

The task force eyed the Department of Defense's remote power needs and pinpointed the department as a guaranteed market. Solar cells would replace only those generators whose operating and maintenance costs resulted in a per-kilowatt-hour cost higher than the generators' proposed replacements. The end result would be, according to the task force, "a significant cost benefit of about $1 billion" for the government.[32] The Department of Defense supported the initiative, finding that "photovoltaics offer a number of advantages," which included "potential cost and fuel savings" and their ability to "provide a highly reliable, silent power source at remote sites, in portable equipment and offshore applications."[33]

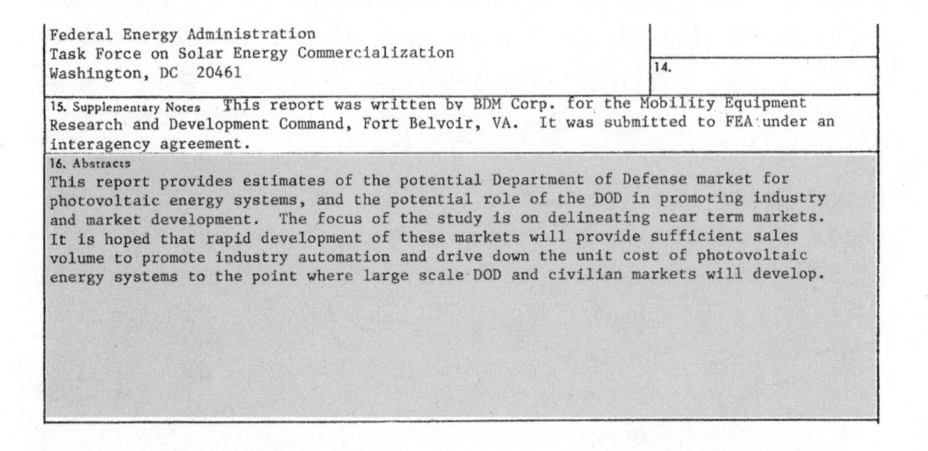

Federal Energy Administration
Task Force on Solar Energy Commercialization
Washington, DC 20461

14.

15. Supplementary Notes This report was written by BDM Corp. for the Mobility Equipment Research and Development Command, Fort Belvoir, VA. It was submitted to FEA under an interagency agreement.

16. Abstracts
This report provides estimates of the potential Department of Defense market for photovoltaic energy systems, and the potential role of the DOD in promoting industry and market development. The focus of the study is on delineating near term markets. It is hoped that rapid development of these markets will provide sufficient sales volume to promote industry automation and drive down the unit cost of photovoltaic energy systems to the point where large scale DOD and civilian markets will develop.

Figure 28.13. Abstract of the findings of the *DOD Photovoltaic Energy Conversion Systems Market Inventory and Analysis*.

The proposed initiative had precedents. The military had jump-started the market for integrated circuits. In 1962 the private sector showed no interest in them, since their cost was prohibitive. The Department of Defense, seeing the benefits of the technology for itself, took on the role of sole purchaser. Through various substantial buys, it created a large enough market to initiate industrial production. In six years, output increased from 160,000 units to 120 million and the price dropped from $50 a piece to around $2.50, giving birth to the gigantic integrated-circuit industry that has spawned everything digital now in use.[34]

The task force asked, why not do the same for solar cells? It found that solar cells at their current price could economically replace generators with a capacity to produce 152 megawatts of electricity at remote sites run by the Department of Defense. Replacement of these generators by photovoltaics would hasten the development of the industry — whose combined output of products had produced less than a megawatt that year — just as it had for integrated circuits. Adoption of the plan seemed assured since the Carter administration had pledged to make its number one goal the resolution of the national energy crisis. However, the bold initiative came to naught. Carter rejected out of hand the idea to seek congressional action supporting the plan.[35] Environmental leader

Denis Hayes noted Carter's ambivalence about the solar cause, asking, "Will Carter join us and lead us into the solar era, or will we have to drag him along behind us?"[36]

The Reagan administration demonstrated an even lower regard for photovoltaics. A letter to the president from his national security adviser, Richard Allen, reflects the antisolar sentiment that prevailed at that moment in the White House. He wrote to the president that he had found a book that contained "ammunition" to help Reagan in his crusade for nuclear power.[37] The book's author was Samuel McCracken, who three years earlier had written an article, antagonistically titled "Solar Energy: A False Hope," published in *Commentary*, a neocon magazine highly popular with members of the Reagan administration.

The War Against the Atom
(Basic Books, 1982, 206 pp.)
by Samuel McCracken
Assistant to the President of Boston University

-- Explains facts--medical, environmental, economic--about nuclear power.

-- Compares nuclear power to alternative energy sources, such as coal, and says nuclear is the only workable energy source reasonably free of environmental risks.

-- Says fears of potential nuclear accidents are grossly exaggerated.

-- Gives charges against nuclear power (radiation danger, reactor accidents, breeder reactors and plutonium, insurance cost, waste disposal, decommissioning hazards, economic disadvantages, thermal pollution, inefficiency, occupational risk, terrorism, proliferation) and rebuts each.

-- Examines alternatives (fossil fuel, solar) and concludes that nuclear energy is environmentally the most benign of major energy sources except natural gas, the most benign in terms of public health, the safest in terms of major accidents, and the only major source able to give us large amounts of flexible energy over a long period of time.

-- Discusses the anti-nuclear power movement and its leadership (Union of Concerned Scientists, Australian-born Boston pediatrician Helen Caldicott, Ralph Nadar, radiation specialist Ernest Sternglass, scientist John Gofman, Barry Commoner). Concludes that their behavior raises grave questions of professional responsibility.

-- Says the nuclear industry has suffered badly from news media coverage.

-- Places moral issue at center of anti-nuclear thought, saying anti-nuclearism, in its most fashionable form, claims that the nuclear industry is manned by moral monsters.

-- Concludes that society is energy-dependent and must become energy-intensive. Least constrained practial source of energy on a large scale remains nuclear.

Figure 28.14. The abstract of McCracken's book supporting nuclear power and attacking solar prepared for President Reagan.

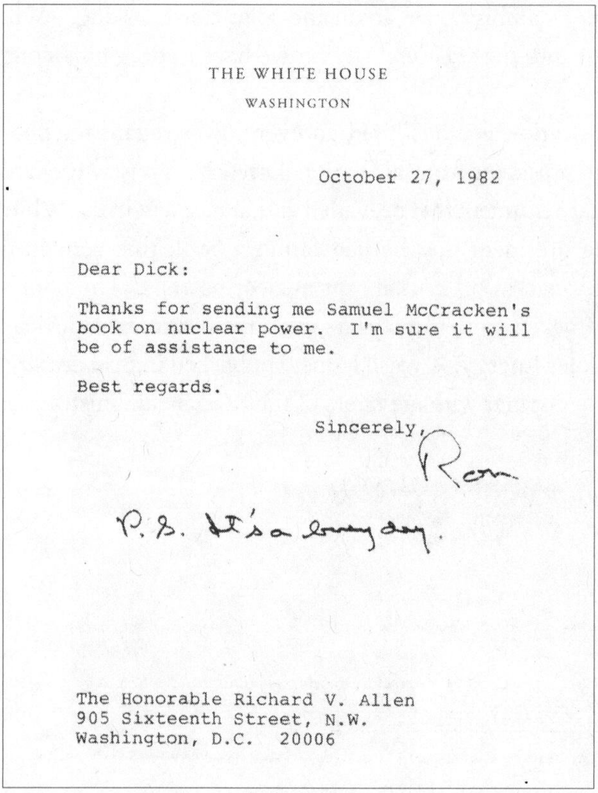

THE WHITE HOUSE

WASHINGTON

October 27, 1982

Dear Dick:

Thanks for sending me Samuel McCracken's book on nuclear power. I'm sure it will be of assistance to me.

Best regards.

Sincerely,

Ron

P.S. It's a lovely day.

The Honorable Richard V. Allen
905 Sixteenth Street, N.W.
Washington, D.C. 20006

Figure 28.15. Letter from President Reagan acknowledging receipt of the book *The War against the Atom* by Samuel McCracken.

Throughout the article McCracken carefully chose his words and phrases to push buttons that would arouse the ire of people like Reagan. It began contentiously, stating, "Solar power is the ideal source of energy for the Me Generation," and then continued by claiming that the advocacy of photovoltaics was a "feature of the New Barbarianism," a catchphrase used by the neocons to vilify the counterculture movement of the 1960s, which they hated more than the "Evil Empire" ruled by the Soviets.[38] Just to make sure his right-wing neocon readership, so influential in the Reagan years, got the message about whom they were dealing with in the solar movement, McCracken concluded his article describing the solar "energy crusade ... as little more than a continuation of the political wars of a decade ago [the anti-Vietnam movement] by other means.... Where salvation was once to be gotten from the Revolution, now it will come from everyone's best friend, the great and simplistic cure of all energy ills, the sun."[39]

McCracken's new book used his *Commentary* article as the basis for his chapter denigrating the solar choice. With the copy that Allen provided for the president came a one-page summary of its contents prepared by the publisher. One of the bullet points in the synopsis stated that the book "examines alternatives (...solar) and concludes that nuclear energy is environmentally the most benign...[and] the most benign in terms of public health, the safest in terms of major accidents, and the only major source able to give us large amounts of flexible energy over a long period of time."[40]

Unsurprisingly, Reagan slashed the meager Carter outlay for photovoltaics by two-thirds. The head of the government's photovoltaic program, Dr. Morton Prince, lamented the consequences of the new administration's war on solar cells. "I was losing all my best people," Prince complained.[41] The lead solar investigator from the National Science Foundation, Lloyd Herwig, analyzed the end result of the carnage: "We yelled from our offices that Japan would be doing it, Germany would be doing it instead."[42]

Herwig's prediction has come to pass. Starting in 1994, the Japanese government offered a generous subsidy to encourage individuals to place solar cells on their rooftops, resulting in the placement of photovoltaic systems with a capacity of 3 kilowatts on 539 rooftops. By the time the program ended in 2004 almost 400,000 rooftops had photovoltaics.[43] The initiative helped Japan become, by 2002, the world leader in the installation of grid-connected systems and the manufacture of solar cells.[44] Rooftop capacity after the first year of the program totaled 1.6 megawatts. Eleven years later, more than a gigawatt (more than a billion watts) had been installed.[45]

The Japanese government supported the program because of its interest in assuring energy security and in dominating the worldwide photovoltaic industry. But by 2004, Germany had become the world leader in photovoltaic installations. This number one position had nothing to do with the amount of sunshine Germany receives. For example, Germany gets about 40 percent less solar radiation throughout the year than the western United States does, and yet Germans have installed more than 30,000 megawatts' worth of photovoltaics, compared to the approximately 4,000 megawatts' worth that Americans have set up, most of it in California.[46] In fact, on May 26, 2012, photovoltaic-generated electricity accounted for nearly 50 percent of Germany's electrical supply. As Norbert Allnoch, director of the Institute of the Renewable Energy Industry in Muenster, remarked, "Never before anywhere has a country produced as much photovoltaic electricity."[49]

Germany's vigor stems from strong government support for photovoltaics, which is part of a program initiated in 2000 that committed the nation "to facilitate a sustainable development of energy supply in the interest of managing global warming."[48] The emphasis on increasing the market for solar cells lay in the government's belief that "in the long term, the use of solar radiation energy holds the greatest potential for providing energy supply which does not have an adverse impact on the climate."[49]

Events in the 1980s shaped German interest in promoting clean energy. The disaster at Chernobyl began to move people away from embracing policies that would have transformed the nation into a nuclear-based society. And watching the nation's beloved forests die off from acid rain emitted by coal plants prompted more citizens to join the base supporting the use of cleaner sources of energy. The general acceptance early on, even by conservative elements in German society, that burning fossil fuels causes climate change strengthened the political will to change energy sources.[50] In this political climate, renewables became a major societal issue. It took a decade, though, for the Germans to successfully promote solar cells.

In 1989, Germany launched a "thousand-rooftop" program designed to enroll more citizens in the production of solar power. To attract participants the government offered to pay 70 percent of the cost of the solar modules, which would range from 1 to 5 kilowatts per roof. The program generated a greater response than anticipated, resulting in around twenty-five hundred systems installed, for a total generating capacity of over 5 megawatts.[51] The program abruptly stopped in 1995, jeopardizing the solar industry. Various municipal utilities sympathetic to photovoltaic development initiated their own rooftop programs to keep the national photovoltaic industry from collapsing. Greenpeace also joined the effort, gathering thousands of orders for photovoltaic rooftop installations. Meanwhile, Eurosolar, a broad-based sociocultural movement in support of solar energy, proposed in 1994 a rooftop program ten times the size of the German effort. By 1996 many German environmental activists and the German Solar Energy Industries Association were pressing the national government to adopt this goal.[52]

The photovoltaics industry in Germany experienced a sea change following the Social Democratic Party–Green Party coalition's victory in the German national elections of 1998. The coalition wanted to really grow the photovoltaic market, so it gave its support to the German Solar Energy Industries Association's proposal and in 1999 launched the "German 100,000-Roof Program," marking the coalition's first step in generating a truly large photovoltaic market. The new ruling parties offered low-interest loans to all those wishing to participate. Participants would also receive an amount equivalent to about ten cents for every kilowatt-hour their systems fed into the power lines. The compensation was based on Germany's 1990 Feed-In Law. The government hoped that, by 2003, 300 megawatts of photovoltaic power would be collectively generated on a hundred thousand roofs.[53] But few responded. The reason for the tepid response, the government believed, was that "compensation rates [provided by the 1990 law had] not been sufficient to stimulate a large-scale market introduction of electricity generated from photovoltaic cells."[54] So when it passed the Renewable Energy Sources Act the following year, it required utilities to purchase photovoltaic-produced electricity at a rate equivalent to about fifty cents per kilowatt-hour, which would be guaranteed for a twenty-year period. The generous offering got the rooftop program moving. The amount of photovoltaics installed on German rooftops doubled between the date the bill was enacted, in 2000, and the following year. By June 2003 a total of one hundred thousand German roofs were collectively producing 300,000 megawatts of power.[55]

Wanting to encourage even greater growth, the German government in 2004 opened the floodgates by removing the incentive program's limit on the number of photovoltaic installations that would receive government subsidies and by further increasing the compensation, to more than fifty-seven euro cents, that small consumers received for generating electricity with solar cells. The industry and the public reacted by increasing the number of multimegawatt ground-mounted systems (solar parks) and installations on private roofs by a factor of six over the next four years.[56]

Many multiple-megawatt solar parks have gone up in recent decades, pioneered by the Spanish and now emulated throughout the world. Germany's feed-in tariffs, which give the generator of photovoltaic power a premium for every kilowatt-hour put into the grid, deserve much of the credit for expanding the amount of photovoltaics worldwide to a cumulative total of around 100 gigawatts — more than two million times the amount that existed in 1975. In summing up the impact of the German program, the investment group Motley Fool wrote, "More than a decade ago, Germany decided it was time to

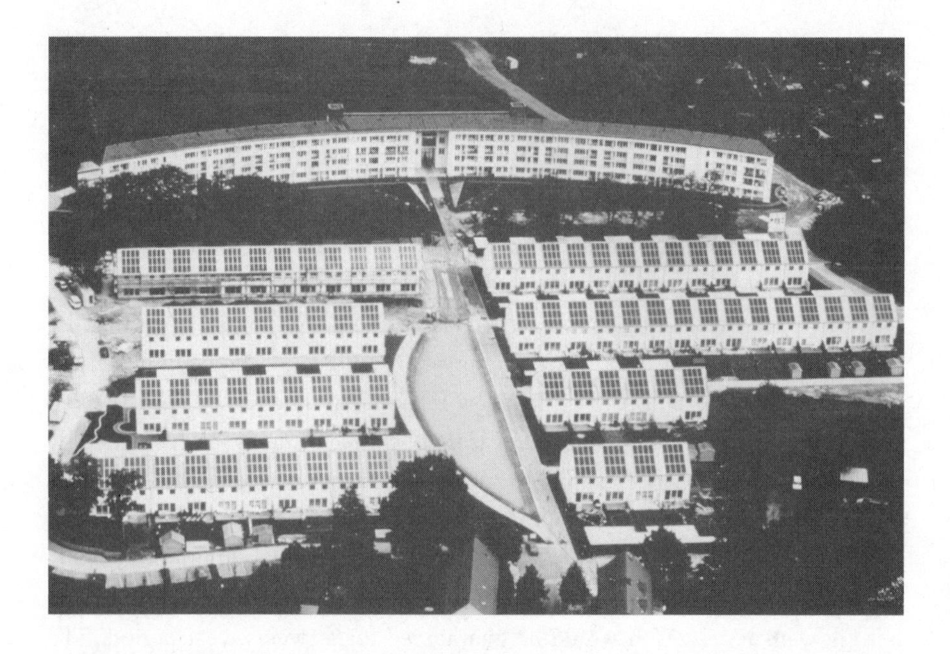

Figure 28.16. A familiar site in Germany: an apartment complex with solar cells mounted on every rooftop.

Figure 28.17. Ground-mounted photovoltaics as seen in this picture allows a German farmer to both raise sheep and sell electricity.

make a big bet on solar. The country heavily subsidized the industry to make it economically viable long-term. The bet paid off."[57] The Chinese government, too, via aggressive support for its photovoltaic industry, has also greatly contributed to widespread adoption of solar cells by bringing down the price of photovoltaic power through the massive production of solar modules. As one expert observed in 2012, "Things that weren't feasible have suddenly opened up. As prices drop, you see huge markets open up."[58] No wonder the International Energy Agency, often biased toward fossil fuels and nuclear energy, expects 230 gigawatts of photovoltaics to be installed throughout the globe by 2017.[59]

View looking Northwest along 42nd Street

Figure 29.1. Advertisement for 4 Times Square, built in the 1990s with the first application of building-integrated photovoltaics in a skyscraper located in a major North American city.

Better Solar Cells, Cheaper Solar Cells

Crystalline silicon remains the dominant photovoltaic material, basically the same substance that was discovered at Bell Laboratories in the early 1950s. Its continued widespread use rests on the fact that, as one expert explained, "it works and it works for a long time."[1] Crystals for single-crystal silicon cells continue to be grown as they were in Daryl Chapin's time: by melting silicon at 2,570°F in a rotating vat and then drawing it up and out of the melt using a puller spinning in the opposite direction. This produces a single-crystalline cylinder 7 to 8 feet in height.

Multicrystalline Solar Cells

Attempting to bring down the price of solar cells, Chapin tried to find new ways to make crystalline silicon. For instance, he tested cells fabricated as multicrystalline silicon. Multicrystalline silicon is formed by casting the silicon material into ingots. The process is less complex than growing a single crystal, and the result is smaller multiple crystals rather than one large one. But the activated charges in the polycrystalline material had great difficulty crossing from crystal to crystal, finding themselves trapped in the boundaries between them. Consequently, relatively few charges made it through this obstacle course to the metal contacts. This meant that Chapin obtained dismal results — an efficiency of only 1 percent.[2] Over the next two decades, researchers increased the efficiency

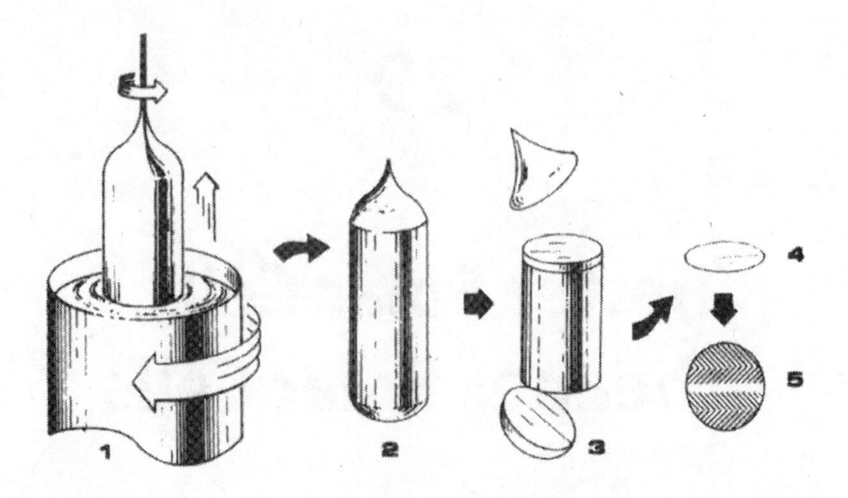

Figure 29.2. The steps involved in preparing single-crystalline silicon cells. (Material loss due to cutting is high: 16.4 grams of silicon are required for 1 peak watt). (1) Silicon is melted at 2,570°F in a rotating vat. As it begins to crystallize, it is drawn upward from the vat by a counter-rotating assembly. The resulting solid cylinder (2) is rounded and the top and bottom are cut off (3). The cylinder is sliced into wafers (4), which are processed into solar cells (5).

of polycrystalline silicon cells by a few percentage points, but never beyond the 6 percent mark. The failure to reach higher efficiencies led experts to conclude that good solar cells could not be made with this material. "This was disproved, though, by Dr. Joe Lindmayer," commented Fritz Wald, who had been involved in the solar-cell industry for many years.

Lindmayer was regarded as one of the best technical people in the solar-cell business during the 1970s: if anyone could make polycrystalline silicon work, he could. He found that past efforts had failed because the crystals produced were too small and, as a result, the area of the boundaries was very large. So he set out to grow relatively large crystallites. In the process, Lindmayer discovered that crystal size depended on the speed at which the ingot cooled: fast cooling meant small crystals. By slowing down the cooling process, he came up with the size needed to make an efficient solar cell, approximating the performance of single-crystalline silicon.[3] High cost still remains an issue.[4] For one thing, the casting procedure still begins with expensive, highly refined silicon, and the cast blocks, which weigh 120 pounds each, still must be sliced into extremely thin wafers to make solar cells. Single-crystalline cylinders also have to be reduced to segments 300 to 400 microns thick ($^1/_{100}$ to $^2/_{100}$ of an inch). Cutting such thin pieces from very large materials creates a lot of waste. "You eventually make as much sawdust as you do cells," Wald observed.[5] Half of the very expensive silicon ends up as trash.

Focusing Devices for Solar Cells

Another way to lower the price of solar cells is to direct more light onto them than they would ordinarily receive. Antonio Luque, one of the pioneers of such a strategy, explains how he approached the problem:

> My first unique contribution to the [photovoltaics] field came in this early period with my invention of the bifacial silicon solar cell, in which the module gains light from not only the top but also the bottom. We tried several prototypes: in one case, we painted the floor white below the installed modules. In this case, we got a net efficiency gain of 20 to 30% over regular silicon. We also tried concentration without the

Figure 29.3. Solar cells at the focus of a parabolic trough, looking very similar to Shuman's device yet producing electricity, not heat, for power. This plant is located in Tenerife, Spain.

> complexity of high-concentration[,] and moving the system to follow the sun by using concentrators developed by Roland Winston. In this case, the collector funnels extra photons to the cells while remaining still. Efficiencies also climbed. Here began a common thread in my work — seeking optical tricks that would get the most out of solar cells.[6]

Currently, most concentrating photovoltaic systems use a lens or concave mirror. Focusing the sun's energy onto a photovoltaic panel enables the system to produce the same amount of electricity with much less photovoltaic material than previously necessary. This is especially important when using the very expensive but highly efficient multijunction cells.[7] The secret of their superior performance is that they consist of layers of different semiconducting materials, with each responding optimally to a different part of the spectrum, allowing more absorption of the sun's energy than with other types of solar cells.[8]

Thin-Film Solar Cells

Yet another avenue for bringing down the cost involves depositing a small amount of a photovoltaically active substance onto an inexpensive supporting material, such as glass or plastic. Solar cells made in this fashion are called thin films. It has long been believed that this process would consume much less of the expensive photovoltaic ingredients, and that it would also lend itself to automation. In the mid-1950s, while at Bell Laboratories, Chapin supported this view, hoping that fellow technologists would devise "a new process whereby a really large surface could be sensitized [photovoltaically] without the preparation of [large] single crystals."[9]

A decade later, Dr. Fred Shirland, a scientist at the Clevite Corporation in Cleveland, thought he had found a way to make a cheap photovoltaic device in exactly the way Chapin imagined. Shirland substituted cadmium sulfide, another substance that produces the photovoltaic effect, for silicon. It appeared to be easy to evaporate or spray a thin film of cadmium sulfide onto plastic or glass. Shirland assured his peers, "There is nothing apparent in the materials required or in the cell design or the fabrication methods to indicate that [cadmium sulfide cells] could not be made in real mass production at very low cost."[10] In 1967, an affordable solar cell ready for mainstream use seemed so near that the

April 1 issue of *Chemical Week* announced that Shirland's firm "has developed a new ... solar cell that can be mass produced at low cost, by... deposition of [cadmium sulfide] onto... plastic."[11]

Despite such sunny expectations, a usable low-cost cadmium sulfide cell has never materialized. Some Clevite cells worked well outdoors for seven or eight years; the majority, however, lost all power when exposed to the elements. Why a few lasted indefinitely and the rest fell apart remains a mystery. NASA concluded that cadmium sulfide cells "are very stable [only] if you keep them from seeing light, oxygen and water vapor."[12]

Figure 29.4. An air force scientist discovered and built cadmium sulfide cells at about the same time Bell scientists were developing the silicon solar cell. This 1955 air force display "shows" cadmium sulfide cells driving a miniature merry-go-round. However, the air force failed to mention that it had surreptitiously attached the carousel to two small wires that had been carefully buried underground and connected to a well-hidden power outlet, as the cells had degraded very quickly when exposed to the elements.

The possibility of discovering the solar device that would power America's grid with the free rays of the sun continued to lure researchers. While most laboratories interested in finding a low-cost photovoltaic device concentrated on cadmium sulfide during the 1960s and 1970s, a few maverick scientists considered other materials. Elliot Berman had initially interested Exxon in the possibility of making a low-cost solar cell from dyes that convert sunlight into electricity, an idea that has resurfaced in a modified form. Berman envisioned depositing the dyes onto a continuous roll-to-roll material, much like the production techniques used in the photographic industry. Exxon gave up on this approach when its researchers failed to achieve conversion efficiencies high enough to be commercially attractive, although some researchers have revisited the idea.

A group at RCA, under the direction of David Carlson, also tried making a cheaper solar cell in the same fashion Shirland employed, but with an alternative material. The oil embargo of 1973 and the subsequent quadrupling of petroleum prices had triggered Carlson's interest in solar electricity. Initially he tried depositing a thin layer of polycrystalline silicon onto common materials such as glass, as Shirland had done with cadmium sulfide. When exposed to light, some of the films he made showed a small but significant photovoltaic effect. The RCA group was so sure that Carlson had successfully made

Figure 29.5. David Carlson at work in his laboratory.

a thin-film form of polycrystalline silicon that Christopher Wronski, the scientist in charge of measurements, based the calculations for his conclusions on the properties of polycrystalline silicon. This led to "getting ridiculous results[,]... impossible numbers," Wronski recalled.[13] When the X-ray analysis of the supposed polycrystalline film came back to the lab, Wronski discovered why he had erred. "The X-rays showed no evidence of crystallinity," Carlson said. "Instead, it revealed that I had made an amorphous silicon solar cell [silicon that lacks the ordered internal structure inherent in crystalline silicon]. I just stumbled into it."[14]

Two years later, in 1976, after much fine-tuning Carlson and his coworkers had upped the efficiency of their new solar device, from under 0.2 percent to a significant 5.5 percent. They reported the accomplishment in the journal *Applied Physics*. The prestigious weekly *Science* called the work at RCA "perhaps the most intriguing recent development" in the solar-cell field. The announcement stirred much excitement throughout the world, because, as David Carlson wrote, "these cells have the potential of producing low-cost power since inexpensive materials such as steel and glass can be used as substrates."

The ensuing flurry of amorphous-silicon research activity worldwide revealed why Carlson and his group had succeeded where earlier investigators had failed. Prior to RCA's initial work, most experimenters investigated only pure amorphous silicon, whose totally disordered composition makes the material useless as an electronic device. By inadvertently contaminating the amorphous silicon with hydrogen, Carlson and his team had greatly improved its photovoltaic capabilities. The hydrogen acted like a molecular repairman. It cleaned up the disheveled internal structure of the pure amorphous silicon by forming chemical bonds with the dangling silicon atoms. "The 'best' amorphous silicon" turned out to be "a silicon-hydrogen alloy," commonly called hydrogenated amorphous silicon.[15]

As the RCA group delved deeper into hydrogenated amorphous silicon's behavior, it appeared that the material had a self-destructive trait. This was discovered after Wronski measured a sample exposed to the sun, his coworker David Staebler retested the sample a month or so later, and their measurements did not agree. "Then we got into an argument as to who measured correctly," Wronski remembered. "We finally discovered that both of us did the measurements right[,] and we found that there was something very wrong with this device — it degraded in sunlight." This revelation gave the RCA people a good scare. "We were very concerned about it," Wronski admitted.[16]

Further studies eased their fears. The researchers found that, although hydrogenated amorphous silicon cells degrade during their first few months in the sun, they then stabilize. "It was not a catastrophic failure. It was something predictable," Carlson recalled. Even better news came from the lab of Joe Hanak, a member of Carlson's research team. "Joe discovered that if you make the cells very thin, the degradation became less," Carlson said.[17] A very thin cell, a hundred or so times thinner than crystalline solar cells, also significantly reduces the amount of starting material needed.

The loss of efficiency of amorphous silicon cells when placed in sunlight changed the course of Harold McMaster and Norman Nitschke's research. They scrapped their work in amorphous silicon to try another material, cadmium telluride, as a thin-film device. They learned how to vaporize powdered particles of cadmium telluride and then have it solidify as a thin layer on a large slab of glass. Success in depositing cadmium telluride vapor on glass meant that the industry now had another photovoltaic material available.[18]

Silicon Solar Cells Revisited

Great strides in making silicon solar cells cheaper and more efficient over the years have maintained silicon's dominance of the photovoltaic industry since the discovery of photovoltaics in 1953. For example, to house the photovoltaically active silicon, scientists at Spectrolab, one of the first photovoltaic companies, introduced in the 1970s some low-cost, readily available, long-lived materials that lend themselves to mass production. Tempered glass was chosen for a top cover. It was robust, it was manufactured everywhere, and it would self-clean after a rain. This same group also came up with a better way to apply contacts to cells — by screen printing, the same method used to put designs on T-shirts.[19]

Meanwhile, the price of the silicon feedstock material dropped, owing to increasing production to satisfy the growing needs of the semiconductor industry. The introduction of automated multiple saws to cut the crystal-grown cylinders or multicrystalline ingots into cells also has reduced their cost.

Light trapping — by texturing the silicon or by the addition of a reflecting device at the bottom, as well as by finding better antireflective material at the surface of the cells — has also improved performance and reduced the amount of silicon needed for modules. Such strategies enhance the amount of light — the fuel for photovoltaics — inside. Silicon's efficiency has also been increased

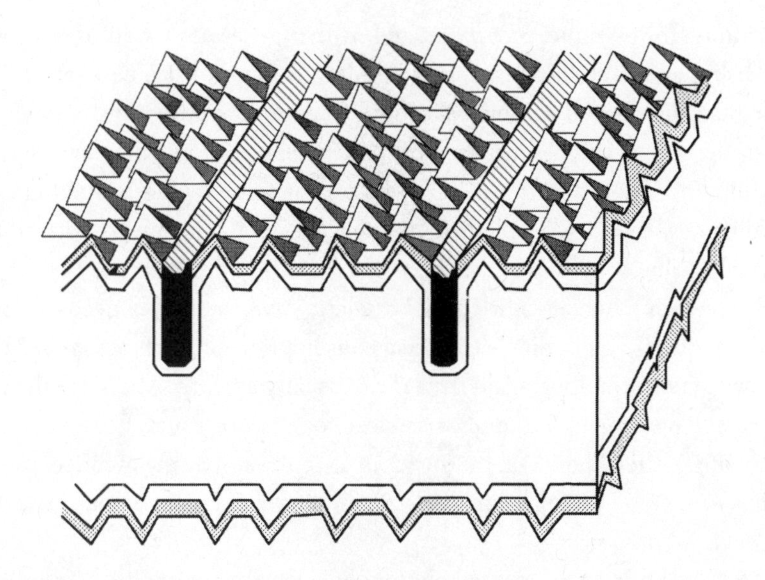

Figure 29.6. The little pyramids on the top help to trap incoming sunlight in the solar cell.

by adding amorphous silicon to silicon in a minute and ordered fashion, which purifies the silicon so that more of the electrons freed by photons reach the contacts. The metal contact lines on the silicon have been thinned to reduce shading so the modules receive more sunlight; some manufacturers have even done away with all metallization on the front, having devised a way to put the contacts on the back so the cells get full sunlight. Cells have also become much thinner — so they are made with less material — and their processing has become much more automated. All the improvements described here have increased silicon solar-cell efficiencies by a factor of four over what Bell Labs had achieved when the company went public with the Bell Solar Battery in April 1954. At the same time, these improvements have reduced the price, from $286 a watt to under $1 a watt.[20]

Building-Integrated Solar Photovoltaics

Using photovoltaics as a building skin completely changes the way of calculating the economics of photovoltaics: The old method, considering electrical generating costs, no longer holds. Instead, the expense of photovoltaic building material must be compared with the price of other facade coverings. Marble and granite, for example, cost more, and neither generates 1 watt of electricity. Purchasing an electricity-generating building skin at a price equivalent to or lower than that of conventional materials is good business. This is especially true in densely developed urban locations, where most commercial buildings are built, and where "power tends to cost a lot more owing to the high cost of adding new capacity," explained Gregory Kiss, a Manhattan-based architect. "Here the value of photovoltaics becomes quite high."[21] There is also a good match between the availability of solar energy and the power needs of office buildings. Most of a building's electrical consumption occurs between 9 AM and 5 PM, when both employees and the sun are usually at work. And since the solar cells can be positioned in the glass as close together or as far apart as desired, their configuration can block out or let in as much sun as the architect wishes, helping to save on illumination, air-conditioning and heating, and the energy such devices consume.

The first building-integrated photovoltaic facade was constructed in 1991; it covered the south side of a utility company's administrative offices in the small town of Aachen, Germany. Since then, many more have been built, including

Figure 29.7. In this very cleverly designed building-integrated photovoltaic structure the eaves not only keep the summer sun out of the building but also use the sun they block to run the air-cooling units, as also do the photovoltaic roof shades.

the facade on a forty-eight-story skyscraper in New York City. Named 4 Times Square and situated on the corner of Broadway and Forty-Second Street, this building is "the opposite extreme of the traditional remote photovoltaic installation," according to Kiss, the architect responsible for its photovoltaic facade. "Here we [have planted] the photovoltaics flag in Times Square, the crossroads of the commercial world. 4 Times Square leads the way to large-scale generation of clean, silent solar electricity at the point of greatest use, in urban centers like New York City."[22]

The number of photovoltaic materials now on the market is dazzling. These materials include crystalline silicon, amorphous silicon, cadmium telluride, gallium arsenide, multijunction devices, and copper indium gallium diselenide. Other new photovoltaic materials entering the market include dyes that imitate photosynthesis yet are more efficient than greenery, and photovoltaic materials that can be mixed in solution to become ink or paint, lending themselves to printing or to being brushed or sprayed onto a surface. Novel concepts are being considered that could one day dramatically increase the efficiency of solar cells and lower their costs. Ideas include increasing a photon's performance once it enters a photovoltaically active material or using cells based on nanowires — structures just a bit bigger than nine water molecules — combined with graphene, a type of carbon many believe will revolutionize the twenty-first century.[23] Only time will tell which ideas and materials will dominate. Or perhaps they will all share a segment of the growing market for electricity that is generated without emitting greenhouse gases. One point, though, seems beyond dispute: in the 1980s, people argued whether the price of photovoltaics would ever go low enough to be competitive with grid electricity, a circumstance now called grid parity. By the second decade of the twenty-first century, everyone agrees that grid parity has already been reached in a few sunny spots where the price of electricity is extremely high, and that grid parity will extend to the rest of the world in the next few years, depending on the amount of solar radiation received, the cost of electricity coming from central power plants, and societies' concern for increased pollution and carbon emissions.

Figure 30.1. Combining solar and wind helps to cancel out the intermittency issue of the two renewables. Often when the sun is not shining, it is more likely to be windy, and when the sun is out, there is usually less wind. Add biogas to the mix and you can be assured of electricity 24-7.

Epilogue

When research began in 1977 on *A Golden Thread: 2,500 Years of Solar Architecture and Technology*, the sire of *Let It Shine*, only 360 watts of photovoltaics operated in the world.[1] At this writing, May 2013, that installed number has increased by a factor of 250,000 to 100 billion watts, or 100 gigawatts.[2] Likewise, when I wrote *From Space to Earth: The Story of Solar Electricity* fourteen years ago, the largest photovoltaic panel factory had a capacity to build 10 megawatts of product per year.[3] Currently, a typical factory will have the ability to manufacture hundreds of megawatts. The world's solar water-heating market, too, has grown more than twentyfold over the past thirteen years.[4]

Concentrating Solar

In 1909, H. E. Willsie, builder of one of the first solar power plants, stated, "Our experience indicated the sun power plants should be built in large units to distribute their power in the form of electricity."[5] Yet in 1976, only a few kilowatts of solar-concentrating power existed in the world, compared with 2 million kilowatts currently at work. The vast majority of solar concentrators resemble the plant Frank Shuman built, although they are far more sophisticated in design and operation. Power towers are also being built. Their ancestry can be traced to Buffon's series of flexible flat mirrors, which he used to focus the sun's rays onto an object at a distance. Instead of trying to set something on fire as Buffon

Figure 30.2. Flat mirrors focus on a power tower.

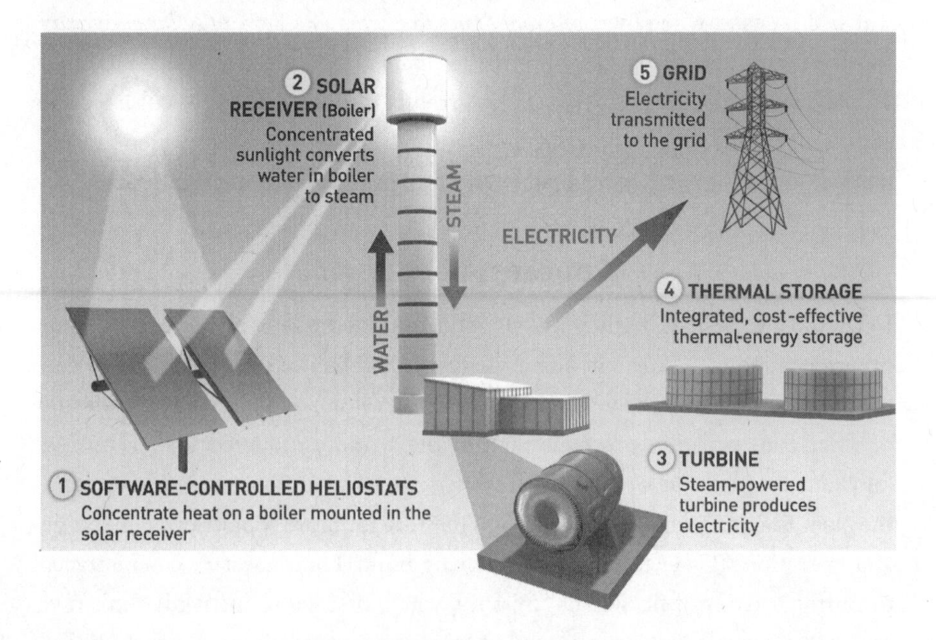

Figure 30.3. A schematic shows how power towers make power.

did, power towers — flat mirrors surrounding a tower — concentrate sunlight onto a boiler atop a tower to produce steam and run a conventional turbo generator for making electricity. One possible limiting factor for the development of such large solar-concentrating plants is that they require building transmission lines from their remote locations to distant urban areas. The lines add greatly to the expense of the solar plant, take a lot of time to build, and are steeped in controversy — resulting in protests from environmental groups — which also adds to their cost.[6]

While solar-concentrating power plants available today generate electricity from solar heat, they may have other uses, too. The kingdom of Saudi Arabia has also been considering the technology to help satisfy its growing need for drinkable water. Overconsumption is threatening the depletion of its aquifers — the country's sole source of natural freshwater — and the heat produced by solar-concentrating plants could turn the ocean off the coast of Saudi Arabia into a source of freshwater for its growing population. The kingdom has plans for multiple solar plants doing just this in the near future.[7]

And in a twist not without historical irony, the oil fields in the sunny parts of the world, whose development quashed the promising work of Frank Shuman, could very well significantly contribute to the revolution in solar-concentrating

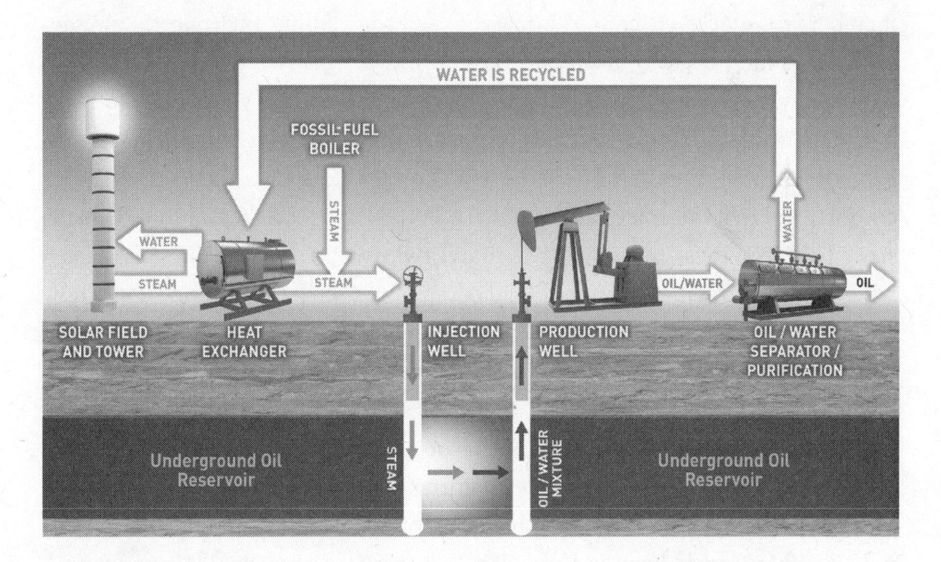

Figure 30.4. How concentrating solar power can thin heavy oil for recovery. Solar-generated steam could transform oil too viscous for current recovery into a lighter mixture capable of extraction.

thermal technology that Shuman and his predecessors, such as Ericsson and Mouchot, dreamed would occur.[8]

Few realize the value of solar energy today. The value of the global photovoltaic market alone climbed to over $82 billion in 2010.[9] The International Energy Agency's annual report, *World Energy Outlook 2012*, which came out during the second week of November 2012, announced that "the rapid expansion of wind and solar power has cemented the position of renewables as an indispensable part of the global energy mix."[10]

U.S. Military as Solar's Ally

The U.S. military's success with solar-powered satellites since the late 1950s has made it particularly bullish on solar energy. Major General Anthony Jackson, for example, stated in 2012, "There is an urgent need for our nation to lead the world in renewables and conservation. For every fifty trucks [carrying fuel] we put on the road [for our forward forces], someone is killed or loses a limb." He added, "I know the cost of oil. I know it up close and personal. If you have never seen the mixture of blood and sand, it's a harsh purple on the desert floor."[11]

In the past decade, the United States has lost more than a thousand service members to attacks on convoys, mainly used for hauling fuel that renewables are increasingly used to displace. One-third of the army's wartime oil is used to generate electricity that is generally cheaper and certainly less risky to make with renewables. The present use of photovoltaics for powering field generators, which was suggested by the Federal Energy Administration's Task Force on Solar Energy Commercialization back in the 1970s, has begun to reduce the size of the convoys transporting fuel and the number of casualties associated with them. A growing amount of field equipment used by the marine corps now runs on solar cells, as do many of the military runway lights for airfields and helipads. And integrated solar arrays–fuel cell generators are currently being developed or field-tested that will reduce overall fuel consumption for tactical electrical generation by 50 percent or more.[12] "The military is keen on photovoltaics," according to Kelly Trauetz, an engineer at the Naval Research Laboratories, "because combined with batteries you will never run out of power. If you are stuck somewhere and you don't know when your resupply is coming, photovoltaics, unlike nonrenewables, will keep going. This is fundamental in situations such as we in the military face."[13]

Figure 30.5. The U.S. Signal Corps celebrated its one hundreth anniversary in 1960 by making the first solar-powered transcontinental radio broadcast from its headquarter in Fort Monmouth, New Jersey, to Hoffman Electronics, its supplier of solar cells, located in El Monte, California.

Figure 30.6. Airfield in Afghanistan using photovoltaic-powered runway lighting. Each light is a self-contained unit, making the entire system nearly invulnerable to destruction by terrorist attack.

Subsidies

Despite solar's resurgence, most governments continue to back fossil fuels with disproportionate subsidies. Oil received thirty times more in subsidies from the federal government than solar between 1950 and 2010.[14] The International Energy Agency reported that in 2012 alone thirty-seven governments spent more than $523 billion subsidizing fossil fuels while assisting renewables with almost one-sixth the funding.[15] And if the nuclear-energy industry did not receive governmental indemnity to shelter it from the burden of carrying third-party liability insurance, the true price of nuclear electricity would skyrocket; without this unique protection, the nuclear industry admits, it couldn't operate.[16]

Congress also continues to give the fossil-fuel industry easier access to capital than it gives to the solar industry and to other renewable-energy industries, by leaving in the tax code investment advantages solely for those exploiting "any resource that is depletable."[17] Ironically, this loophole was inserted one year after Congress had cut the tax credit to solar devices in the name of "deficit reduction and tax reform."[18]

The Private Sector's Support for Solar Energy

Even with unequal government subsidies, the private sector is increasingly embracing solar. Walmart, for example, has a total of 65 megawatts of solar panels on the roofs of 144 of its stores. All told, the power generated by the rooftop panels of corporate buildings in the United States is equivalent to the amount of power used by almost four hundred thousand homes. The Bloomberg Corporation stands with Walmart in converting rooftops to solar-energy production. "We see solar playing an increasingly important role in our energy mix. We will add several megawatts to our portfolio in the next couple of years," said Curtis Ravenel, global head of sustainability at Bloomberg. "The environmental and financial benefits are attractive on multiple levels for us as a firm."[19] Meanwhile, Apple announced in early 2013 that all of its data centers and 75 percent of its worldwide corporate facilities run on renewable energy; its efforts include a 20 megawatt solar farm next to the company's Maiden, North Carolina, data center.[20]

Climate Change and Other Environmental Disasters

The Fukushima Daiichi nuclear disaster in 2011 succinctly demonstrated what's wrong with nuclear power, just as Chernobyl did in 1986. And the allure of attaining energy independence by exploiting oil, shale gas, and tar sands blinds many to the inherent danger of increased carbon dioxide in the atmosphere. Despite the fact that natural gas does emit less carbon dioxide than other fossil fuels, it still contributes significant amounts of greenhouse gases, both through carbon emissions from combustion and through the possibility of escaped methane, a far more potent greenhouse gas than carbon dioxide. As National Oceanic and Atmospheric Administration (NOAA) scientists plainly state, "Carbon dioxide (CO_2), emitted by fossil-fuel combustion and other human activities, is the most significant greenhouse gas contributing to climate change."[21] Or as Nobel Laureate and former secretary of energy Dr. Steven Chu succinctly puts it, "Global warming is happening now. Humans are causing it."[22]

Meanwhile, NOAA scientists reported in May 2013 that humanity has just passed a long-feared milestone, with carbon dioxide levels reaching their highest concentration in human history, exceeding 400 parts per million. The last time this level was reached was about three million years ago. In that period, scientists know through geological data, it was much hotter and sea levels may have been as much as 60 to 80 feet higher than they are now. "Humanity may be precipitating a return to such conditions," climate-change experts suggest, with the bone-chilling caveat that now billions of people are endangered."[23]

The devastation wrought by Hurricane Sandy in parts of the Eastern Seaboard in late October 2012 brought the issue of global warming to the forefront of Americans' attention once more. Some directly blame climate change for the hurricane's intensity, while others see its fury and its flooding as a foretaste of future disasters that will arise as the world warms. More and more people are finding that the arguments put forth by the world's best scientists are compelling. U.S. intelligence agencies, including the CIA, recently published a study warning that, when the world reaches a certain point in global warming, "states will fail, large populations subjected to famine, flood or disease will migrate across international borders, and national and international agencies will not have the resources to cope."[24] The investment guru Jim Jubak, too, has become a convert after Sandy, stating, "Personally I find the data on climate change convincing."[25] Even oil and gas executives are finally admitting that burning fossil fuels contributes to global warming. Rex Tillerson, CEO of the Exxon Mobil Corporation, now says that burning fossil fuels contributes to global warming, but he sees adaptation, such as moving cities to new locations, as a way to mitigate the damage.[26] An executive of BHP Billiton, the world's largest coal-mining company, also sees changes ahead for the fossil-fuel industry because of climate change. "In a carbon constrained world where energy coal is the biggest contributor to a carbon problem,...I suspect the usage of thermal coal is going to decline. And frankly it should," he says. As a consequence of the company's recognition of global warming, it is rebuilding one of its facilities whose vulnerability to cyclones is quite high.[27]

Shouldn't the price of adapting be factored into the end cost of fossil fuels to show their true expense? George Shultz, who served in the cabinets of both Richard Nixon and Ronald Reagan, believes so, stating,

> We have to have a system where all forms of energy bear their full costs. For some, their costs are the costs of producing the energy, but many other forms of energy produce side effects, like pollution, that are a cost of society. The producers don't bear that cost, society does. There has to be a way to level the playing field and cause those forms of energy to bear their true costs. That means putting a price on carbon.[28]

Although discussion of carbon taxes is increasing as this book goes to press, we seem a long way from demonstrating the political will to follow through. Reality will eventually intrude. According to investment expert Jubak, "the real world will deliver enough disasters, storms and data to convince everyone that global climate change is real and that the world needs to reduce its burning of all forms of hydrocarbons, even natural gas."[29] The International Energy Agency concludes that to maintain a livable environment, "no more than one-third of proven reserves of fossil fuels can be consumed prior to 2050."[30] Noncarbon fuels like solar, in combination with renewed efforts to radically increase our energy efficiency, will have to make up the difference.

In addition to limiting how much carbon we can safely emit, a warming climate will continue to wreak havoc with our current energy systems. Thermal-electric power generation from fossil fuels, nuclear energy, or even solar-concentrator plants, for example, require massive amounts of water for cooling and the production of steam. Droughts and the rising water temperatures predicted by global warming models will "make it harder to generate electricity by these means in decades to come," a 2012 scientific study suggests.[31] Perhaps Professor Robert Henry Thurston had it right when he wrote in a report for the Smithsonian in 1901:

> It may be not found that this threatened exhaustion of our fuel supplies is not the first limit likely to be set to the progress of humanity on our globe.... The mingling with the air of the products of combustion of our fuels while they still last, [and] the pollution of our waters[,]... will have their effects[;]... and our most serious problems are quite likely to be found at an earlier date than that of the loss of our fuels.[32]

Demand for More Solar around the World

Ultimately, how the world responds to climate change, and the quick deployment of renewables, will largely hinge on rapidly developing countries like India and China. Energy authority Daniel Yergin believes that countries like China and India will give a boost to renewables such as solar because of their growing power needs.[33] Stock advisers at the Motley Fool concur, writing in a report to investors: "Energy-hungry countries around the world are seeing solar as the next [energy] frontier."[34] The Chinese government plans to have 40 gigawatts (40 billion watts) of photovoltaics installed by 2015. Record levels of air pollution in major Chinese cities have, according to one official, created more pressure "to save energy and reduce emissions as smog worsens."[35] Analyst Wang Xiaoting of Bloomberg New Energy Finance says that Chinese government policies "are signaling obvious support [for solar-energy production], stimulating speculation that solar companies will benefit."[36]

Meanwhile, India is seeking to install 20 gigawatts by 2020.[37] India, like much of the world outside of the United States, relies largely on diesel for its electrical generation. According to Shubjra Mohanka, an executive at an Indian photovoltaic company, with the "rise in diesel prices [in India]...[it] makes sense for companies to switch to solar."[38]

The resurgence of photovoltaics in Japan, owing to the closing of its nuclear power plants, will most likely turn another Asian powerhouse into a world leader in solar power. The Japanese government, in the wake of Fukushima Daiichi, offers one of the most generous incentive programs in the world to those making the changeover.

Interest in both solar concentrators and photovoltaics in Saudi Arabia stems from the country's desire to maximize its national income by exporting every possible barrel of oil while relying on the power of the sun at home. To achieve this goal the Saudi government recently announced its intention to install

Figure 30.7. The Carlisle House, built by Steve Strong, combines photovoltaics with passive-solar design to approach net-zero-energy consumption.

enough solar devices to generate more than 40 gigawatts over the next two decades.[39]

In California, the state's revised Title 24 building standards for 2013 will also move solar further into the mainstream. The new code requires that by 2020 all new residential housing and by 2030 all commercial buildings produce as much energy as they consume, a designation called net-zero energy. The new building rules require photovoltaics on all rooftops.[40]

To meet the demand for net-zero energy, many architects are combining older solar technologies — solar architecture and solar water heaters — with the newest, photovoltaics. Solar pioneer Steve Strong built the first net-zero-energy house back in 1979 by using such a strategy. Back then, people called this type of structure "an energy independent house."[41]

Moving beyond the Grid

As energy produced onsite becomes more prevalent, how buildings are tied to the electrical grid becomes more of an issue. Hurricane Sandy demonstrated the necessity of wiring solar houses in a way that lets them continue to deliver electricity to their inhabitants when the power lines fail. Presently, when the power goes out, it goes out for everyone, including those homes with photovoltaics installed. While virtually everyone in the affected region went without electricity for days, if not weeks, the lights stayed on at Midtown Community School in Bayonne, New Jersey, thanks to the prescience of local officials, who eight years ago installed a hybrid solar backup system. Unlike other photovoltaic systems installed at some New Jersey school districts, which generate extra power to cushion operation costs, this system was designed to operate independently of the grid during times when it failed — the same as the Pentagon's default design for new renewables at military bases.

When Sandy came ashore and utility poles went down, the school's 232-kilowatt photovoltaic system shunted electricity generated by the solar panels away from the nonfunctional grid to a stand-by generator. Working in tandem with diesel fuel, sun-generated electricity saved enough energy to keep the generator running throughout the aftermath of the superstorm, and its smooth functioning provided a well-lit shelter for evacuees. Richard Schaeffer, who is in charge of the school's facilities management, explained how solar saved the day: "Without our solar system on the roof of the school, we would have needed even more fuel, which would have been difficult to find because it was

Figure 30.8. A passive-solar New Jersey house with photovoltaics and solar water heating rode out Hurricane Sandy with no discomfort for those living inside.

needed for all the repair trucks operating around the state."[42] Lyle Rawlings, the engineer who oversaw the installation of the Bayonne system, observed, "Sandy created a lot of implications for the future. The fragility of the electric grid was highlighted, people were knocked back to the Stone Age almost, and it taught us lessons about how deep the need for emergency power can go."[43] But even though most of the solar-power panels in the service territory of New Jersey's utility, Public Service Electric and Gas Company (PSE&G), survived the storm, only those specially wired to work without the grid were able to deliver their vital power to those in need.

A nearby net-zero-energy house that included orientation, water heating, and photovoltaics demonstrated the advantages of an autonomous house. While all other houses in the neighborhood had no power for more than a week and a half, a battery backup allowed the occupants of this house to keep their lights on. Since the house was situated and designed to also take in solar heat and effectively hold it during nighttime and inclement weather, the occupants kept warm while others froze.[44]

The Changing Role of Electrical Generation

Solar-generated electricity offers a singular advantage: it peaks on hot summer days when demand surges. Because the cost of photovoltaics has continued to drop, rooftop arrays now produce electricity as cheaply as, or more cheaply than, central power plants in many areas of the United States and the world. Using these arrays eliminates the cost of financing the construction of large power plants and transmission lines.[45] As a consequence, "residential solar has tremendous potential for growth and will become a preferred energy choice for mainstream consumers throughout the U.S.," according to Lori Neuman, spokesperson for the utility giant NRG.[46]

As solar becomes dominant, the role of utilities will inevitably change. According to Lazlo Varro, economist for the International Energy Agency, "the electricity network company could essentially be an insurance company, providing insurance against not having sunshine when you need power."[47] Or they can enter the photovoltaic rooftop business themselves, as Southern California Edison has, helping to lead the transition from centralized to distributed generation of electricity in much the same way as large telecommunication companies have evolved from being solely owners of landlines to also being purveyors of cellular services.[48] Today, most rooftop systems interact with the grid, putting electricity into utility lines during sunny periods and extracting electricity from the lines during sunless periods. But a new German program provides incentives for homeowners to combine photovoltaics with electricity storage, allowing homes to actually cut themselves from the grid. The program will reduce the twenty-year cost of a PV system with storage to 10 percent less than one without.[49] Once again, Germany leads the way in photovoltaics, this time toward autonomous living with solar electricity.

The CEO of NRG Energy, David Crane, foresees a similar situation in which the homeowner connects the home's rooftop photovoltaics with a generator, a battery or a fuel cell, and a natural-gas line. Verizon, the telecommunications giant, actually plans to combine fuel cells with solar panels.[50] In such cases, Crane believes, customers "don't need the power industry at all.... They can say to the [utility], 'Disconnect that [electrical] line.'" The utility executive projects that such a scenario "is ultimately where big parts of the country [will] go."[51]

A report from the consulting firm Navigant supports Crane's prediction. It has told clients that by 2018 there will be over 200 gigawatts of localized photovoltaic installations.[52] Even the utilities themselves recognize what's ahead. A recent report by the Edison Electric Institute, the association of investor-owned utilities, flatly recognized the threat rooftop photovoltaics presents to the current utility model, stating, "One prominent example is in the area of distributed solar PV, where the threats to the centralized utility business model have accelerated."[53] The realization of these forecasts will be a major victory for national security in light of the increased threat of terrorism. "It wouldn't take that much to take the bulk of the power system down," according to the chairman of the U.S. Federal Energy Regulatory Commission (FERC), Jon Wellinghoff. "If you took down the transformers and the substations so they're out permanently, we could be out for a long, long time.... A more distributed system [such as rooftop solar] is much more resilient. Millions of distributed generators can't be taken down at once"[54]

Tom Friedman, Pulitzer Prize–winning columnist for the *New York Times*, contends that in the poorer countries of the world, "every problem [people face] is an energy problem. The school that has no light, that's an energy problem. A clinic in a remote part of Africa that doesn't have the capacity to refrigerate medicines, that's an energy problem." Polluted drinking water, the scourge of the developing world, is also an energy problem because, without electricity, people must drink and wash from bodies of water full of parasites as well as animal and human excrement. Most people in these poorer regions have no access to utilities; only a modular power source relying on locally available fuel such as the sun can provide the necessities we in the developed world take for granted. That is why, Friedman says, as "we, the developed countries, take the lead in driving down the cost of distributed energy [that is, the cost of photovoltaics], we are solving both climate change and poverty."[55]

Figure 30.9. Children watch as a photovoltaic-powered pump draws clean drinking water into a cistern.

As energy visionary Amory Lovins wrote in the foreword to this book, challenges remain, many created by a deeply entrenched fossil-fuel industry dedicated to maintaining the energy status quo. But all indicators point toward renewable energy finally taking its place in the sun. Perhaps "Let it shine" should replace "Drill, baby, drill" in the national parlance, and we should be looking skyward for our energy solutions rather than digging up ancient carbon and dumping it in our atmosphere. "Alternative Energy Will No Longer Be Alternative," trumpeted a recent headline in *The Economist*. The article finished

Figure 30.10. A young Malian girl drinks from the tap clean water drawn by a photovoltaic-powered pump.

by saying that in this great shift "the alternatives become normal, and what was once normal becomes quaintly old-fashioned."[56] As Amory Lovins has said, "Rediscovering our solar rootstock, and grafting modern efficiency technologies onto it, offers good news for all people and for the earth."[57]

Acknowledgments

To thoroughly cover a story that spans six thousand years and five continents has required the help of many groups and people. I would like to thank them for generously providing me with the assistance that made *Let It Shine* the definitive story of solar energy, from the alpha to the omega.

I would like to thank Jan Corazza for spending many hours tightening up my manuscript before it went to the publisher. Her work brought order and discipline to my writing. Jan's husband, Dr. Luciano Corazza, lent his keen eye to the manuscript, too, eliminating redundancies and inconsistencies that had inadvertently appeared. I also have nothing but praise for help I have received from the staff of New World Library in transforming the rough draft I submitted into a final book. My editor Jason Gardner's enthusiasm for the project translated into many hours of collaborative improvements to both the writing and the choice and placement of images. Copyeditor Bonita Hurd offered many valuable insights and suggestions that greatly improved the text. Tracy Cunningham produced a very handsome cover and helped prepare the images. Tona Pearce Myers lovingly designed and typeset the book, aligning the images with the text so as to make them synergistic. Proofreader Karen Stough did the last read-through, making sure all errors disappeared. I thank them all for their interest in my work and the time and energy they spent on my manuscript.

Images are essential to this book's success. I would like to thank Roberta Bloom and Edgar Peralta of the University of California, Santa Barbara, Art Works for their tireless work in cleaning up and digitalizing all the graphics.

I am indebted to the Solar Energy Institute (IES) of Spain, one of the finest photovoltaic research centers worldwide, housed at the Polytechnic University of Madrid, for its kind hospitality in showing me the photovoltaic achievements in a country where almost one-third of the electric consumption is supplied with renewable energy.

I would like to express my appreciation to the following people who generously shared their expertise and time to help make *Let It Shine* the definitive story of solar energy: Richard Acker; Edward Arthur; William J. Bailey, Jr.; Dr. Douglas Balcomb; Larry Beil; Dr. Elliot Berman; Robert Black; Arthur Brown; Dr. Dominique Campana; Dr. David Carlson; William Crandall; Dr. Albert G. Dietz; Freeman Ford; Lisa Frantzis; Dr. J. Walter Graham; Dr. Martin Green; Professor Quinghua Guo; John Hammond; Bill Hampton; Mark Hankins; Harold Heath; Dr. Lloyd Herwig; Arnold Holderness; Dr. Hoyt Hottel; James Husbands; Dr. Borimir Jordan; Patrick Jourde; George Fred Keck; Dr. Charles Keeling; Gregory Kiss; Dr. Ronald Knapp; Panos Lamanis; Dr. George Löf; Dr. Joseph Loferski; Captain Lloyd Lomer; Michael Mack; Mike McCulley; Bernard McNelis; Dr. Ionnes Michaelides; John Oades; Dr. Michael Ornetzeder; Donald Osborn; Dr. Wolfgang Palz; Dr. Morton Prince; Markus Real; Arthur Rudin; Frank Shuman; Steven Strong; Dr. Ichimatsu Tanashita; Dr. Maria Telkes; Steve Trenchard; Walter van Rossem; Dr. Stuart Wenham; Dr. Tom Wigley; Dr. Martin Wolf; David Wright; Henry N. Wright; Dr. Christopher Wronsk; Bill Yerkes; Gonen Yissar; Dr. Hans Ziegler.

The following archives provided invaluable material: Niedersachsischen Staatsarchiv Buckeburg; Ronald Reagan Presidential Foundation and Library; Special Collections, Davidson Library, University of California, Santa Barbara; Special Collections, Shields Library, University of California, Davis; Solar Energy Collections, Architectural and Design Library, Arizona State University; Stadtarchiv Muenchen; Wisconsin Historical Society.

For a book like *Let It Shine* that covers many areas of the world and is based on primary documents, translators are required. I would like to thank Brian Jones, Wolf Kittler, Grete Kolstad, Craig Morris, Susanne Tejeda (German); Allen Huang (Chinese); and Itzhik ben Sassom (Hebrew).

I also wish to thank my employers which include the California Space Grant Consortium, whose vision is to inspire and educate the next generation of aerospace scientists, engineers, and managers; and the University of California Student Affairs team of Michael Young, Bill McTague, Bill Shelor, and Gary Jurich for their personal and professional support while I was writing *Let It Shine*. Sean Henderson also provided generous financial support.

Notes

Foreword

1. Amory B. Lovins and Jon Creyts, "Hot Air about Cheap Natural Gas," *RMI Outlet*, September 6, 2012, http://www.rmi.org/blog_hot_air_about_cheap_natural_gas.
2. For information about this publication, go to http://www.smallisprofitable.org.
3. For information about this publication, go to http://www.reinventingfire.com.
4. See http://www.reinventingfire.com.
5. National Renewable Energy Laboratory, "Renewable Energy Futures Study," 2012, http://www.nrel.gov/analysis/re_futures/.
6. Amory B. Lovins, "Asia's Accelerating Energy Revolution," *RMI Outlet*, March 26, 2013, http://blog.rmi.org/blog_2013_03_26_2013_Asias_Accelerating_Energy_Revolution.
7. Amory B. Lovins, "Germany's Renewables Revolution," *RMI Outlet*, April 17, 2013, http://blog.rmi.org/blog_2013_04_17_germanys_renewables_revolution.

Introduction

1. Charles Pope, *Solar Heat: Its Applications* (Boston: Charles H. Pope, 1903), 22–23.

Chapter 1. Chinese Solar Architecture

1. Yu Li, *Xian qing ou ji* (Beijing: Zuo jia chu ban she…, 1995). Unless otherwise noted, translations of quotations from Chinese works are by unknown translators. These translations have been reviewed by Dr. Qinghua Guo, the world's foremost authority on the history of Chinese architecture, who teaches in Australia. All translations of quotations in other languages are my own.

2. Ibid.
3. David W. Pankenier et al., "The Xiangfen Taosi Site," *Archaeologica Baltica* 10 (2009): 145–46.
4. Thomas Kuhn, *The Copernican Revolution* (Cambridge, MA: Harvard University Press, 1957), 8.
5. Edouard Biot, "Of an Ancient Chinese Work Entitled Tcheou-Pei," *Journal Asiatique* (June 1841): 612.
6. Édouard Biot, *Le Tcheou-Li; ou, Rites de Tcheou* (Paris: Impremière Nationale, 1851), 1, bk. 9, fols. 15–17, 201–203 and note 3, 201–203; 2, bk. 33, fols. 60, 279 and note 5, 279.
7. James Legge, trans., *The Chinese Classics*, vol. 4, *She King* (Taipei, Republic of China: Southern Materials Center, 1994), 81.
8. Ban Gu quoted in David Knechtges, trans., *Wen Xuan; or, Selections of Refined Literature*, vol. 1 (Princeton, NJ: Princeton Library of Asian Translations, 1982), 174.
9. James Legge, trans., *The Chinese Classics*, vol. 3, *The Shoo King* (Taipei, Republic of China: Southern Materials Center, 1985), 21.
10. Ban Gu quoted in Knechtges, *Wen Xuan*, 133.
11. Ibid., 175.
12. Arthur Waley, *Translations from the Chinese* (New York: Alfred Knopf, 1941), 186.
13. James Legge, trans., *Mencius* (New York: Dover Publications, 1970), 250n7.
14. Ibid., 250–51.
15. Ibid.
16. Ibid.
17. David Pankenier, "The Cosmo-Political Background of Heaven's Mandate," *Early China* 20 (1995): 121.
18. Qinghua Guo, *Chinese Architecture and Planning: Ideas, Methods, Techniques* (Stuttgart: Edition Axel Menges, 2005), 11.
19. "Earliest Palace City Discovered in Henan," July 25, 2004, People's Daily Online, english.peopledaily.com.cn/200407/25/eng20040725_150754.html.
20. Gabriel de Magalhaes, *A New History of China: Containing a Description of the Most Considerable Particulars of the Vast Empire* (London: printed for Thomas Newborough, 1688), 266.
21. Ibid, 268.
22. Legge, *The Chinese Classics*, vol. 4, *She King*, 440.
23. Ibid., 304.
24. Ibid., 230.
25. Bi Yuan, *Sanfu Huangto* (Shanghai: Shangwu yinshuguan, 1938), 3, 48.
26. Henri Maspero, "La vie privé en Chine à l'époque des Han," *Revue des Arts Asiatiques* 7 (1931–32): 187.
27. Qinghua Guo, "The Chinese Domestic Heating System [Kang]: Origins, Applications and Techniques," *Architectural History* 45 (2002): 32.
28. Legge, *The Chinese Classics*, vol. 3, *Shoo King*, 21.
29. *Erya*, "Shigong [Explaining dwellings]," chap. 5.

30. James Legge, trans., *The Chinese Classics*, vol. 1, *Confucian Analects, The Great Learning, The Doctrine of the Means* (Oxford: Clarendon Press, 1883), 184n1.

Chapter 2. Solar Architecture in Ancient Greece

1. Xenophon, *Memorabilia* 3.8.8f.
2. David M. Robinson and J. Walter Graham, *Excavations at Olynthus: Domestic and Public Architecture*, Testimonia Selecta 12 (Baltimore: Johns Hopkins University Press, 1946), 411.
3. Homer Thompson and R. E. Wycherley, *The Agora of Athens: The History, Shape and Uses of an Ancient City Center*, The Athenian Agora 14 (Princeton, NJ: American School of Classical Studies at Athens, 1972), 174–76.
4. J. Jones, L. Sacket, and A. Graham, "The Dema House in Attica," *Annual of the British School at Athens* 57 (1962): 103–5.
5. For the most authoritative analysis of Olynthian houses, see David M. Robinson and J. Walter Graham, *Excavations at Olynthus: 8, Domestic Architecture*, Part III, "The Houses: Plan and Rooms" (Baltimore: Johns Hopkins University Press, 1938), 141–248.
6. Aristotle's quote about "the modern fashion" is from his *Politics*, 7.11.6. Aristotle credits the famous urban planner Hippodamus with introducing the "checkerboard" street plan.
7. Theodor Wiegand and H. Schrader, *Priene* (Berlin: Georg Reimer, 1904). See especially chap. 10, "Die Privathäuser."
8. A discussion of the original excavations of Delos appeared in *École française d'Athènes: Exploration archéologique de Délos*, 21 vols. (Paris: Éditions de Boccard, 1909–59). Of these works, see especially Joseph Chamonard, *Le quartier du théâtre: Étude sur l'habitation delienne à l'époque hellenistique* (1922–24), vol. 8, pt. 1.
9. J. Perlin, *A Forest Journey: The Story of Wood and Civilization* (Woodstock, VT: Countryman Press, 2005), 85–86, 98.
10. Plato, *Critias* 3.b.d.
11. Perlin, *A Forest Journey*, 96–97.
12. Ibid., 78 and map on 79.
13. Xenophon, *Economics* 9.4
14. Edwin D. Thatcher, "The Open Rooms of the Terme del Foro," *Memoirs of the American Academy in Rome* 24 (1956): 167–264.
15. Aeschylus, *Prometheus Bound* 447f.

Chapter 3. Ancient Roman Solar Architecture

1. Homer Thompson and R. E. Wycherley, *The Agora of Athens: The History, Shape and Uses of an Ancient City Center*, The Athenian Agora, 14 (Princeton, NJ: American School of Classical Studies at Athens, 1972), 181–82.
2. Vitruvius, *De architectura* 6.1.2.
3. Ibid., 6.4.1–2.

4. Ibid.
5. *Varro on Farming, M. Terenti Varronis Rerum Rusticarum Libri Tres*, trans. with introduction, commentary, and excursus by Lloyd Storr-Best (London: G. Bell and Sons, 1912).
6. Pliny, *Epistulae*, "To Gallus," 2.17 (Laurentum) and 5.6 (Tuscany).
7. Ibid.
8. Ibid.
9. Aegidio Forcellini, *Lexicon totius latinitatis*, vol. 2 (Patavi: A. Forni, 1965), 645.
10. Robert James Forbes, *Studies in Ancient Technologies* (Leiden, Holland: Brill, 1964), 5, 185–87; D. B. Harden, "Domestic Window Glass: Roman, Saxon and Medieval," in *Studies in Building History*, ed. E. M. Jope (London: Odhams Press, 1961), 41–42; Pliny, *Natural History* 36.45.160.
11. In *Epistulae morales* (90.25), Seneca discusses the invention of windowpanes but does not specifically say that they were made from transparent stone or glass. But it can be safely conjectured that he was talking about glass. He referred to the substance from which the panes were made as *testa*, Latin for "baked material." Glass is the only known "baked material" that is transparent and used for windowpanes. It was not until three hundred years later that the Romans differentiated windows made from stone from those made with glass, when Lactantius used the term *vitrum*, specifically "glass," instead of the more general term, *specularium*, which could refer to window coverings of either glass or thin stone. For this discussion see Lactantius, *De opificio Dei* 8; and Hugo Blumner, *Technologie und Terminologie der Gewerbe und Künste bei Griechen und Römern* (Leipzig, Germany: B. G. Teubner, 1912), 66. Pliny, though, in his *Natural History*, 30.46.163, states that the "specularium" allowed sunlight to enter the building. Seneca also says that it "admits clear light."
12. Pliny, *Natural History* 28.28.64.
13. Martial, *Epigrams* 8.14. Martial, in *Epigrams* 8.68, writes that his friend Entelles shut his vineyard in transparent glass so "that jealous winter may not sear the purple clusters nor chilly frost consume the gifts of Bacchus."
14. Seneca, *Epistulae morales* 56.1–2.
15. Ibid., 86.6–11.
16. Seneca, *Epistulae morales* 86.6–11.
17. Vitruvius, *De architectura* 5.10.1 and 6.4.1.
18. August Mau provides a comparison of the old and new baths at Pompeii, in *Pompeii* (London: Macmillan, 1902), 208–21.
19. James Ring, "Windows, Baths, and Solar Energy," *American Journal of Archaeology* 100, no. 4 (1996): 722.
20. John Perlin, *A Forest Journey: The Story of Wood and Civilization* (Woodstock, VT: Countryman Press, 2005), 128. See also Quintus Aurelius Symmachus, *Q. Aurelii Symmachi Relationes, Recensuit*, ed. Wilhelm Meyer (Leipzig, Germany: B. G. Teubner, 1872), relato 44; and J. P. Waltzing, *Étude historique sur les corporations professionnelles chez les Romains* (Bologna: Froni Editore, 1968), II, 55, 125–26.
21. William H. Plommer discusses the role that Faventinus and Palladius fulfilled for their

rural aristocratic clients, in *Vitruvius and Later Roman Building Manuals* (London: Cambridge University Press, 1973), 2–3.

22. Ulpian, *Digestia* 8.2.17.

Chapter 4. Burning Mirrors

1. James Legge, trans., *The Sacred Books of China: The Texts of Confucianism* in *The Sacred Books of the East* 27 (Delhi: Motilal Banarsilass, 1986), 449.
2. Lu Diming and Zhai Keyong, "Chemical Composition and Microstructure of the Zhouyuan Solar Speculum for Ignition," *Archaeology*, no. 5 (2000).
3. Junchang Yang and Duan Yanli, "Thoughts on Yang Sui in Early Period of Ancient China," *Southeast Culture*, no. 8 (2000).
4. Lu Diming and Zhai Keyong, "Chemical Composition and Microstructure of the Zhouyuan Solar Speculum for Ignition."
5. Theophrastus, *De igne* (Assen, Holland: Van Gortum, 1971), 73.
6. Diocles, *Diocles: On Burning Mirrors; The Arabic Translation of the Lost Greek Original*, ed. with English translation and commentary by G. I. Toomer (Berlin: Springer Verlag, 1976), 12.
7. Ibid., 6.
8. Plutarch, *The Lives of the Noble Grecians and Romans*, trans. John Dryden; revised by Arthur Hugh Clough (New York: Modern Library, 1932), 82.
9. Édouard Biot, *Le Tcheou-Li; ou, Rites de Tcheou* 2 (Paris: Impremière Nationale, 1851), bk. 37, fols. 27, 381.
10. Alpheus Hyatt Verill, *America's Ancient Civilizations* (New York: Putnam, 1953), 177.
11. Theophrastus, *De igne*, 73.
12. Wang Chong, "Miscellaneous Essays," in *Lun-Hêng*, trans. from the Chinese and annotated by Alfred Forke (New York: Paragon Book Gallery, 1972), pt. 2, p. 412.
13. Ibn al-Haytham, *Discourse on the Paraboloidal Focussing Mirror*, trans. H. J. Winter and W. Arafat, *Journal of the Royal Asiatic Society of Bengal*, series 3, 15, no. 1–2 (1949): 25.
14. Roger Bacon, *Opus Majus of Roger Bacon*, trans. Robert Burke (Philadelphia: University of Pennsylvania Press, 1928), 135.
15. Ibid.
16. Roger Bacon, *Fr. Rogeri Bacon Opera*, ed. J. S. Brewer (London: Longham, Green, Longham, and Roberts, 1959), Opus Tertium, XXXVI.
17. Probably no myth has persisted as fact for so long as the story of Archimedes burning the Roman fleet with the aid of burning mirrors. For an excellent discussion of the Archimedes legend, see Hugo Blumner, *Technologie und Terminologie der Gewerbe und Künste bei Griechen und Römern* (Leipzig, Germany: B. G. Teubner, 1912), 36.
18. Giovanni Magini, *Breve Instruttione sopra l'Apparenze et Mirabili Effeti dello Specchio concavo sferico* (Bologna: Presso Clementi Ferroni, 1628), chap. 3.
19. Galileo Galilei, *Dialogue Concerning the Two New Sciences*, trans. Henry Crew and Alfonso de Salvio (Evanston, IL: Northwestern University Press, 1950), 43.

20. Giambattista della Porta, *Natural Magick* (London: for Thomas Young and Samuel Speed, 1658), 355.
21. Ibid., 362.
22. Ibid.
23. Adam Lonicer quoted in Augustin Mouchot, *La chaleur solaire*, 2nd ed. (Paris: Gauthier-Villars, 1879), 94–95.
24. Leonardo da Vinci, *Prophecies, Notebooks of Leonardo da Vinci*, 2, compiled and edited from the original manuscript by J. Richter (New York: Dover, 1970), 365.
25. *Il Codice Atlantico di Leonardo da Vinci nella Biblioteca Ambrosiana Milano* (Milan: U. Hoepli, 1894–1904), 371 v-a (blue paper) and 277 r-a; Carlo Pedretti, *The Literary Works of Leonardo da Vinci* (Berkeley: University of California Press, 1973), 120.
26. Leonardo da Vinci, *The Notebooks of Leonardo da Vinci*, vol. 2, arranged and rendered into English...by Edward McCurdy (New York: G. Braziller, 1958), 538.
27. Ibid.
28. Martin Kemp, *Leonardo da Vinci: Experience, Experiment and Design* (Princeton, NJ: Princeton University Press, 2006).
29. Athanasius Kircher, *Ars Magna Lucis et Umbrae* (Rome: Sumtibus Hermanni Scheus, 1646), pt. 3, bk. 10, corollary 2.
30. Ibid.
31. "An Account of a Not Ordinary Burning Concave, Lately Made at Lyons, and Compared with Several Others Made Formerly," *Philosophical Transactions of the Royal Society of London* 1 (1665): 95–98; and "An Account from Paris Concerning a Great Metallin Burning Concave, and Some of the Most Considerable Effects of It: Communicated by Severall Persons upon the Place, Where Tryals Have Been Made of It," *Philosophical Transactions of the Royal Society of London* 4 (1669): 986–87.
32. Peter Hoesen, *Kurze Nachricht von der Beschaffenheit und Wirkung derer parabolischen Brennspiegel* (Dresden: Bey F. Hetel, 1755): chap. 1.
33. "A Relation of the Great Effects of a New Sort of Burning Speculum Lately Made in Germany: Taken from the Acta Eruditorum of the Month of January Last; Being a Letter from the Inuentor to the Authors of That Journal," *Philosophical Transactions of the Royal Society of London* 16, no. 188 (1687): 352–53.
34. Hoesen, *Kurze Nachricht von der Beschaffenheit und Wirkung derer Parabolischen Brennspiegel* (Dresden: Bey F. Hetel, 1755), 1–15.
35. Ibid.
36. Ibid.
37. D. L. Simms and P. L. Hinkley, "Brighter than How Many Stars? Sir Isaac Newton's Burning Mirror," *Notes and Records of the Royal Society of London* 43 (January 1989): 31–50.
38. Diocles, *Diocles: On Burning Mirrors*, 36–37.
39. Aristophanes, *The Clouds* 765–70.
40. James Augustus St. John, *The History of the Manners and Customs of Ancient Greece*, vol. 3 (Port Washington, NY: Kennikat Press, 2007).
41. Pliny, *Natural History* 37.10.28.

42. Joseph Needham, *Science and Civilization in China*, vol. 4, pt. 1 (Physics) (Cambridge: Cambridge University Press, 1954), 115.
43. "Viking Lenses from Visby, Sweden," Viking Rune: All Things Scandinavian, undated, vikingrune.com/2009/01/viking-lenses-visby.
44. E. W. von Tschirnhaus[en], "The Effects of Burning Glasses of Three or Four Feet in Diameter," *Histoire de l'Académie des Sciences* (1699), in *The Philosophical History and Memoirs of the Royal Academy of Sciences at Paris Abridged*, 1699–1720, trans. and abridged by J. Martyn and E. Chambers (London: n.p., 1742), vol. 1, p. 26.
45. "Solar Cannon of the Palais Royal," *Scientific American* (December 30, 1882): 420; Frank Richard Stockton, "The Cannon of the Palais Royal," in *Round-About Rambles in Lands of Fact and Fancy* (New York: Scribner, Armstrong, 1872), 97, available at Futility Closet, www.futilitycloset.com/2006/02/26/the-cannon-of-the-palais-royal.
46. Extract of a "Letter from the Marquis Nicolini F. R. S. to the President, concerning the Same Mirror Burning at 150 Feet Distance," *Philosophical Transactions of the Royal Society* 44 (1746–47): 495–96.
47. Jacques Roger, *Buffon, un Philosophe au Jardin du Roi* (Paris: Fayard, 1989), 53.
48. Edward Gibbon, *The History of the Decline and Fall of the Roman Empire* (London: Methuen, 1898), vol. 4, p. 243.

Chapter 5. Heat for Horticulture

1. In England, for example, glass windows continued to be a luxury until the latter part of the sixteenth century. Eleanor Godfrey, *The Development of English Glassmaking, 1560–1640* (Chapel Hill: University of North Carolina Press, 1975), 207. Likewise, residents in many important cities of sixteenth-century Europe were still using linen, paper, and parchment panes. R. J. Forbes, *Studies in Ancient Technology* (Leiden, Holland: E. J. Brill, 1966), vol. 5, p. 185.
2. John C. Loudon, *Remarks on the Construction of Hot Houses* (London: for J. Taylor, 1817), 3.
3. The greatest advances of glaciers in the Northern Hemisphere since the Ice Age occurred from 1550 to 1850. H. H. Lamb, *The English Climate* (London: English Universities Press, 1964), 162–64.
4. John Laurence, *The Clergyman's Recreation* (London: for John Lintott, 1726), 72.
5. Joseph Carpenter, *The Retir'd Gardener* (London: for J. Tonson, 1717).
6. Nicolas Fatio de Duillier, *Fruit Walls Improved* (London: R. Everingham, 1699), 2 (heat-absorbing qualities of brick); 3 (shortcomings of vertical fruit walls).
7. Stephen Switzer, *The Practical Gardener* (London: for T. Woodward, 1724), 295; and Stephen Switzer, *Icongraphia Rustica* (London: D. Browne, 1718), 240.
8. Switzer, *Icongraphia Rustica*, 240.
9. Fatio de Duillier, *Fruit Walls Improved*, 4–6.
10. Hugh Plat, *The Garden of Eden* (London: for William Leake, 1659–60), 41.
11. Cold frames continued to be used for centuries in Great Britain. They were definitely common in England during Victorian times, since they are mentioned in Lewis

Carroll's *Alice's Adventures in Wonderland*, where, in chapter 4, Alice puts her arm through the window of the house she finds herself crammed into: "she heard a... crash of broken glass, from which she concluded that it was just possible [the rabbit] had fallen into a cucumber frame." In chapter 4, note 2, in the World Library edition of *Alice*, Lynne Vallone comments on the incident, stating that the "*Cucumber Frame* [is a] small greenhouselike structure used to maximize solar energy." Vallone, Explanatory notes for *Alice's Adventures in Wonderland and Through the Looking-Glass and What Alice Found There*, by Lewis Carroll (New York: Modern Library Classics, Random House, 2002), 245–62.

12. Francis Rogers and Alice Beard, *5,000 Years of Glass* (New York: Fredrick A. Stokes, 1937), 140–50; R. W. Douglas and Susan Frank, *A History of Glass Making* (Henley-on-Thames, Oxfordshire: Foulis, 1972), 137–48.
13. James Anderson, *A Description of a Patent Hot-House* (London: for J. Cumming, 1803), vi–vii.
14. Hermann Boerhaave, *Element Chemiae* (London: Sumtibus S. K. et J. K., 1732), vol. 1, p. 213.
15. Thomas Andrew Knight, "Description of a Forcing House of Grapes," *Transactions of the Horticultural Society of London* 1 (1821): 99.
16. Michel Adanson, *Familles des plantes* (Paris: Vincent, 1763), vol. 1, p. 132.
17. Anderson, *A Description of a Patent Hot-House*, 23–42.
18. Robert Kerr, *The Gentleman's House* (London: J. Murray, 1864), 353.
19. John C. Loudon, *Encyclopedia of Cottage, Farm, and Villa Architecture* (London: F. Warne and Company, 1846), 974.
20. *The English Gardener* quote is found in John Hix, *The Glass House* (Cambridge, MA: MIT Press, 1974), 89.
21. Ibid., 87.
22. W. Bridges Adams quoted in *The Gardener's Chronicle* (January 14, 1860): 29.

Chapter 6. Solar Hot Boxes

1. Horace de Saussure, letter to the editor, *Journal de Paris*, supplement, 108 (April 17, 1784): 475–78.
2. Saussure's entire account can be found in ibid.
3. William Adams, *Solar Heat: A Substitute for Fuel in Tropical Countries* (Bombay: Bombay Education Society's Press, 1878), 30.
4. Horace de Saussure, *Voyages dans les Alpes*, vol. 2 (Geneva: Barde, Manget et Compagnie, 1786), chap. 35, para. 933.
5. Ibid.
6. Joseph Fourier, *Mémoire sur les températures du globe terrestre et des espaces planétaires*, trans. W. M. Connolley (1827), 584–85, available on the translator's website, www.wmconnolley.org.uk/sci/fourier_1827/fourier_1827.html#cl2.
7. John F. W. Herschel, *Results of Astronomical Observations Made during the Years 1834... at the Cape of Good Hope* (London: Elder and Company, 1847), 443–44.
8. Ibid.

9. Adams, *Solar Heat*, 33–34.
10. Samuel Pierpont Langley, "The Mt. Whitney Expedition," *Nature* (August 3, 1882): 315.
11. Adams, *Solar Heat*, 30.
12. Horace de Saussure, "Letter to the Editors," *Le Journal de Paris*, Supplement #108 (April 17, 1784) 475–478.
13. Fourier, *Mémoire sur les températures*, 586–87.

Chapter 7. The First Solar Motors

1. Augustin Mouchot, *La chaleur solaire et ses applications industrielles*, 2nd ed. (Paris: Gauthier-Villars, 1879), 256.
2. Ibid., 154.
3. Mouchot, *La chaleur solaire*, 256; Hero of Alexandria, *The Pneumatics of Hero of Alexandria*, introduction by Marie B. Hull (London: MacDonald, 1971), 33.
4. Mouchot, *La chaleur solaire*, 170–78.
5. Isaac de Caus, *New and Rare Inventions of Water Works* (London: Joseph Moxon, 1659), 33.
6. Ibid.
7. Mouchot, *La chaleur solaire*, 261.
8. Ibid., 267; Augustin Mouchot, "Sur les effets mécaniques de l'air Confiné échauffé par les rayons du soleil," *Comptes Rendus de l'Académie des Sciences* 59 (1864): 527.
9. Mouchot, *La chaleur solaire*, 123.
10. Mouchot, *La chaleur solaire*, 142–43.
11. Ibid.
12. Ibid., chap. 6.
13. Ibid., 224.
14. Ibid., 225.
15. Ibid.
16. Leon Simonin, "L'emploi industriel de la chaleur solaire," *Revue des Deux Mondes* (May 1, 1876).
17. Ibid.; Augustin Mouchot, "Résultats obtenus dans les essais d'applications industrielles de la chaleur solaire," *Comptes Rendus de l'Académie des Sciences* 81 (1875): 571–74; and Mouchot, *La chaleur solaire*, 222–30.
18. Mouchot, *La chaleur solaire*, 257.
19. Ibid., 263.
20. Ibid., 263–64.
21. Ibid., 235–36.
22. Examples of the many solar devices Mouchot tested in Algeria are found in Augustin Mouchot, "Résultats d'expériences faites en divers points de l'Algérie pour l'emploi industriel de la chaleur solaire," *Comptes Rendus de l'Académie des Sciences* 86 (1878): 1019–21.
23. Mouchot, *La chaleur solaire*, 243–46.
24. Ibid.
25. Ibid., 263. Apparently Mouchot was the first person to use solar energy to provide the

heat needed in Carré's apparatus, which relied on the absorption of heat as the cooling mechanism.

26. Ibid., 248–49. The problem with thermocouples is their low efficiency in transforming heat into electricity. Scientists currently are working on ways to split water with solar energy to generate hydrogen.
27. Ibid.
28. Abel Pifre, "Nouveaux résultats d'utilization de la chaleur solaire obtenus à Paris," *Comptes Rendus de l'Académie des Sciences* 9 (1880): 388–89.
29. A. Crova, "Étude des appareils solaires," *Comptes Rendus de l'Académie des Sciences* 94 (1882): 943–45.

Chapter 8. Two American Pioneers

1. William C. Church, *The Life of John Ericsson* (New York: Scribners, 1890), 2, 266. This book is the most complete account of Ericsson's life and accomplishments, including his work in the field of solar energy.
2. William Dennis, "Salt — Its Uses and Manufacture — Salt Meats," *De Bow's Review*, 23 (1857): 133–63.
3. *Syracuse Courier*, August 17, 1861.
4. John Perlin, *A Forest Journey: The Story of Wood and Civilization* (Woodstock, VT: Countryman Press, 2005), 173.
5. J. Leander Bishop, *A History of American Manufacturers: 1608 to 1860* (Philadelphia: Edward Young and Co., 1866), 1, 289.
6. Ibid., 293; Onondaga Historical Association, *Annual Volume of the Onondaga Historical Association*, ed. W. M. Beauchamp (Syracuse NY: Dehler Press, 1914), para. 158, available at www.archive.org/stream/annualvolumeofon1914onon/annualvolume ofon1914onon_djvu.txt.
7. "The Salt Business," *Syracuse Daily Standard*, September 13, 1858.
8. Ibid.; and Charles Sprague Sargent, *Report on the Forests of North America* (Washington, DC: Government Printing Office, 1884), 502.
9. "The Salt Manufacture in '48 and '49," *Onondaga Standard*, May 2, 1849.
10. Ebenezer Merriam, "Report of the Superintendent of the Onondaga Salt Springs for 1849," *[New York State] Assembly Report* no. 36 (January 12, 1850): 38.
11. "The Salt Business."
12. "Annual Report of the Superintendent of the Onondaga Salt Springs for 1863," *[New York State] Assembly Report* no. 70 (February 12, 1862): 8.
13. Charles Jolly Werner, *A History and Description of the Manufacture and Mining of Salt in New York State* (Huntington, NY: self-published, 1917), 24–25.
14. "Annual Report of the Superintendent of the Onondaga Salt Springs for 1864," *[New York State] Assembly Report* no. 35 (January 20, 1865): 5–6.
15. Mohinder Singh, "The Story of Salt," *Gandhi Marg* 24, no. 3 (October–December 2002), available at Gandhian Institute Bombay, Sarvodaya Mandal and Gandhi Research Foundation, Mahatma Gandhi's Writings, Philosophy, Audio, Video and Photographs, www.mkgandhi.org/articles/the%20story%20of%20Salt.htm.

16. Church, *The Life of John Ericsson*, 265–66.
17. Ibid.
18. John Ericsson, *Contributions to the Centennial Exhibition* (New York: "The Nation" Press, 1876), 564.
19. Ibid., 559.
20. Ibid., 571–74.
21. Church, *The Life of John Ericsson*, 268.
22. Ibid., 271; Ericsson, *Contributions to the Centennial Exhibition*, 563.
23. John Ericsson, "The Sun Motor," *Nature* 29 (January 3, 1884): 217–18; and 38 (August 2, 1888): 319–21.
24. Ericsson, "The Sun Motor" (August 2, 1888): 321.
25. Church, *The Life of John Ericsson*, 276.
26. "Power from Sunlight," *Harper's Weekly* 47 (February 4, 1903): 256.
27. Charles H. Pope, *Solar Heat: Its Applications* (Boston: Charles H. Pope, 1903), 13.
28. Robert H. Thurston, "Utilizing the Sun's Energy," *Smithsonian Institution Annual Report* (1901): 265.
29. Pope, *Solar Heat*, 33.
30. Correspondence between Aubrey Eneas and Samuel Pierpont Langley, Secretary of the Smithsonian Institution, Incoming Correspondence, Box 23, Record Unit 31 (Office of the Secretary: 1891–1906), Smithsonian Institution Archives.
31. Ibid.
32. U.S. Patent 670,916, March 26, 1901.
33. Anna Laura Meyers, interview by author.
34. Arthur Inkersley, "Sunshine as Power," *Sunset* magazine 10 (April 1903): 515; Frank Millard, "Harnessing the Sun," *World's Work* 1 (April 1901): 600.
35. Millard, "Harnessing the Sun," 600.
36. Alfred Davenport, "The Solar Motor at Pasadena, California," *Engineering News* 45 (May 9, 1901): 330; C. Holder, "Solar Motors," *Scientific American* 84 (March 16, 1901): 169.
37. Pete Estrada, interview by author.
38. Ruth Mellenbruch, "Solar Energy Experiment Was Conducted Here in 1906–1907," *Arizona Range News* (September 12, 1941). Mellenbruch's article provides the background to Eneas's solar work in Arizona.
39. Ibid.
40. Aubrey Eneas to Samuel Pierpont Langley, letter, February 10, 1896, Smithsonian Archives.

Chapter 9. Low-Temperature Solar Engines

1. "The Utilization of Solar Heat for the Elevation of Water," *Scientific American* 53, (October 3, 1885): 214
2. The primary information on Tellier's solar work comes from "The Utilization of Solar Heat for the Elevation of Water," *Scientific American* 53 (October 3, 1885): 214; and *Kaiserliches Patentamt*, Patentschrift #34749, Klasse 46, Tellier (August 15, 1885): chap. 15.

3. A. S. E. Ackermann, "The Utilization of Solar Energy," *Smithsonian Institution Annual Report* (1915): 158.
4. The primary sources for information about Willsie and Boyle's work are Henry Elmer Willsie, "Experiments in the Development of Power from the Sun's Heat," *Engineering News* 61, no. 19 (May 1909): 512–14; U.S. Patent 1,101,001 (June 23, 1914); and U.S. Patent 1,130,871 (March 9, 1915).
5. Ibid.
6. Ibid.
7. Ibid.
8. Ibid.
9. Ibid.
10. Ibid.

Chapter 10. The First Practical Solar Engine

1. Frank Shuman Jr., son of Frank Shuman, interview by author.
2. Frank Shuman, "Feasibility of Utilizing Power from the Sun," *Scientific American* 110 (February 28, 1914): 179.
3. "Power from the Sun's Heat," *Engineering News* 61, no. 19 (May 13, 1909): 509.
4. Frank Shuman, "The Generation of Mechanical Power by the Absorption of the Sun's Rays," *Mechanical Engineering* 31 (December 19, 1911): 1.
5. Frank Shuman, "Sun Power Plants Not Visionary," *Scientific American* 110 (June 27, 1914): 60.
6. Frank Shuman, "Solar Power," *Scientific American* 71 (February 4, 1911): 78.
7. "Power from the Sun's Heat," 509.
8. Frank Shuman to Judge Thomas South, August 13, 1907, Philadelphia, letter, Historical Society of Tacony, Philadelphia, Pennsylvania. Courtesy Louis M. Iaerola.
9. Shuman, "The Generation of Mechanical Power by the Absorption of the Sun's Rays," 1.
10. "Power from the Sun's Heat," 509.
11. Shuman, "The Generation of Mechanical Power by the Absorption of the Sun's Rays," 2.
12. Ibid., 6.
13. A .S. E. Ackermann, "Shuman Sun-Heat Absorber," *Nature* 89 (April 4, 1912): 122–23; Frank Shuman, "Power from Sunshine, a Pioneer Solar Power Plant," *Scientific American* 105 (September 30, 1911): 291–92.
14. R. C. Carpenter, "Tests of a Simple Engine," *Scientific American*, supplement, 74 (July 13, 1912): 28–29.
15. Shuman, interview.
16. "Sun Power Plant: A Comparative Estimate of the Cost of Power from Coal and Solar Radiation," *Scientific American*, supplement, 77 (January 17, 1914): 37.
17. Shuman, "Feasibility of Utilizing Power from the Sun," 179.
18. Ibid.
19. Shuman, interview.
20. Frank Shuman, "The Most Rational Source of Power: Tapping the Sun's Radiant Energy Directly," *Scientific American* 109 (November 1, 1913): 350.

Chapter 11. The First Commercial Solar Water Heaters

1. The homesteader is the author's mother, who grew up on a ranch in the Soledad Canyon, California.
2. Theodore Hotchkiss, interview by author, Monrovia, California.
3. Anna Laura Myers, interview by author.
4. William Ingalls, retired Santa Barbara plumber, interviewed by author.
5. Federal Reserve Bank of Minneapolis, Consumer Price Index (Estimate) 1800–, www.minneapolisfed.org/community_education/teacher/calc/hist1800.cfm; U.S. Energy Information Administration, Natural Gas Monthly, Data for December 2012, Release Date: February 28, 2013, www.eia.gov/naturalgas/.
6. James Harrison, interview by author. Performance data can be obtained from Frederick Augustus Brooks, *Use of Solar Energy for Heating Water in California*, Bulletin no. 602 (Berkeley: University of California Agricultural Experiment Station, 1936).
7. Samuel Pierpont Langley, "The Mt. Whitney Expedition," *Nature* (August 3, 1882): 315.
8. The Kemp Climax brochure can be accessed at the Lawrence B. Romaine Trade Catalog Collection (Mss. 107), "Heating and Ventilating Systems," Box 17: "Kemp, Clarence," Davidson Library, Special Collections, University of California, Santa Barbara.
9. Lee McCrae, "Utilizing Sun Rays in a Home," *Countryside* magazine 23 (October 1916): 203.
10. "1600...," *Los Angeles Times*, June 3, 1900.
11. "The Solar Water Heater," *Pasadena Daily Evening Star*, August 17, 1896.
12. Walter van Rossem, interview by author. All van Rossem quotes in the following paragraphs are from this source.
13. U.S. Patent 705,167, April 19, 1898.
14. Jim Bailey, installer, interview by author.
15. J. J. Backus, "Praise for Solar Heater," *Architect and Engineer of California* 8 (March 1907): 98.

Chapter 12. Hot Water, Day and Night

1. Edward "Ned" Arthur, one of the company's five original employees, interview by author.
2. William Crandall, Day and Night installer, interview by author.
3. "Solar Heater — A Monrovian Perfects a New and Improved Heater," *Monrovia Daily News*, July 3, 1909.
4. Arthur, interview.
5. Day and Night advertisement.
6. Paul Squibb, Santa Ynez rancher, interview by author.
7. William J. Bailey Jr., son of the inventor, interview by author.
8. U.S. Patent 1,242,511, October 9, 1917; Frederick Augustus Brooks, *Use of Solar Energy for Heating Water in California*, Bulletin no. 602 (Berkeley: University of California Agricultural Experiment Station, 1936), 52–54.
9. "Let the Sun Do the Work," *Arizona* magazine (October 1913): 9.

10. Arthur, interview.
11. Ibid.
12. "Old Sol Harnessed for Domestic Service," *Metal Worker, Plumber and Steam Fitter* 81 (May 29, 1914): 711–12.
13. Arthur, interview.
14. Corporate Records at BDP [Bryant, Day and Night and Payne] Corporation.
15. Robert Carroll, Santa Barbara plumber, interview by author.
16. Arthur, interview.
17. Bailey Jr., interview.

Chapter 13. A Flourishing Solar Industry

1. William J. Bailey Jr., interview by author.
2. Harold Heath, interview by author.
3. Ibid.
4. L. J. Ursem Company advertisement, *Miami Herald*, April 4, 1926.
5. William D. Munroe, interview by author.
6. Ibid.
7. "Firms Keep Pace with the Rapid Growth of Miami," *Miami Herald*, April 9, 1925.
8. Heath, interview.
9. Ibid.
10. Ibid.; Walter Morrow, later head of the Solar Water Heater Company, interview by author. For printed information, see Harold Hawkins, *Domestic Solar Water Heating in Florida*, Bulletin no. 18 (Gainesville: Florida Engineering and Industrial Experimental Station, University of Florida, September 1947).
11. Heath, interview.
12. Ibid.
13. "Solar Water Heater Demand Is Increased," *Miami Herald*, August 5, 1935.
14. Heath, interview.
15. Jerome E. Scott, *Solar Water Heater Industry in South Florida, 1923–1974* (Washington, DC: National Science Foundation, 1975), 49.
16. "Let Sol Do the Work," *Domestic Engineering* 157 (May 1941): 48–49, 124–27; Federal Housing Authority, *Public Housing Design: A Review of Experience in Low-Rent Housing* (Washington, DC: Government Printing Office, June 1946), 234–35.
17. Heath, interview.
18. Harold Alt, "Sun Effect and the Design of Solar Water Heaters, American Society of Heating and Ventilating Engineers (ASHVE)," *Transactions* 41 (1935): 131–48.

Chapter 14. Solar Water Heating Worldwide, Part 1

1. *Jerusalem Post*, January 18, 1951.
2. Levi Yissar, "The Electricity Crisis and an Alternative Solution," *Haaretz*, June 20, 1951 (in Hebrew).

3. Gonen Yissar, Levi Yissar's son, interview by author.
4. Ibid.
5. "Do You Have Hot Water?", *Hadavar Hashuvah*, February 7, 1952 (in Hebrew).
6. Ibid.
7. Ibid.
8. "The Commercialization of Solar Water Heating Begins," *Maariv*, August 19, 1953 (in Hebrew).
9. "Mrs. Ben-Gurion Becomes Acquainted with Her Solar Water Heater," *Maariv*, December 14, 1953 (in Hebrew).
10. Gonen Yissar, interview.
11. Harry Tabor, scientific pioneer of solar applications in Israel, interview by author. For further information on Miromit and the improvements it made, see John I. Yellot and Rainier Sobotka, "An Investigation of Solar Water Heater Performance" (paper presented to the American Society of Heating, Refrigeration and Air Conditioning Engineers (ASHRAE), 71st Annual Meeting, June 29–July 1, 1964).
12. Rainier Sobotka, "Solar Water Heaters from Katmandu to Bamako," *Sun at Work* 10 (second quarter, 1966): 10.
13. Keith Jenkins, Solahart Company, interview by author.
14. Roger N. Morse, "Solar Energy Research: Some Australian Investigations," *C.S.I.R.O.* (October 1959): 26–27.
15. Roger N. Morse, "Solar Energy in Australia," *Ambio: Royal Swedish Academy of Sciences* (1977): 213.
16. R. G. Collander quoted in Austin Wheeler, "The Prospects for Engineering Utilization of Solar Energy," *South African Mechanical Engineer* 5, (October 1956): 88.
17. The story of the solar water-heater industry in South Africa comes from Lewis Rome, interviews by author.
18. Ichimatsu Tanashita, "Recent Developments of Solar Water Heaters in Japan" (paper presented to the United Nations Conference on New Sources of Energy: Solar Energy, Wind Power and Geothermal Energy, May 1, 1961), 1.
19. Ichimatsu Tanashita, "Present Status of Solar Water Heaters in Japan," *Transactions on the Use of Solar Energy: The Scientific Basis* 3, no. 2 (October 31–November 1, 1955): 67–68.
20. Ichimatsu Tanashita, correspondence with author.
21. Tanashita, "Recent Developments of Solar Water Heaters in Japan," 104–5; Ichimatsu Tanashita, "Present Situation of Commercial Solar Water Heaters in Japan" (paper presented to the International Solar Energy Conference, 1970, Australia).
22. Tanashita, "Recent Development of Solar Water Heaters in Japan," 105–6; Tanashita, "Present Situation of Commercial Solar Water Heaters in Japan," 2–4.
23. G. Löf and Dr. Susamu Karaki, "U.S. Solar-Thermal Systems," in *Japanese/United States Symposium on Solar Energy Systems, Washington, D.C., June 3–5, 1974*, vol. 2, *Summaries of Technical Presentations*, ed. Mitre Corporation, National Science Foundation, Research Applied to National Needs Program (McLean, VA: Mitre Corporation, June 3–5, 1974), 4-1.

Chapter 15. Saving Airmen with the Sun

1. Countess Stella Andrassy, "Solar Stills Transform Sea Water into Fresh," *New York Herald Tribune*, April 8, 1955.
2. Maria Telkes, "Solar Distillers Tested under Operating Conditions off Miami," *Section 11.2 of the National Defense Research Committee of the Office of the Scientific Research and Development*, Cambridge, MA: Massachusetts Institute of Technology, February 1–7, 1944, 3, in Maria Telkes Papers, Accession #2007-04062, Box 20, Folder 11, Arizona State University Libraries: Architecture and Environmental Design Library Archives and Special Collections, Phoenix, Arizona.
3. Science News Letter, "Still converts salt water into safe drinking water," February 10, 1945, in Maria Telkes Papers, Accession #2007-04062, Box 18, Folder 7, Arizona State University Libraries: Architecture and Environmental Design Library Archives and Special Collections, Phoenix, Arizona.
4. Maria Telkes, "Solar Distillers Tested under Operating Conditions off Miami," 3.
5. Ibid., 5.
6. "How Downed Navy Fliers Are Being Rescued from Pacific Waters," *Herald Tribune*, December 10, 1944.
7. "Rickenbacker Urges Air-Sea Rescue Expansion, *Air Sea Rescue Bulletin* 11, no. 4 (December 4, 1945): 2.
8. "*Life* Tries Out a New Toy: Chicago Artists Play with Giant Plastic Balls Reconverted from War Use," *Life* (February 4, 1946): 100, in Maria Telkes Papers, Accession #2007-04062, Box 20, Folder 13, Arizona State University Libraries: Architecture and Environmental Design Library Archives and Special Collections, Phoenix, Arizona.
9. Ibid.

Chapter 16. Solar Building during the Enlightenment

1. Theodor Wiegand and Hans Schrader, *Priene* (Berlin: G. Reimer, 1904), 290.
2. Thomas Tusser, *FIUE HUNDR* (London: published for the *GOOD* English Dialect Society by Trubner and Co., 1878), 141, www.archive.org/stream/fivehundredpointo8tussuoft/fivehundredpointo8tussuoft_djvu.txt.
3. Ernst Gladbach, *Die Holz-Architektur der Schweiz*, 2nd ed. (Zurich: Orell Füssli, 1885), 8.
4. Humphry Repton, *Observations on the Theory and Practice of Landscape Gardening* (London: T. Bensley, 1803), 195, 196n1.
5. "What Success May Germany, and Bavaria in Particular, Hope for from Previous Efforts in Land Improvement," *Monatsblatt für Bauwesen und Landesverschönerung in Bayern* (from this point on, abbreviated as *Monatsblatt*) 10, no. 1, article 2 (January 1830): 7.
6. Bernhard Christoph Faust, *Zur Sonne nach Mittag sollten alle Häuser der Menschen gerichtet seyn* [All people should face their houses to the midday sun] (n.p.: Bruchstücke als Handschrift gedruckt, 1824), introduction.
7. Ibid.

8. Ibid., 1.
9. Ibid., 2–3.
10. "Dr. Faust's Desired Ideals Regarding the Rebuilding of the City of Hof after the Fire," *Monatsblatt* 3, no. 10, article 53 (October 1823): 62, 63.
11. "On Positioning Domestic Buildings According to the Sun," *Monatsblatt* 2, no. 10, article 54 (October 1822): 58–59, and "The House and the Lawn in the Front," *Monatsblatt* 7, no. 2, article 4 (February 1827): 9n.
12. Faust, *Zur Sonne nach Mittag*, sec. 46.
13. Ibid., sec. 36.
14. Ibid., tables, secs. 21–22.
15. Ibid., sec. 66a; and "Dr. Faust's Response to Some Objections against Building Houses and Cities toward the Sun," *Monatsblatt* 9, no. 6 (June 1829): 29–32.
16. Ibid., sec. 61.
17. Ibid., sec. 71.
18. "Laws and Duties of the Brotherhood of Masons, by Prince Edwin, 926," *Monatsblatt* 4, no. 12, article 41 (December 1824): 66–67, note d.
19. Faust, *Zur Sonne nach Mittag*, sec. 76.
20. Ibid., sec. 72.
21. Ibid., sec. 68.
22. Ibid., sec. 72.
23. Ibid., and "The House and the Lawn Areas in Front," 9–11.
24. Faust, *Zur Sonne nach Mittag*, sec. 72; and "Building with Adobe," *Monatsblatt* 10, no. 12, article 42 (December 1830): 67.
25. Aeschylus quoted in "Dr. Faust's Desired Ideals Regarding the Rebuilding of the City of Hof after the Fire," 62–63.
26. "What Success May Germany, and Bavaria in Particular, Hope for from Previous Efforts in Land Improvement," 7.
27. Ibid.
28. Faust, *Zur Sonne nach Mittag*, introduction.
29. "Laws and Duties of the Brotherhood of Masons"; and "Report of the Royal School of Building Arts for Winter 1825/26," *Monatsblatt* 6, no. 3, article 14 (March 1826): 14.
30. "News from England Pertaining to Civil Engineering and National Beautification," *Monatsblatt* 4, no. 9, article 30 (September 1825): 47–48.
31. Anton Ritter Camerloher, "Useful Undertakings and Suggestions: Response to the Query on Solar Architecture, *Allgemeiner Anzeiger*, no. 131 (May 15, 1829): 1546.
32. Königlich Bayerisches, "Was Ist unter Sonnenbau zu verstehen?" *Intelligenzblatt für den Isarkreis*, XXII Stück, München (May 6, 1836): 54; "Report of the Royal School of Building Arts for the Winter 1825/1826," 14.
33. "Renewed Call for Solar Building," *Monatsblatt* 6, no. 7, article 30 (July 1826): 34.
34. Ibid.
35. Hans Plessner, "Bavarian School House Construction One Hundred Years Ago," *Der Baumeister* (August 1932): 44.

36. Ibid., 43–44.
37. Ibid.
38. Faust, *Zur Sonne nach Mittag*, sec. 4, note; "Baupolizei," *Monatsblatt* 5, no. 7, article 23 (July 1825): 35, note.
39. "What May Be Understood as Solar Architecture?," *Allgemeine Bauzeitung mit Abbildungen* 1 (1836): 26–34.
40. "Bernhard Christoph Faust, http://de.wikipedia.org/wiki/Bernhard_Christoph_Faust.
41. "Some Questions and Answers Regarding the Solar Building and Theory," *Monatsblatt* 10, no. 2, article 5 (February 1830): 9–11.
42. Marc Emery, "Faust et Le Corbusier," *Nouvelle Revue Neuchâteloise* 1, no. 3 (Autumn 1984): 3–48.
43. World Heritage Site, "Ideal City," n.d., www.worldheritagesite.org/tag.php?id=233.
44. Faust, *Zur Sonne nach Mittag*, sec. 73.

Chapter 17. Solar Architecture in Europe after Faust and Vorherr

1. "News from England Pertaining to Civil Engineering and National Beautification," *Monatsblatt* 4, no. 9, article 30 (September 1825): 47–48; "Excerpt from Minutes of the Engineering Deputation in Munich," *Monatsblatt* 5, no. 12, article 39 (December 25, 1825): 61–62.
2. Friedrich Engels, *The Condition of the Working Class in England*, in vol. 4 of *Collected Works of Karl Marx and Friedrich Engels* (New York: International Publishers, 1975), 355; and Friedrich Engels, "The Condition of the Working Class in England," in *The Sociology and Politics of Health: A Reader*, ed. Michael Purdy and David Banks (London: Routledge, 2001), 10.
3. *The Artisan* (October 1843), quoted in Engels, *The Condition of the Working Class in England*, 339.
4. I. Bird, "Notes on Old Edinburgh," in "Letter no. 40" (1879), in *The Works of John Ruskin*, ed. Edward Tyas Cook and Alexander Wedderburn (London: George Allen, 1907), vol. 28, p. 73.
5. John Ruskin, *Fors Clavigera* (1874), vol. 4 of *The Works of John Ruskin*, ed. Edward Tyas Cook and Alexander Wedderburn (London: George Allen, 1907), vol. 28, p. 73.
6. Walter L. George, *Labour and Housing in Port Sunlight* (New York: G. P. Putnam's Sons, 1916), 23.
7. Raymond Unwin, "Cottage Plans and Common Sense," *Fabian Tracts*, no. 109 (1908): 3.
8. Ibid., 5.
9. Sigfried Giedion quoted in Jose Luis Sert, *Can Our Cities Survive? An ABC of Urban Problems, Their Analysis, Their Solutions* (Cambridge, MA: Harvard University Press; London: H. Milford, Oxford University Press, 1942), lx.
10. Ibid., 242.
11. Ibid., 26.
12. H. Plessner, "History of Solar Architecture," *Architect*, supplement to *Der Baumeister* 9 (September 1931).

13. Ferdinand Kramer, "The Bauhaus and New Architecture," in *Bauhaus and Bauhaus People: Personal Opinions and Recollections of Former Bauhaus Members and Their Contemporaries*, ed. Eckhard Neumann, trans. Eva Richter and Alba Lorman (New York: Van Nostrand Reinhold, 1970), 79.
14. Marcel Breuer, interview by author.
15. Wilhelm von Moltke, interview by author.
16. Walter Gropius, *Scope of Total Architecture* (New York: Collier Books, 1962), 110.
17. J. Mueller, "City Planning in Frankfurt, Germany," *Journal of Urban Planning* 4 (1977): 17–18.
18. Lewis Mumford, "Machines for Living," *Fortune* 7 (February 1933): 87.
19. Adolf Behne, "Dammerstock," *Die Form* 6 (1930): 163–64.
20. Breuer, interview.
21. Behne, "Dammerstock," 163–66.
22. Teo Otto quoted in Eckhard Neumann, ed., *Bauhaus and Bauhaus People: Personal Opinions and Recollections of Former Bauhaus Members and Their Contemporaries*, trans. Eva Richter and Alba Lorman (New York: Van Nostrand Reinhold, 1970), 290–91.
23. Hannes Meyer, "Bauen," *Bauhaus* 2, no. 1 (1928): 13.
24. Hannes Meyer, "Notes for Lectures in Vienna and Basel" (March 22, 1929), in *Bauen und Gesellschaft: Schriften, Briefe, Projekte*, ed. Lena Meyer-Bergner (Dresden: Verlag der Kunst, 1980), 54.
25. Ibid.
26. Hannes Meyer, "Erläuterungen zum Schulprojekt," *Bauhaus* 2, nos. 2–3 (1928): 15.
27. Hannes Meyer, "Laubenganghäuser der Siedlung Dessau-Torten, 1929–1930," in *Hannes Meyer, Architekt, Urbanist, Lehrer* (Berlin: Ernst, 1989).
28. Philipp Tolziner quoted in Frank Whitford, *Bauhaus* (London: Thames and Hudson, 1984), 262.
29. Ibid., 263.
30. Hugo Häring, "Bemerkingen zum Flachbau," *Moderne Bauforme* 33, pt. 2 (November 1934): 619–27.
31. "Insolation and Natural and Artificial Lighting in Relation to Housing and Town Planning," *Bulletin of the Health Organization of the League of Nations* 7 (June 1938): 590.
32. Sert, *Can Our Cities Survive?*, 6.
33. Alfred Roth, *Die neue Architektur* (Zurich: Les Éditions d'Architecture, 1948), 71–90.

Chapter 18. Solar Heating in Early America

1. Michael Zeilik, interview by author.
2. Mrs. William T. Sedgwick, *Acoma, the Sky City* (Cambridge, MA: Harvard University Press, 1926).
3. Ralph Knowles, *Energy and Form* (Cambridge, MA: MIT Press, 1974), 28–33.
4. Dennis R. Hannaford and Revel Edwards, *Spanish Colonial or Adobe Architecture of California, 1800–1850* (New York: Architectural Book Publishing, 1931), 1.

5. "Two Craftsman Houses: A Plaster Dwelling That Is Suitable for Either Town or Country Dwelling," *The Craftsman* 15 (January 1909): 472–80.
6. Bruce Price, *A Large Country House* (New York: William T. Comstock, 1887).
7. William Atkinson, *The Orientation of Buildings; or, Planning for Sunlight* (New York: John Wiley and Sons, 1912), 62–78.

Chapter 19. An American Revival

1. *The Orientation of Buildings, Being the Report, with Appendices of the R.I.B.A. Joint Committee on the Orientation of Buildings* (London: Royal Institute of British Architects, 1933), 3.
2. Ibid., 4.
3. Ibid.
4. Ibid., 42.
5. W. W. Shaver, "Heat Absorbing Glass Windows," *Transactions of the American Society of Heating and Ventilating Engineers* 41, no. 1018 (1955): 297.
6. George Nelson and Henry Wright, *Tomorrow's House: A Complete Guide for the Home-Builder* (New York: Simon and Schuster, 1945), 178.
7. George Fred Keck to Earl Aiken, letter, May 27, 1938, Box 5, LOF main folder 5291, M73-026, George Fred Keck and William Keck Papers, 1905–95, Wisconsin Historical Society Archives, Madison.
8. Ibid.
9. Ibid.
10. George Fred Keck, interviews by author.
11. H. Sloan, "We Call Ours a Solar Home," *Pencil Points* (February 1944): 82.
12. "Solar House, Glenview, Ill.," *Architectural Forum* 79, pt. 1 (September 1943): 5–6; and "Two Developments in Glenview, Ill.," *Architectural Forum* 80 (March 1944): 86.
13. George Fred Keck to Earl Aiken, letter, May 1943, Box 5, LOF main folder 5291, M73-026, George Fred Keck and William Keck Papers, 1905–95, Wisconsin Historical Society Archives, Madison.
14. *Solar Heating for Post-War Dwellings* (Toledo, OH: Libbey-Owens-Ford Glass Company, 1943); and "Solar Heating," *Architectural Forum* 79 (August 1943): 6–8, 114.
15. "Solar Heating," 6.
16. Ibid.
17. Sloan, "We Call Ours a Solar Home," 82.
18. Ralph Wallace, "The Proven Merits of a Solar Home," *Reader's Digest* 44 (January 1944): 10.
19. "Dream Houses — U.S. and British," *Newsweek* (September 6, 1943): 103.
20. Wallace, "The Proven Merits of a Solar Home," 1.
21. Earl Aiken to George Fred Keck, letter, October 16, 1944, Box 5, LOF main folder 5291, M73-026, George Fred Keck and William Keck Papers, 1905–95, Wisconsin Historical Society Archives, Madison.
22. George Fred Keck to Kenneth Reid, editor of *Pencil Points*, letter, M89-424, Box 26,

George Fred Keck and William Keck Papers, 1905–95, Wisconsin Historical Society Archives, Madison.

23. Ibid.
24. Ibid.
25. Ibid.
26. George Fred Keck to Earl Aiken, letter, May 6, 1943, Box 5, LOF main folder 5291, M73-026, George Fred Keck and William Keck Papers, 1905–95, Wisconsin Historical Society Archives, Madison.
27. Charles Huchins, Box 4, folder "Huchins, Charles," M73-026, George Fred Keck and William Keck Papers, 1905–95, Wisconsin Historical Society Archives, Madison.
28. Green Ready-Built folder, Box 27, M73-026, George Fred Keck and William Keck Papers, 1905–95, Wisconsin Historical Society Archives, Madison.
29. "It's Here—Solar Heating for Post-War Homes," *American Builder* 65 (September 1943): 34–36.
30. George Fred Keck to Earl Aiken, letter, May 6, 1943, Box 5, LOF main folder 5291, M73-026, George Fred Keck and William Keck Papers, 1905–95, Wisconsin Historical Society Archives, Madison.
31. Nelson and Wright, *Tomorrow's House*, 178.
32. Arthur Brown, interviews by author; "Black Wall Stores Winter Sun Heat," *House Beautiful* 92 (April 1950): 150; "They Heat Their House with Sunshine," *Better Homes and Gardens* 31 (February 1953): 175, 199.
33. Nelson and Wright, *Tomorrow's House*, 178.
34. George Fred Keck quoted in Richard Windfield Hamilton, ed., *Space Heating with Solar Energy: Proceedings of a Course Symposium Held at the Massachusetts Institute of Technology, August 21–26, 1950* (Cambridge, MA: MIT Press, 1954), 91.
35. F. W. Hutchinson, "Solar House: A Full Scale Experimental Study," *Heating and Ventilating* 42 (September 1945): 96–97; F. W. Hutchinson, "Solar House: Research Progress Report," *Heating and Ventilating* 43 (March 1946): 53–57; F. W. Hutchinson and W. P. Chapman, "A Rational Basis for Heating Analysis," *Heating, Piping and Air Conditioning* 18 (July 1946): 109–17.
36. Maron J. Simon, *Your Solar House: A Book of Practical Homes for All Parts of the Country* (New York: Simon and Schuster, 1947), 10.
37. Ibid., 7.
38. Ibid., cover and 7.
39. Vice president of LOF to George Fred Keck, August 25, 1945; C. Dean Lowry to George Fred Keck, August 19, 1946; and "Illinois Solar Home" (the set of plans and drawing that George Keck supplied for the book *Your Solar House*), Box 5, LOF main folder 5291, M73-026, George Fred Keck and William Keck Papers, 1905–95, Wisconsin Historical Society Archives, Madison.
40. Earl Aiken to architects who submitted plans and drawings for the book *Your Solar House*, January 7, 1948, Box 5, LOF main folder 5291, M73-026, George Fred Keck and William Keck Papers, 1905–95, Wisconsin Historical Society Archives, Madison.
41. Patricia Moore, "Let the Sun Shine In…," *Chicago Sun-Times*, January 7, 1979, 24–25.

Chapter 20. Solar Collectors for House Heating

1. Edward S. Morse, "Warming and Ventilating Apartments by the Sun's Rays," p. 2, U.S. Patent 246,626, issued September 6, 1881.
2. Edward S. Morse, "The Utilization of the Sun's Rays in Heating and Ventilating Apartments," in *The Proceedings of the Society of Arts of the Massachusetts Institute of Technology* (Cambridge, MA: MIT Press, 1885), 116.
3. Ibid., 116–17.
4. Ibid., 117.
5. Ibid., 118.
6. Ibid., 119 and 120.
7. "Heating by Sunshine," *Scientific American* 46 (May 13, 1882): 288.
8. Dorothy Wayman, *Edward Sylvester Morse: A Biography* (Cambridge, MA: Harvard University Press, 1942), 37–372.
9. Ibid.
10. "Solar Attack," *Time* 31 (June 16, 1938): 20.
11. Hoyt Hottel, interviews by author; Hoyt Hottel, "Converters of Solar Energy," in *Annual Report of the Smithsonian Institution* (Washington, DC: Smithsonian Institution, 1941), 152–56; H. C. Hottel and B. B. Woertz, "The Performance of Flat-Plate Collectors," in *Transactions of the American Society of Mechanical Engineers* (February 1943): 91–104.
12. Georg Löf, interviews by author; Office of Production, Research and Development, U.S. War Production Board, and University of Colorado Engineering Experiment Station, *Solar Energy Utilization for House Heating*, project no. 459, contract no. 100 (Washington, DC: Office of Production, Research and Development, War Production Board, May 18, 1946).
13. Albert Dietz and Hoyt Hottel, interviews by author; Albert Dietz and Edmund Czapek, "Solar Heating of Houses by Vertical South Wall Storage Panels," *Heating, Piping and Air Conditioning* 22 (March 1950): 118–25.
14. A. L. Hesselschwerdt, "Performance of the MIT Solar House," in *Solar Heating with Solar Energy: Proceedings of a Course Symposium Held at the Massachusetts Institute of Technology, August 21–26, 1950* (Cambridge, MA: MIT Press, 1954).
15. Dietz and Hottel, interviews.
16. "Test House Heated Only by Solar Heat," *Architectural Record* 105 (March 1949): 136–37.
17. Maria Telkes, "Solar House Heating — a Problem of Heat Storage," *Heating and Ventilating* 44 (May 1947): 71–73.
18. Dietz and Hottel, interviews.
19. Esther Nemethy, interview by author.
20. Arthur Brown, interview by author.
21. "Structural Components for Schools," *Architectural Record* (August 1956): 164–65, images 19, 20, 21.
22. Dietz and Hottel, interviews; C. D. Engbretson and N. G. Asher, *Progress in Space Heating with Solar Energy*, report no. 60-WA-88 (New York: American Society of Mechanical Engineers, July 28, 1960).

Chapter 21. From Selenium to Silicon

1. James Clerk Maxwell to Peter Guthrie Tate, letter, late April 1874, in *The Scientific Letters and Papers of James Clerk Maxwell*, vol. 3, 1874–1879 (Cambridge: Cambridge University Press, 1995), 67.
2. Willoughby Smith, *The Rise and Extension of Submarine Telegraphy* (New York: Arno Press, 1974), 310.
3. Willoughby Smith, "The Action of Light on Selenium," *Journal of the Society of Telegraph Engineers* 2 (1873): 32.
4. Ibid., 32.
5. William Grylls Adams and Richard Evans Day, "The Action of Light on Selenium," *Philosophical Transactions of the Royal Society of London* 168 (1877): 115.
6. William Grylls Adams to Professor Stokes, director of the Royal Society, November 27, 1875, papers of Sir George Gabriel Stokes, Add. 7656 RS 1116, Cambridge Manuscript Collection, Cambridge University Library.
7. Werner von Siemens, "On the Electromotive Action of Illuminated Selenium Discovered by Mr. Fritts of New York," *Van Nostrand's Engineering Magazine* 32 (1885): 392n.
8. Ibid.
9. Ibid., 515.
10. James Clerk Maxwell to George G. Stokes, letter, October 31, 1876, RR 429–31, Referees' Reports: reports on scientific papers submitted for publication to the Royal Society from 1832 to date (peer review), Archives of the Royal Society, London.
11. Von Siemens, "On the Electromotive Action of Illuminated Selenium Discovered by Mr. Fritts of New York," 515.
12. George Minchin to Oliver Lodge, letter, May 28, 1887, Ms. Add. 89/172, University College London, Library Services, Special Collections. Minchin expressed similar thoughts in a scientific poem: "Tis time that each successive age / Brings forth its crowd of theories and rules / What yesterday was 'law' to-day the sage / Holds up for scorn or laughter in schools." George Minchin, *Naturae Veritas* (London: Macmillan, 1887), 57.
13. George Minchin, "Experiments in Photoelectricity," in *Proceedings of the Physical Society* (London: Physical Society, 1890), vol. 11, pp. 108–9.
14. George Minchin quoted in Rollo Appleyard, "Photo-Electric Cells," *Telegraphic Journal and Electric Review* 28 (January 31, 1891): 126.
15. Minchin, "Experiments in Photoelectricity," 101–2.
16. Abraham Pais, *Subtle Is the Lord: The Life and Science of Albert Einstein* (New York: Oxford University Press, 1982), 357, 358, 414.
17. Daryl Chapin, *Energy from the Sun* (New York: Bell Laboratories, 1962), 24.
18. Pais, *Subtle Is the Lord*, 359.
19. Lee Dubridge, *Photoelectric Phenomena* (New York: McGraw-Hill, 1932), v.
20. "Magic Plates Tap Sun for Power," *Popular Science Monthly* 118 (June 1931): 41.
21. Appleyard, "Photo-Electric Cells," 125.
22. Thomas Benson, *Selenium Cells* (New York: Spon and Chamberlain, 1919), 60.
23. Dr. Maria Telkes to Dan Trivich, letter, August 31, 1952, private collection of Dr. Richard Komp, a graduate student of Dan Trivich.

24. E. D. Wilson, "Power from the Sun," *Power* 28 (October 1935): 517.
25. Vladimir Zworkin and Edward Ramberg, *Photoelectricity and Its Applications* (New York: J. Wiley, 1949), 470.
26. Gerald Pearson, Laboratory Book no. 233588, Location no. 121-07091, January 20, 1953, and March 6, 1953, 146–47, 154–55, AT&T Archives, Warren, NJ.
27. Daryl Chapin, Laboratory Book no. 28161, Location no. 124-11-02, October 2, 1952, 1, AT&T Archives, Warren, NJ.
28. Ibid., February 12, 1953, 23.
29. Daryl Chapin to Robert Ford, letter, AT&T Media Relations, Bell Laboratories, Murray Hill, NJ, provided by Mrs. Aubrey Chapin Svensson; Gerald Pearson, Book no. 23588, Location no. 121-0007-01, March 6, 1953, 155, AT&T Archives, Warren, NJ.
30. Daryl Chapin, Laboratory Book no. 28161, Location no. 124-11-02, March 23, 1953, 25, AT&T Archives, Warren, NJ.
31. Ibid.
32. Daryl Chapin, "Progress Report for July and August," September 4, 1953, Department 1730, provided by Mrs. Audrey Chapin Svensson.
33. Daryl Chapin, "Progress Report for January and February," March 1, 1954, Department 1314, provided by Mrs. Audrey Chapin Svensson.
34. Daryl Chapin, "Construction of Power Photocells," coversheet for Technical Memorandum, March 2, 1954, MM-54-131-9, provided by Mrs. Audrey Chapin Svensson.
35. "Sun's Energy: Fuel Unlimited," *U.S. News and World Report* 36 (May 7, 1954): 18.
36. "Vast Power Is Tapped by Battery Using Sand Ingredient," *New York Times*, April 26, 1954.

Chapter 22. Saved by the Space Race

1. J. Yellott, "Progress Report: Solar Energy in 1960," *Mechanical Engineering* 82 (December 1960): 41.
2. "Power by Sunlight," *Newsweek* 46 (October 10, 1955): 108.
3. Daryl Chapin, "The Conversion of Solar Energy to Electrical Power," May 18, 1956, UCLA lecture notes, 8, courtesy of Mrs. Audrey Chapin Svensson.
4. H. Leslie Hoffman, "Harnessing the Sun's Energy," *Trusts and Estates* 9 (October 1997): 1021.
5. "Converting Atomic Energy Direct into Electricity," *London News* 224 (June 12, 1954): 1009.
6. Director of RCA quoted by Dr. Joseph Loferski, research scientist at RCA Laboratories, interview by author.
7. "Converting Atomic Energy Direct into Electricity," 1009.
8. Martin Wolf, "Notes on the Early History of Solar Cell Development Particularly at National Semiconductor and Hoffman Semiconductor Division," October 18, 1995, 4, courtesy of Martin Wolf.
9. Daryl Chapin to Robert Ford, letter, AT&T Media Relations, Bell Laboratories, Murray Hill, NJ, December 4, 1994, courtesy of Mrs. Aubrey Chapin Svensson.

10. Gordon Raisbeck, "The Solar Battery," *Scientific American* (December 1955): 110.
11. Hans Ziegler, "Utilization of Solar Energy," briefing to Lieutenant General J. O'Connell, September 1955, 9, provided by Dr. Hans Ziegler. The signal corps oversees all power needs of the U.S. Army.
12. Signal Corps Engineering Laboratories, "Proposal for Satellite Program" 1 (top secret report, September 9, 1955), 25–26, courtesy of Dr. Hans Ziegler.
13. "U.S. to Launch History's First Man-Made-Earth-Circling Satellite," *New York Times*, July 30, 1955, 1, 7.
14. "Report of the Working Group on Internal Instrumentation of the USNG/IGY Technical Panel on the Earth Satellite Program" (paper presented at the Conference on Use of Solar Batteries for Powering Satellite Apparatus, Washington, DC, April 30, 1955), James van Allen Papers, Box 251, Folder 7, University of Iowa Archives, Iowa City.
15. Hans Ziegler, "Solar-Power Sources for Satellite Applications," *Annals of the International Geophysical Year* (1958): pts. 1–5, 3.3.3, 302–3.
16. Hans Ziegler, "Solar Power for Satellite Applications" (paper presented at the International Geophysical Year Rocket and Science Conference, Washington, DC, September 30–October 5, 1957), provided by Dr. Hans Ziegler.
17. U.S. Army Signal Corps Garrison, Public and Technical Information Division, press release, June 18, 1957, ECOM/Historical Research Collection, Fort Monmouth, NJ.
18. "Radio Fails as Chemical Battery Is Exhausted: Solar-Powered Radio Still Functioning," *New York Times*, April 6, 1958, 37.
19. Office of Technical Liaison, Office of the Chief Signal Corps Office, Department of the Army, "Sun-Powered Vanguard I Satellite Marks First Anniversary on March 17, 1959" (Washington, DC, undated), 1, ECOM/Historical Research Collection, Fort Monmouth, NJ.
20. "Reds Boast of Solar Battery," *Electronics*, Business Edition, 31 (July 11, 1958): 33.
21. National Research Council, Ad Hoc Panel on Solar Cell Efficiency, *Solar Cells' Outlook for Improved Efficiency* (Washington, DC: NRC, 1972), 3.
22. Rawson Bennett, "Navy Energy Needs, Present and Future," in *Proceedings of a Seminar on Advanced Energy Sources and Conversion Techniques...3-7 November, 1958. Sponsored by the U.S. Dept. of Defense, in Cooperation with California Institute of Technology [and] University of California, Los Angeles, under the Auspices of the U.S. Army Signal Corps* (Washington, DC: U.S. Department of Commerce, Office of Technical Services, [1961?]), 15.
23. Owen Bowen, Fred Kennedy, and Wade Pulliam, *DARPA's Space History*, www.monkeypuppet.net/truth/usgovernment/DARPASpaceHistoryFinal.pdf.
24. Loferski, interview.
25. Wolf, "Notes on the Early History of Solar Cell Development Particularly at National Semiconductor and Hoffman Semiconductor Division," 7.

Chapter 23. The First Large-Scale Photovoltaic Applications on Earth

1. Paul Rappaport, "Photoelectricity," *Proceedings of the National Academy of Sciences* 47, no. 8 (August 1961): 1303.

2. Harry Tabor, "Power for Remote Areas," *International Science and Technology* (May 1967): 54.
3. Lou Shrier, Exxon Mobil Corporation, interview by author.
4. B. P. Kelly et al., "Investigations of Photovoltaic Applications," in *Photovoltaic Power and Its Applications in Space and on Earth* (Bretigny-sur-Orge, France: Centre National d'Études Spatiales, 1973), 514.
5. D. Nolan, "Solar Energy Used for Production Applications," *Oil and Gas Journal* 76 (March 6, 1978): 82.
6. Steven Trenchard, president, Automatic Power, interview by author.
7. President Ronald Reagan to Captain Lloyd Lomer, United States Coast Guard, April 8, 1981, courtesy of Captain Lloyd Lomer.
8. C. Guy, "Experience of the French Lighthouse Service in Powering Aids to Navigation Either on Land or Floating Aids with Photovoltaic Generators," in *Proceedings of the 2nd International Photovoltaic Science and Engineering Conference, Beijing, China* (Hong Kong: Adfield Advertising Company, 1986), 670, 673.
9. K. Soras and V. Makios, "Feasible Stand-Alone Photovoltaic Systems in Greece," in *Fifth E.C. Photovoltaic Solar Energy Conference, Athens, Greece* (Dordrecht, Holland: Kluwer Academic Publishers, 1984), 487.
10. Larry Beil, interview by author.
11. John Oades, interview by author.
12. Bill Hampton, interview by author.
13. Oades, interview.
14. Ibid.
15. William Hunter, telecommunications executive knowledgeable about the Hunts Mesa project, interview by author.
16. J. Shepherd, "Navajoland's Solar Powered RF Repeater Is First in the Nation," *Telephone and Engineer Management* (November 15, 1976), reprint.
17. Bill Hampton, interview by author.
18. Michael Mack, interview by author.
19. Arnold Holderness, interview by author.
20. Australia, Parliament, Senate, Standing Committee on Natural Resources, *Reference, Solar Energy: Official Hansard Report* (Government Printer of Australia, 1976), 2.1706, 2.1708, 2.1730.
21. Ibid., 2.1706, 2.1730.
22. Michael Mack, "Solar Power for Telecommunications: The Last Decade," in *International Telecommunications Energy Conference, New Orleans* (New York: IEEE, 1984), 274; Michael Mack, "Solar Power for Communications," *Telecommunications Journal of Australia* 29, no. 1 (1979): 390.
23. Australia, Parliament, Senate, Standing Committee on Natural Resources, *Reference, Solar Energy*, 2.1730.
24. Mack, "Solar Power for Communications," 390.
25. *NSF/NASA Solar Energy Panel, An Assessment of Solar Energy as a National Resource* (College Park, MD: University of Maryland, December 1972), 1.
26. D. Eskenazi, D. Kerner, and L. Slominski, *Evaluation of International Photovoltaic Projects*, SAND 85-7018/2 DE 87 002943, B-10, (Albuquerque, NM: Sandia National Laboratories, 1986).

27. Michael Mack and George Lee, "Telecom Australia's Experience with Photovoltaic Systems in the Australian Outback," *Telecommunications Journal* 56 (1989): 514–15.
28. Mack, "Solar Power for Telecommunications: The Last Decade," 279.
29. Ibid., 274.
30. Arnold Holderness and Bill Yerkes, interviews by author; L. Permenter, "In Touch in the Outback," *Sunworld* 12, no. 3 (1980): 80.
31. I. Garner, "Design Optimization of PV Power Systems for Telecommunication Applications," in *The Conference Record of the Eighteenth IEEE Photovoltaic Specialists Conference — 1985: Riviera Hotel, Las Vegas, Nevada, October 21–25, 1985* (New York: IEEE, 1985), 275.

Chapter 24. Prelude to the Embargo

1. *Statistical Abstract of the United States* (Washington, DC: Department of Labor, 1965), 366, 369; Bassam Fattauh, "An Anatomy of the Crude Oil Pricing System" (Oxford: Oxford Institute for Energy Studies, January 2011): 6.
2. *Statistical Abstract of the United States* (Washington, DC: Department of Labor, 1955), 790; *Statistical Abstract of the United States* (Washington, DC: Department of Labor, 1965), 723.
3. Gas Company official, interview by author.
4. Executive of Southern California Edison who requested anonymity, interview by author.
5. Employee of the Southern California Gas Company who requested anonymity, interview by author.
6. *Statistical Abstract of the United States* (Washington, DC: Department of Labor, 1969), 511; *Statistical Abstract of the United States* (Washington, DC: Department of Labor, 1965), 672; and *Historical Statistics of the United States* (Washington, DC: Department of Commerce, 1975), pt. 2, p. 819.
7. Eric Hodgins, "Power from the Sun," *Fortune* 148 (September 1953): 130.
8. Charles A. Scarlott, interview by author. Scarlott is coauthor of *Energy Sources: The Wealth of the World* (New York: McGraw-Hill, 1952).
9. Ibid.
10. President's Materials Commission, "The Promise of Technology," in *Resources for Freedom* (Washington, DC: June 1952), 220, 127.
11. Thomas Kidder, "Tinkering with Sunshine," *Atlantic Monthly* (October 1977): 71; S. David Freeman, "Is There an Energy Crisis? An Overview," *Annals of the American Academy* 410 (November 1973).
12. Memorandum regarding "Operation Candor," July 22, 1953, White House Office, National Security Council Papers, PSB Central Files Series, Box 17, PSB 091.4 U.S. (2).
13. *President Eisenhower's "Atoms for Peace" Speech, December 8, 1953, Before the General Assembly of the United Nations on Peaceful Uses of Atomic Energy*, Atomic Archive, www.atomicarchive.com/Docs/Deterrence/Atomsforpeace.shtml.
14. "Only the Soviets Say 'No,'" *New York Times*, December 11, 1953; "U.S. to Disregard Initial Soviet 'No' on an Atomic Pool," *New York Times*, December 11, 1953.

15. Charles E. Egan, "Atomic Program for Peace Hailed," *New York Times*, April 13, 1955, 9.
16. "The Atom: Age of Nuclear Power Has Arrived," *Life* (January 4, 1954): 88.
17. The Atomic Energy Act of 1954, 42 U.S.C. § 2011 et seq.
18. H. Alfvén, "Fission Energy and Other Sources of Energy," *Bulletin of the Atomic Scientists* 30, no. 1 (January 1974): 4.
19. William L. Laurence, "Nuclear History Due to Be Made in Geneva Parley," *New York Times*, August 7, 1955.
20. Admiral Lewis Strauss, "Record of Proceedings of Session 1, Monday Morning, 8 August 1955," in *Proceedings of the International Conference on the Peaceful Uses of Atomic Energy* (New York: United Nations, 1956), 16.31–35.
21. "Atom Power Curb Seen: Seaborg Says Waste Disposal May Limit Development," *New York Times*, April 6, 1955.
22. "Senate Hearings on Atomic Energy," *Bulletin of the Atomic Scientists* 1–2 (December 24, 1945): 3.
23. "Random Notes from Washington," *New York Times*, October 24, 1955.
24. Ralph Eugene Lapp, *The New Force: The Story of Atoms and People* (New York: Harper Brothers, 1956), 103.
25. *New York Times*, editorial, April 24, 1954, 24.
26. NSF/NASA Solar Energy Panel, An Assessment of Solar Energy as a National Resource (College Park, MD: University of Maryland, December 1972), 1.
27. Paul Donovan and William Woodward, *An Assessment of Solar Energy as a National Energy Resource: Prepared by the NSF/NASA Solar Energy Panel* (Washington, DC: NASA/NSF, 1972), 1; Farrington Daniels, *The Direct Use of the Sun's Energy* (New Haven, CT: Yale University Press, 1964), 3.
28. "Toward a Captive Sun," *Newsweek* (November 14, 1955): 66; John Desmond Bernal, *World without War* (White Fish, MT: Kessinger Publishing, 2010), 46.
29. Leonard Engel, "Harnessing the Electron," *New York Times*, June 13, 1954; "The Sun: Prophets Study Rays for Far-Off Needs," *Life* (January 4, 1954): 93.
30. M. K. Hubbert, "Nuclear Energy and the Fossil Fuels" (paper presented before the Spring Meeting of the Southern District, Division of Production, American Petroleum Institute, San Antonio, Texas, March 7–9, 1956), available at www.hubbertpeak.com/hubbert/1956/1956.pdf.

Chapter 25. Solar in the 1970s and 1980s

1. "Sun Day — 30 Million Americans Celebrate," *Solar Age* (June 1978): 10.
2. William Metz and Allen Hammond, *Solar Energy in America* (Washington, DC: American Association for the Advancement of Science, 1978), xi–xii.
3. Solar Energy Industries Association, *Development and Status of the Renewable Energy Industries* (Washington, DC: Solar Energy Industries Association, 1985), 16.
4. Freeman Ford, interview by author.
5. E. Kiester, "How to Make Sun Power Work for You," *Skeptic* magazine (March–April 1977): 53.

6. Peter Rabbit, *Drop City* (Paris: Olympia Press, 1971), 27.
7. "Steve Baer and Holly Baer: Dome Home Enthusiasts," *Mother Earth News* (July–August 1973): 1, www.motherearthnews.com/nature-community/steve-baer-holly-baer-dome-home-zmaz73jazraw.aspx#axzz2OQ5df6fN.
8. Ibid., 3.
9. Ibid., 7.
10. Steve Baer, *Sunspots: Collected Facts and Solar Fiction* (Corrales, NM: Zomeworks, 1977), 24.
11. John Yellot and Harold Hay, "A Naturally Air-Conditioned Building," *Mechanical Engineering* 92 (January 1970): 19.
12. "Harold R. Hay Talks about Solar Energy," *Mother Earth News* (September–October 1976): 8, www.motherearthnews.com/nature-community/harold-r-hay-zmaz76soztak.aspx#axzz2OQ5df6fN.
13. Baer, *Sunspot*, 24.
14. John Yellott, "Passive Solar Heating and Cooling Systems," *ASHRAE Journal* (January 1978): 64.
15. A. D. Cohen to Harold Hay, letter, September 23, 1973, John I. Yellott Papers, 2007-04075, Preliminary Inventory of the John Yellot Papers, Folder 1: "Harold Hay 1972–1974," Box 29, Arizona State University Libraries: Architecture and Environmental Design Library Archives and Special Collections, Phoenix, www.azarchivesonline.org/xtf/view?docId=ead/asu/yellott_acc.xml;query=;brand=default.
16. "Steve Baer and Holly Baer," 13.
17. David Wright, interview by author.
18. Travis Brock, "David Wright: Passive Design," *Mother Earth News* (September–October 1977): 10, 11, www.motherearthnews.com/Green-Homes/1977-09-01/The-Plowboy-Interview.aspx?page=10#axzz2OQ5df6fN.
19. Ibid., 12.
20. David Wright, "Passive: Adobe," *Solar Age* (July 1976): 15.
21. David Wright, *Natural Solar Architecture: A Passive Solar Primer* (New York: Van Nostrand, 1978), x.
22. Noel Grove, "Oil, the Dwindling Treasure," *National Geographic* (June 1974): 792–93.
23. Ibid., 794.
24. Ibid., 823.
25. P. E. Glaser, O. E. Maynard, J. Mackovciak, and E. L. Ralph, Arthur D. Little, Inc., "Feasibility Study of a Satellite Solar Power Station," NASA CR-2357, NTIS N74-17784, February 1974.
26. David Godolphin, "Solar Power in the Desert," *Solar Age* (April 1985): 7.
27. J. Grosskveutz, "High Temperature Technology" (1988), in *Implementing Solar Thermal Technology*, ed. Ronald Larsen and Ronald E. West (Cambridge, MA: MIT Press, 2003), 287.
28. Martin McPhillips, *New Energy-Conserving Passive Solar Single-Family Homes: Cycle 5, Category 2, HUD Solar Heating and Cooling Demonstration Program* (Washington, DC:

U.S. Department of Housing and Urban Development in cooperation with the U.S. Department of Energy, 1981), 3–4.

29. W. Metz and A. Hammond, *The Science Report on Solar Energy in America* (Washington, DC: American Association for the Advancement of Science, 1978), ix, 5.
30. Douglas Balcomb, interview by author.
31. Ibid.
32. Douglas Balcomb, "Summary of the Passive Solar Heating and Cooling Conference," in *Passive Solar Heating and Cooling: Conference and Workshop Proceedings, May 18–19, 1976, University of New Mexico, Albuquerque, New Mexico* (Washington, DC: Energy Research and Development Administration, 1977).
33. Balcomb, interview.
34. "The 25 Most Important Houses in America," in "Houses," special issue, *Fine Homebuilding*, no. 107 (Summer 2006): 66, www.vanderryn.com/Docs/article-fh.pdf.
35. C. Blankton, "Significant Results of the Federal Solar R and D Programs," in *History and Overview of Solar Heat Technologies*, ed. Donald Beattie (Cambridge, MA: MIT Press, 1989), 209.
36. "Solar — Here Is a Serious Look at Tomorrow," *Sunset* magazine (November 1976): 80; "Passive Solar: Yesterday Is Tomorrow," *Sunset* magazine (February 1979): 77.
37. "Sun-Conscious Houses: Primer on Passive Solar Heating," *Life* (January 1, 1980): 49.
38. H. Antolini, ed., *Sunset Homeowner's Guide to Solar Heating and Cooling* (Menlo Park, CA: Lane Publishing, 1978), 2.
39. Balcomb, interview.
40. Robert Besant et al., "The Passive Performance of the Saskatchewan Conservation House," in *Proceedings of the 3rd National Passive Solar Conference* (Newark, DE: International Solar Energy Society, 1979), 713–15.
41. M. Jenior and R. Lorand, "Passive Technologies," in *History and Overview of Solar Heat Technologies*, ed. Donald Beattie (Cambridge, MA: MIT Press, 1997), 10, 208.
42. McPhillips, *New Energy-Conserving Passive Solar Single-Family Homes*, 3–4.
43. Douglas Balcomb, *Passive Solar Building* (Cambridge, MA: MIT Press, 2008), 5.
44. Freeman Ford, "Solar Business: A Case Study," in *Finance and Utilization of Solar Energy: Proceedings of a Conference at University of California, Santa Cruz, California, October 8–9, 1981* (Santa Cruz, CA: Center for Innovation and Entrepreneurial Development, 1982), 110.
45. Freeman Ford, interview by author.
46. R. Bliss, "Why Not Just Build It Right?," *Bulletin of the Atomic Scientists* (March 1976).

Chapter 26. America's First Solar City

1. Frederick Augustus Brooks, "Space Heating and Domestic Uses," in *Solar Energy Research*, ed. Farrington Daniels and John Duffie (Madison: University of Wisconsin Press, 1955), 76.

2. Arthur Farral, *Solar Heat* (Berkeley, California Experiment Station, 1929), 3.
3. Ibid.
4. Frederick Augustus Brooks, *Solar Energy and Its Use in Heating Water in California* (Berkeley, California Experiment Station, 1936), 62.
5. Ibid., 17.
6. Frederick Augustus Brooks, "Storage of Solar Energy in the Ground," in *Solar Energy Research*, ed. Farrington Daniels and John Duffie (Madison: University of Wisconsin Press, 1955), 245.
7. Brooks, "Space Heating and Domestic Uses," 76.
8. Frederick Augustus Brooks, *Introduction to Physical Microclimatology* (Berkeley: University of California Press, 1960), 2–13; Brooks, "Storage of Solar Energy in the Ground," 245.
9. Loren Wenzel Neubauer, FSC-1955, D-109, Box 1, Folder "The Solaranger," Manuscripts, Neubauer, Loren W. (1904–91), Agricultural Technology, University of California, Davis, Special Collections.
10. Loren Wenzel Neubauer and Harry Bruce Walker, *Farm Building Design* (Englewood Cliffs, NJ: Prentice-Hall, 1961), 91–92.
11. Ibid.
12. Neubauer, FSC-1955.
13. Brooks, *Solar Energy and Its Use in Heating Water in California*, 7.
14. Frederick Augustus Brooks, "The Availability of Solar Energy," in *Introduction to the Utilization of Solar Energy*, ed. Abe Mordechai Zarem and Duane D. Erway (New York: McGraw-Hill, 1963), 38.
15. Brooks, *Introduction to Physical Microclimatology*, 2.
16. Frederick Augustus Brooks, "Agricultural Needs for Special and Extensive Observations of Solar Radiation," *Botanical Review* 30 (April–June 1964): 265.
17. R. Deering, "A Study for the Moderation of Extremes of Heat, Cold and Undesirable Winds," 12–14, in Frederick Augustus Brooks, D-100, Box 1, Folder: "Project 1536," Manuscripts, Brooks, Frederick Augustus, Agricultural Technology, University of California, Davis, Special Collections.
18. Ibid., 12.
19. Loren Wenzel Neubauer, *Ag Eng 105 Syllabus*: "Discussion #9, 9.2," 1946, D-109, Box 1, Folder "Binder 1 of 2," Manuscripts, Neubauer, Loren W. (1904–91), Agricultural Technology, University of California, Davis, Special Collections.
20. Loren Wenzel Neubauer, *Ag Eng 105 Syllabus* (1946): "Discussion No. 9 and No. 2-2 and 2-9," 1946, D-109, Box 1, Folder "Binder 1 of 2," Manuscripts, Neubauer, Loren W. (1904–91), Agricultural Technology, University of California, Davis, Special Collections.
21. Loren Neubauer, "The Solaranger," in *Solar Heating Pamphlets*, by Frederick A. Brooks, Arthur William Farrall, and L. W. Neubauer (1955).
22. Loren Neubauer, FSC-1951 and FSC-1950, D-109, Box 1, Folder "Farm Structures Conference 1951," in Folder: "Binder 1 of 2," Manuscripts, Neubauer, Loren W. (1904–91), Agricultural Technology, University of California, Davis, Special Collections.

23. Loren Neubauer, "Farm Structures Conference," February 1, 1951, D-109, Box 1, Folder: "Binder 1 of 2," Manuscripts, Neubauer, Loren W. (1904–91), Agricultural Technology, University of California, Davis, Special Collections.
24. Loren Neubauer, "The Semi-Solar Low Cost House Saves Energy through Environmental Orientation," in *Proceedings [of the] IAHS International Symposium on Housing Problems, 1976: Atlanta, Georgia, May 24–28, 1976* (Clemson, SC: Continuing Engineering Education, Clemson University, 1976), 1417.
25. H.E. Wickes to Tod Neubauer, letter, June 1, 1959, and Tod Neubauer to Miss Faye C. Jones, letter, February 12, 1958, D-109, Box 1, Folder: "Binder 1 of 2," Manuscripts, Neubauer, Loren W. (1904–91), Agricultural Technology, University of California, Davis, Special Collections.
26. Loren Neubauer, "Southermation: Heat Control for Farm Building," paper no. PCR-69-117 (paper presented to the ASAE, Pacific Coast Region, Phoenix, Arizona, 1969), 8, reprint.
27. Neubauer and Walker, *Farm Building Design*, 307.
28. Jon Hammond, interview by author.
29. Neubauer, "The Semi-Solar Low Cost House Saves Energy through Environmental Orientation," 1416–18.
30. Bob Black, interview by author.
31. Neubauer, "The Semi-Solar Low Cost House Saves Energy through Environmental Orientation," 1418.
32. Ibid., 1416–18.
33. Ibid., 1419.
34. Ibid., 1420.
35. Jon Hammond, interview by the author.
36. The Elements, *The Davis Experiment: One City's Plan to Save Energy* (Washington, DC: Public Resource Center, 1977), 9.
37. Ibid., 11–12.
38. Living Systems, *Davis Energy Conservation Report: Practical Use of the Sun* (Davis, CA: Living Systems, 1977), 35.
39. Ibid., 36.
40. Ibid., 43.
41. David A. Bainbridge, Judy Corbett, and John Hofacre, *Village Homes' Solar House Designs* (Emmaus, PA: Rodale Press, 1977), 20.
42. Rob Thayer, *Gray World, Green Heart: Technology, Nature and the Sustainable Landscape* (New York: John Wiley and Sons), 287.
43. Bainbridge, Corbett, and Hofacre, *Village Homes' Solar House Designs*, 12.
44. Josephine Carothers, interview by author.
45. Edgar DeMeo, interview by author.
46. Doug Balcomb, interview by author.

Chapter 27. Solar Water Heating Worldwide, Part 2

1. Freeman Ford, "Solar Business: A Case Study," in *Finance and Utilization of Solar Energy: Proceedings of a Conference at University of California, Santa Cruz, California,*

October 8–9, 1981 (Santa Cruz, CA: Center for Innovation and Entrepreneurial Development, 1982), 112.
2. Solar Energy Industries Association, *Development and Status of the Renewable Energy Industries* (Washington, DC: Solar Energy Industries Association, 1985), 16.
3. Ibid., 33.
4. Martin Enowitz, "Investing in Energy," *Solar Today* (January–February 1987): 7.
5. B. D'Allesandro, "Dark Days for Solar," *Sierra* magazine (July–August 1987): 34.
6. R. Kleinman, "Haifa Leads Madison Sixty to One in Solar Use," *Country News* (September 27, 1982): 6.
7. Itzhak Shomron, interview by author.
8. Werner Weiss and Franz Mauthner, *Solar Heat Worldwide: Markets and Contribution to the Energy Supply, 2010* (Gleisdorf, Austria: AEE-Institute for Sustainable Technologies, 2012), 12, http://archive.iea-shc.org/publications/downloads/Solar_Heat_Worldwide-2012.pdf.
9. Dr. Ioannis Michaelides, interview by author. Michaelides is known throughout the world as the foremost authority on the solar water-heater industry in Cyprus.
10. Weiss and Mauthner, *Solar Heat Worldwide*, 12.
11. Panos Lamaris, personal correspondence with author.
12. Weiss and Mauthner, *Solar Heat Worldwide*, 12.
13. All information of the Barbados solar water-heater industry is based, unless specified otherwise, on my personal correspondence with James Husband, and on *Solar Technology and Sustainable Development: Building on the Solar Dynamics Experience: Barbados*,www.eclac.org/iyd/noticias/pais/3/31463/Barbados_Doc_1.pdf.
14. Ibid., 177–78.
15. B. Perlack and W. Hinds, *Evaluation of the Barbados Solar Water Heating Experience* (September 2003), http://solardynamicsltd.com/wp-content/uploads/2010/07/SWH-report1-2.pdf.
16. Weiss and Mauthner, *Solar Heat Worldwide*, 12.
17. Michael Ornetzeder, "Old Technology and Social Innovations. Inside the Austrian Success Story on Solar Water Heaters," *Technology Analysis and Strategic Management* 15, no. 1 (2001): 13, 1.
18. Michael Ornetzeder and Harald Rohracher, "User Led Innovations and Participation Processes: Lessons for Sustainable Energy Technologies," *Energy Policy* 34, no. 2 (2006): 138–50.
19. Ornetzeder, "Old Technology and Social Innovations," 109, 111.
20. The story of solar water heating on Æro Island is from my personal correspondence with Jess Heinemann, structural engineer, Æro Island; and Flemming Ulber, chief consultant, energy production, Æro Island.
21. "Australian Designed Solar Water Heaters for Japan," *Solar Energy Progress in Australia and New Zealand,* International Solar Energy Society 14 (July 1975): 31.
22. Y. B. Ng and C. T. Leung, "Solar Technology in China: A Review," *Sunworld* 6, no. 4 (1982): 115.
23. Weiss and Mauthner, *Solar Heat Worldwide*, 26.
24. "Solar Energy Applications in Remote Areas of Northwestern China," *Sunworld* 10, no. 2 (1986): 47.

25. Xuemei Bai, "China's Solar-Powered City," Renewable Energy World, May 22, 2007, www.renewableenergyworld.com/rea/news/article/2007/05/chinas-solar-powered-city-48605.
26. Barbel Epp, "Solar Thermal Scales New Heights in China," Renewable Energy World, June 27, 2012, www.renewableenergyworld.com/rea/news/article/2012/06/solar-thermal-scales-new-heights-in-china?cmpid=WNL-Friday-June29-2012.
27. Weiss and Mauthner, *Solar Heat Worldwide*, 26.

Chapter 28. Photovoltaics for the World

1. "Energy Problems of Developing Countries — West Africa — The Principal Energy Problems of West Africa," in *Energy Needs/Expectations: World Energy Conference, 13th Congress, Cannes 5/11: Technical Papers*, vol. 2, pt. 4 (London: World Energy Conference, 1986).
2. M. R. Starr et al., "Design of PV Lighting Systems for Developing Countries" (paper presented to Right Light Three, 3rd European Conference on Energy-Efficient Lighting, Newcastle-upon-Tyne, U.K., June 18–21, 1955), reprint.
3. Patrick Jourde, interview by author.
4. Bernard McNelis, interview by author.
5. Jourde, interview.
6. "How to Sell the Sun at the Equator? Solar in Africa," *Ecotec Resources bv* (1994): 4.
7. M. Kimani and Mark Hankins, "Rural PV Lighting Systems," in *Whose Technologies? The Development and Dissemination of Renewable Energy Technologies (RETS) in Sub-Saharan Africa*, ed. Ann Heidenreich Dominic Walubengo and Muiruri J. Kimani (Nairobi: KENGO Regional Wood Energy Programme for Africa, 1993), 91.
8. Mark Hankins, interview by author.
9. Richard Swanson, solar pioneer, interview by author.
10. William Metz and Allen Hammond, *Solar Energy in America* (Washington, DC: American Association for the Advancement of Science, 1978), xiii.
11. John Ericsson quoted in William Adams, *Solar Heat, a Substitute for Fuel in Tropical Countries for Heating Steam Boilers, and Other Purposes* (Bombay: Education Society Press, 1878): 16.
12. Charles Fritts, "Note," in "On the Electromotive Action of Illuminated Selenium, Discovered by Mr. Fritts, of New York," by Werner von Siemens, *Van Nostrand's Engineering Magazine* 32 (1885): 515.
13. Shell Reports, *Solar Energy* (Houston: Shell Oil Company, 1978).
14. Alpha Real, *Alpha Real sucht 333 Kraftwerkbesitzer, MEGAWATT Solarkraftwerke für unsere Umwelt* (Switzerland: Alpha Real), company sales brochure.
15. Markus Real, video interview by Mark Fitzgerald, 1994.
16. Donald Osborn, interview by author.
17. Steve Strong, interview by author. Strong is a renowned building-integrated-photovoltaics architect.
18. Markus Real, video interview by Mark Fitzgerald, 1994.

19. Donald Osborn, interview by author.
20. Jim Barnett, "SMUD Solar Programs for Commercial and Residential Customers," undated, Sacramento Municipal Utility District, PowerPoint presentation, slide no. 1, available at Build It Green, www.builditgreen.org/attachments/files/170/090825_Barnett.pdf.
21. Skip Fralick, interview by author.
22. Osborn, interview.
23. Farrington Daniels, "Conclusions," in *Solar Energy Research*, ed. Farrington Daniels and John Duffie (Madison: University of Wisconsin Press, 1955), 254.
24. Quoted in William Metz and Allen Hammond, *Solar Energy in America* (Washington, DC: American Association for the Advancement of Science, 1978), 35.
25. Hoyt Hottel, quoted in Edmund Faltermayer, "Solar Energy Is Here, But It's Not Yet Utopia," *Fortune* (February 1976, vol. 93): 105; and Hoyt Hottel, "Solar Energy: Cloudy Forecast," *Skeptic* magazine (March–April 1977): 68.
26. *The Nation's Energy Future: A Report to Richard Nixon, President of the United States*, submitted by Dr. Dixy Lee Ray, chairman, United States Atomic Energy Commission, WASH-1281 (Washington, DC: U.S. Atomic Energy Commission, December 1, 1973), 21–24.
27. Barry Commoner, *The Poverty of Power* (New York: Alfred Knopf, 1976), 142.
28. Ibid., 143.
29. Alfred J. Eggers, subpanel chairman, *Subpanel IX: Solar and Other Energy Sources* (Washington, DC: National Science Foundation, October 27, 1973), sec. 7, "Photovoltaic Conversion."
30. Lloyd Herwig, interview by author. Herwig is head of the solar section of the National Science Foundation.
31. Federal Energy Administration, Task Force on Solar Energy Commercialization, *Preliminary Analysis of an Option for the Federal Photovoltaic Utilization Program (FPUP)* (Washington, DC: Federal Energy Administration, July 20, 1977), 1.
32. Ibid., 25.
33. Robert Terry, Clarence P. Carter, Judy Israel, Orrin H. Merrill, and Michael Semmans, *DOD Photovoltaic Energy Conversion Systems Market Inventory and Analysis, Summary Volume* (Vienna, VA: BDM Corp., June 1977), xv.
34. Michael Eckert and Helmut Schuber, *Crystals, Electrons, Transistors: From Scholar's Study to Industrial Research* (Houten, Netherlands: Springer, 1990), 184.
35. Barry Commoner, *The Politics of Energy* (New York: Knopf, 1979), 34–38.
36. Denis Hayes quoted in "Sun Day — 30 Million Americans Celebrate," *Solar Age* (June 1978): 10.
37. Richard V. Allen to Ronald Reagan, October 7, 1982, letter, identification no. 098203SS, Danny Boggs File, Energy, Atomic Energy, Ronald Reagan Foundation and Library, Simi Valley, CA.
38. Samuel McCracken, "Solar Energy: A False Hope," *Commentary* (November 1979): 61.
39. Ibid., 67.
40. Samuel McCracken, *The War against the Atom* (New York: Basic Books, 1982).
41. Dr. Morton Prince, interview by author.

42. Herwig, interview.
43. Arnulf Jager-Waldau, *JRC PV Status Report 2004* (Ispra, Italy: European Commission Joint Research Centre, 2005), 14.
44. Reinhard Haas, "Market Development Strategies for Photovoltaics: An International Review," *Renewable and Sustainable Energy Review* 4, no. 4 (August 2003): 217–315.
45. Assun Lopez Polo, Reinhard Haas, and Demut Suma, *Promotional Drivers for Grid-Connected PV* (Vienna, Austria: Vienna University of Technology, Institute of Power Systems and Energy Economics, Energy Economic Group, March 2009), 12.
46. Eric Wesoff, "22 Gigawatts of Solar in Germany on May 25: A New World Record," *Greentech Media*, May 29, 2012, www.greentechmedia.com/articles/read/22-Gigawatts-of-Solar-in-Germany-on-May-25/.
47. Jonathan Gifford, "Germany: Record 40 Percent Solar Weekend," *PV Magazine* (May 29, 2012), www.pv-magazine.com/news/details/beitrag/germany-record-40-percent-solar-weekend_100006953/#ixzz1wT9IEv8A.
48. Federal Ministry for the Environment, Nature Conservation and Nuclear Safety, *Act on Granting Priority to Renewable Energy Sources* (Renewable Energy Sources Act) (Berlin: Federal Ministry for the Environment, Nature Conservation and Nuclear Safety, March 2000), sec. 1.
49. Ibid., Explanatory Memorandum, B. Provisions, sec. 8, para. 1.
50. Volkmar Lauber and Mez Lutz, "Three Decades of Renewable Electric Policies in Germany," *Energy and Environment* 15, no. 4 (2004): 599–623.
51. T. Erge, V. U. Hoffmann, K. Kiefer, "The German Experience with Grid-Connected PV-Systems," *Solar Energy* 70, no. 6 (2001): 479–87.
52. Volkmar Lauber and Mez Lutz, "Three Decades of Renewable Electric Policies in Germany," *Energy and Environment* 15, no. 4 (2004): 6, www.wind-works.org/cms/uploads/media/Three_decades_of_renewable_electricity_policy_in_Germany.pdf.
53. Ibid., 10.
54. Federal Ministry for the Environment, Nature Conservation and Nuclear Safety, *Act on Granting Priority to Renewble Energy Sources*, sec. 12.
55. Federal Ministry for the Environment, Nature Conservation and Nuclear Safety, *Renewable Energy Sources in Figures: National and International Development* (Berlin: Federal Ministry for the Environment, Nature Conservation and Nuclear Safety, 2012), www.erneuerbare-energien.de/fileadmin/Daten_EE/Dokumente__PDFs/_/broschuere_ee_zahlen_en_bf.pdf.
56. Arnulf Jager Waldau, *PV Status Report 2011: Research, Solar Cell Production and Implementation of Photovoltaics* (Ispra, Italy: European Commission, Joint Research Centre, July 2011), 15.
57. T. Hoium, "2013 Is the Year of Solar," Motley Fool, February 12, 2013, www.fool.com/investing/general/2013/02/12/2013-is-the-year-of-solar.aspx?source=isesitlnk0000001&mrr=1.00.
58. Natalie Obiko Pearson, "Solar Cheaper than Diesel Making India's Mittal Believer: Energy," Bloomberg.com, January 24, 2012, www.bloomberg.com/news/2012-01-25/solar-cheaper-than-diesel-making-india-s-mittal-believer-energy.html.
59. Herman K. Trabish, "Stat of the Day: 40 Percent More Wind, Solar, Hydro and Biopower in 2017," *Greentech Media*, July 9, 2012, www.greentechmedia.com/articles/read/Stat-of-the-Day-40-Percent-More-Wind-Solar-Hydro-and-BioPower-in-2017.

Chapter 29. Better Solar Cells, Cheaper Solar Cells

1. Bob Johnson, longtime solar-cell analyst, interview by author.
2. Daryl Chapin, "Progress Report for May and June 1955," July 1, 1955, Department 1314; and Daryl Chapin, "Progress Report for May and June 1955," July 1, 1954, Department 1314, courtesy of Mrs. Audrey Chapin Svensson.
3. Martin Wolf, "Photovoltaic Solar Energy Conversion Systems," in *Solar Energy Handbook*, ed. Jan Krieder and Frank Krieth (New York: McGraw-Hill, 1981), 24–25.
4. M. Savelli, "Thin-Film Solar Cells and Spray Technology," *Solar Cells* 12 (1984): 192.
5. Fritz Wald, photovoltaic pioneer, interview by author.
6. Personal correspondence with Antonio Luque, one of the most respected scientists in solar-cell concentration.
7. Multijunction cells under solar concentration have experienced major breakthroughs in recent years, reaching a record efficiency of 44 percent. See "Solar Junctions sets CPV Record of 44%," *Semiconductor Today*, October 25, 2012, www.semiconductor-today.com/news_items/2012/OCT/SOLARJUNCTION_151012.html. Future enhancements of the quality of the semiconductor material and the increase of the number of cells in the stack could drive efficiencies beyond 50 percent. See Antonio Luque, "Will We Exceed 50% Efficiency in Photovoltaics," *Journal of Applied Physics* 110 (2011): 31301.
8. Ibid.
9. Daryl Chapin, "The Conversion of Solar Energy to Electrical Energy," UCLA lecture notes (May 18, 1956): 18, courtesy of Mrs. Audrey Chapin Svensson.
10. Fred Shirland, "The History, Design, Fabrication and Performance of CdS Thin-Film Solar Cells," in *Solar Cells*, ed. Charles Backus (New York: IEEE Press, 1966), 66.
11. "Sun-Power Prospects," *Chemical Week* (April 1, 1967): 53.
12. NASA Lewis, *Photovoltaic Conversion of Solar Energy for Terrestrial Applications, NZ4-22704 (23–25 October 1973)* (Cherry Hill, NJ: Jet Propulsion Laboratory, 1974), 23.
13. Christopher Wronski, interview by author.
14. David Carlson, interview by author.
15. A. Robinson, "Amorphous Silicon: A New Direction for Semi-conductors," *Science* 197 (August 25, 1977): 852.
16. Wronski, interview.
17. Carlson, interview.
18. "Solar Energy: Glasstech Builds on Its Pioneering Ideas," Glasstech World, special edition, undated, www.glasstech.com/downloads/Newsletters/GLAS1335 Newsletter7.PDF.
19. Arthur Rudin, interview by author. Rudin worked at Solar Power Corporation and Arco Solar, two of the first solar companies that made modules for the terrestrial market.
20. Personal correspondence with Dr. Richard Swanson, founder of SunPower.
21. Gregory Kiss, interview by author.
22. Ibid.
23. These novel concepts are the object of intense research among scientists throughout the world. They are usually known as third-generation photovoltaics. For detailed information on the subject, consult Antonio Martí and Antonio Luque, eds., *Next Generation*

Photovoltaics: High Efficiency through Full Spectrum Utilization (Bristol, UK: Institute of Physics Publishing, 2003) and Martin Green, *Third Generation Photovoltaics* (Berlin: Verlag Springer, 2003). For more information on nanowires combined with graphene, see Bård Amundsen and Else Lie, "Graphene and Semiconductor Technology: Smaller, Cheaper, Better," *Renewable Energy World*, June 14, 2013, www.renewableenergyworld.com/rea/news/article/2013/06/graphene-and-semiconductor-technology-smaller-cheaper-better.

Epilogue

1. Henry Kelly, "Photovoltaic Power Systems: A Tour through the Alternatives," *Science* 199 (February 10, 1978): 635.
2. Ben Willis, "World PV Capacity Tops 100GW," *PV Tech*, February 11, 2013, www.pv-tech.org/news/world_pv_capacity_tops_100gw?utm_source=PV-Tech&utm_campaign=fof8fob584-PV_Tech_Daily_Newsletter_Monday_11_02_132_11_2013&utm_medium=email.
3. John Perlin, *From Space to Earth: The Story of Solar Electricity* (Ann Arbor, MI: Aatec Publications, 1999), 202.
4. Werner Weiss and Franz Mauthner, *Solar Heat Worldwide: Markets and Contribution to the Energy Supply, 2010* (Gleisdorf, Austria: AEE-Institute for Sustainable Technologies, 2012), 22, www.iea-shc.org/Data/Sites/1/publications/Solar_Heat_Worldwide-2012.pdf.
5. H. E. Wilsie, "Experiments in the Development of Power from the Sun's Heat," *Engineering News* 61 (May 1909): 514.
6. Katie Fehrenbacher, "BrightSource's Cancelled Projects Highlight Hurdles for Desert Solar Thermal Plants," *GigaOM*, April 4, 2013, http://gigaom.com/2013/04/04/brightsources-cancelled-projects-highlight-hurdles-for-desert-solar-thermal-plants.
7. Heba Hashem, "CSP Secures Large Portion of Saudi Solar Market," CSP Today, September 21, 2012, http://social.csptoday.com/emerging-markets/csp-secures-large-portion-saudi%E2%80%99s-solar-market.
8. "The Oil Production Challenge," GlassPoint, undated, www.glasspoint.com/about-market.html.
9. National Renewable Energy World, "Device Tosses Out Unusable PV Wafers," NREL.gov, January 11, 2013, www.nrel.gov/news/features/feature_detail.cfm/feature_id=2076.
10. International Energy Agency, *World Energy Outlook 2012, Executive Summary* (Paris: International Energy Agency, 2012), 6, www.iea.org/publications/freepublications/publication/English.pdf.
11. Felicity Carus, "US Military Takes Solar to the Frontline," Editor's Blog, *PV Tech*, May 8, 2012, www.pv-tech.org/editors_blog/us_military_takes_solar_to_the_frontline.
12. Jessica Lyons Hardcastle, "US Military Support for Solar Energy," Editor's Blog, *Solar Novus Today*, October 3, 2012, www.solarnovus.com/index.php?option=com_content&view=article&id=5763:us-military-support-for-solar-energy&Itemid=352.

13. Kelly Trauetz, interview by author, April 29, 2013.
14. Darrell Issa, chairman, U.S. House of Representatives, Committee on Oversight and Government Reform, *How Obama's Green Energy Agenda Is Killing Jobs*, Staff Report, U.S. House of Representatives, 112th Congress, September 22, 2011, http://oversight.house.gov/wp-content/uploads/2012/02/9-22-2011_Staff_Report_Obamas_Green_Energy_Agenda_Destroys_Jobs.pdf.
15. Howard H. Baker Jr. Center for Public Policy, *Assessment of Incentives and Employment Impacts of Solar Industry Deployment* (Knoxville: Howard H. Baker Jr. Center for Public Policy, University of Tennessee, May 1, 2012), 8, http://bakercenter.utk.edu/wp-content/uploads/2012/04/Solar-incentives-and-benefits-_complete-report_May-1-2012.pdf.
16. Joe Romm, "How Much of a Subsidy Is the Price-Anderson Nuclear Industry Indemnity Act?" Thinkprogress.org, August 7, 2008, http://thinkprogress.org/climate/2008/08/07/202962/how-much-of-a-subsidy-is-the-price-anderson-nuclear-industry-indemnity-act/.
17. International Energy Agency, *World Energy Outlook 2012, Executive Summary*, 1.
18. Ibid.
19. "Solar Means Business: Top 20 US Corporate Solar Customers Are Iconic Brands," Vote Solar Initiative, September 12, 2012, http://votesolar.org/2012/09/solar-means-business-top-20-us-corporate-solar-customers-are-iconic-brands.
20. "Powering Our Facilities with Clean, Renewable Energy," Apple.com, undated, www.apple.com/environment/renewable-energy.
21. "NOAA: Carbon Dioxide Levels Reach Milestone at Arctic Sites," NOAA, May 31, 2012, http://researchmatters.noaa.gov/news/Pages/arcticCO2.aspx.
22. Steven Chu, interview by author, May 1, 2013.
23. Justin Gillis, "Heat-Trapping Gas Passes Milestone, Raising Fears," *New York Times*, May 10, 2013, www.nytimes.com/2013/05/11/science/earth/carbon-dioxide-level-passes-long-feared-milestone.html.
24. John M. Broder, "Climate Change Report Outlines Perils for U.S. Military," *New York Times*, November 9, 2012, www.nytimes.com/2012/11/10/science/earth/climate-change-report-outlines-perils-for-us-military.html?_r=0.
25. Jim Jubak, "Ready for the US Energy Boom?" *Jubak's Journal* (blog), MoneyShow, November 20, 2012, 3, www.moneyshow.com/investing/article/37/Jubak_Journa-29804/Ready-for-the-US-Energy-Boom/.
26. "Climate Change Fears Overblown, Says ExxonMobil Boss," *The Guardian*, June 28, 2012, www.guardian.co.uk/environment/2012/jun/28/exxonmobil-climate-change-rex-tillerson.
27. Rosana Francescato, "World's Largest Mining Company Admits Climate Change Is Real," *The Blog* (contributors' blog), Mosaic, undated, https://joinmosaic.com/blog/worlds-largest-mining-company-admits-climate-change-real.
28. Mark Golden and Mark Shwartz, "Stanford's George Shultz on Energy: It's Personal," *Stanford Report*, July 12, 2012, http://news.stanford.edu/news/2012/july/george-shultz-energy-071212.html.
29. Jubak, "Ready for the US Energy Boom?"

30. International Energy Agency, *World Energy Outlook 2012, Executive Summary*, 3.
31. Matthew L. Wald, "Climate Change Threatens Power Output, Study Says," Green (blog), *New York Times*, June 4, 2012, http://green.blogs.nytimes.com/2012/06/04/climate-change-threatens-power-output-study-says/.
32. Robert H. Thurston, "Utilizing the Sun's Energy," Annual Report of the Board of Regents of the Smithsonian Institution (Washington, DC: Government Printing Office, 1902), 270.
33. Daniel Yergin, "Will Gas Crowd Out Wind and Solar?," *Fortune* (April 30, 2012): 96.
34. Travis Hoium, "First Solar: Analyst Report," Motley Fool, undated, www.fool.com/reports/fslr-q.aspx.
35. "China to Boost Solar Power Goal 67% as Smog Envelops Beijing," Bloomberg.com, January 30, 2013, www.bloomberg.com/news/2013-01-30/china-to-boost-solar-power-goal-67-as-smog-envelops-beijing.html.
36. "Deadly China Pollution Breathes New Life into Solar Debt," Bloomberg.com, February 7, 2013, www.bloomberg.com/news/2013-02-07/deadly-pollution-breathes-new-life-into-solar-debt-china-credit.html.
37. "India Seeks 20,000 MW of Solar Power Capacity by 2020," *Reliable Plant*, undated, www.reliableplant.com/Read/18150/india-seeks-20,000-mw-of-solar-power-capacity-by-2020.
38. Shreya Jai & Mitul Thakkar, ET Bureau, "Diesel Price Hikes Make Solar Power Attractive Option," *Economic Times* (January 23, 2013): http://articles.economictimes.indiatimes.com/2013-01-23/news/36505789_1_diesel-price-solar-power-natural-gas.
39. Danny Kennedy, "Saudi Arabia Makes Big Bet on Solar," *Forbes*, December 10, 2012, www.forbes.com/sites/dannykennedy/2012/12/10/saudis-invest-heavily-in-solar-just-as-the-us-tries-to-catch-up-to-saudis-in-oil/.
40. Justin Gerdes, "Net-Zero Energy Buildings Are Coming — What about the Buildings Already Standing?," *Forbes*, February 28, 2012, www.forbes.com/sites/justingerdes/2012/02/28/net-zero-energy-buildings-are-coming-what-about-the-buildings-already-standing/.
41. "MIT 150 Exhibition, Problem-Solving MIT: Carlisle Solar House," MIT Museum, undated, http://museum.mit.edu/150/138.
42. "Bayonne School Stays in Power after Sandy: Thanks to a Solar System from Flemington Company," *Hunterdon County Democrat*, November 6, 2012, www.nj.com/hunterdon-county-democrat/index.ssf/2012/11/bayonne_school_stays_in_power.html.
43. Lyle K. Rawlings quoted in Felicity Carus, "PV Poised to Light Up NYC after Lessons from Hurricane Sandy," Editor's Blog, *PV Tech*, March 5, 2013, www.pv-tech.org/editors_blog/pv_poised_to_light_up_nyc_after_lessons_from_hurricane_sandy?utm_source=PV-Tech&utm_campaign=92ea8079ce-PV_Tech_Daily_Newsletter_Tuesday_5_March_2013&utm_medium=email.
44. Lyle Rawlings, president and CEO of Advanced Solar Products in New Jersey, interview by author, November 13, 2012.
45. Patrick O'Grady, "Arizona Solar Installation Numbers Poised for a Fall," Phoenix Business (blog), *Phoenix Business Journal*, March 28, 2013, www.bizjournals.com

/phoenix/blog/business/2013/03/arizona-solar-installation-numbers.html?ana=yfcpc.

46. Chris Meehan, "NRG Brings Solar to Rooftops," *Solar Reviews*, March 26, 2013, www.solarreviews.com/blog/nrg-brings-solar-to-rooftops.
47. Geert De Clercq, "Analysis: Renewables Turn Utilities into Dinosaurs of the Energy World," *Reuters*, March 8, 2013, www.reuters.com/article/2013/03/08/us-utilities-threat-idUSBRE92709E20130308.
48. "Chicago's SoCore Energy to Be Purchased by California Utility," *Chicago Tribune*, July 2, 2013, http://www.chicagotribune.com/business/breaking/chi-ore-energy-to-be-purchased-by-california-utility-20130702,0,7907350.story.
49. Ben Willis, "IHS: Germany to Lead 'Explosion' in Global PV Storage Market," *PV Tech*, April 25, 2013, www.pv-tech.org/news/ihs_germany_to_lead_explosion_in_global_pv_storage_market?utm.
50. Jeff St. John, "Verizon's $100M Fuel Cell and Solar Power Play," *Green Tech Media*, April 30, 2013, www.greentechmedia.com/articles/read/verizons-100m-fuel-cell-and-solar-power-play?utm.
51. Christopher Martin and Naureen S. Malik, "NRG Skirts Utilities Taking Solar Panels to US Rooftops," Bloomberg.com, March 25, 2012, www.bloomberg.com/news/2013-03-24/nrg-skirts-utilities-taking-solar-panels-to-u-s-rooftop.html?cmpid=yhoo.
52. Navigant Research, "Distributed Solar Energy Generation: Market Drivers and Barriers, Technology Trends, and Global Market Forecast," undated, www.navigantresearch.com/research/distributed-solar-energy-generation.
53. Peter Kind, "Disruptive Challenges: Financial Implications and Strategic Response to a Changing Retail Electric Business," Edison Electric Institute, January 4, 2013, www.eei.org/ourissues/finance/Documents/disruptivechallenges.pdf.
54. Jim Jenal, "Thoughts on Solar by Run on Sun's Founder & CEO, Jim Jenal: FERC Says Solar Protects Grid from Attack," Run on Sun Founder's Blog, April 27, 2013, http://runonsun.com/~runons5/blogs/blog1.php/solworks/ferc-says-solar-protects-grid.
55. Bob Freling, "Time for a Smarter Approach to Environmental Thinking: Thomas Friedman vs. Bjorn Lomborg," Bob Freling's Solar Blog, December 18, 2009, www.bobfreling.com/2009/12/ive-just-returned-from-the.htm.
56. Geoffrey Carr, "Sunny Uplands: Alternative Energy Will No Longer Be Alternative," *The Economist*, November 21, 2012, www.economist.com/news/21566414-alternative-energy-will-no-longer-be-alternative-sunny-uplands.
57. From remarks by Amory B. Lovins on accepting the Onassis Foundation's Delphi Prize for Man and His Environment, in Athens, Greece. Helena Smith, "Fulbright Gets Onassis Foundation Award," AP News Archive, April 20, 1989, www.apnewsarchive.com/1989/Fulbright-Gets-Onassis-Foundation-Award/id-c18e7d2d460db402957863553e9cf6cd.

Illustration Credits and Permission Acknowledgments

1.1, 2.1, 3.1, 3.3, 3.8, and 14.9: Sarah Bore

1.2: *Bulletin of the Museum of Far Eastern Antiquities*, Stockholm, 1950, 2.1

1.3 and 1.5: Dr. Ronald Knapp

1.4: Dr. Qinhua Guo

2.2: David Moore Robinson, *Excavations at Olynthus, Part XII: Domestic and Public Architecture With Excursus I on Pebble Mosaics with Colored Plates, Excursus II on the Oecus Unit by George E. Mylonas, Testimonia, List of Greek Words, Etc.*, plate 1, Plan of Blocks A iv to A viii. Copyright © 1946 by The John Hopkins Press. Reprinted with permission of The John Hopkins Press.

2.3: David Moore Robinson and J. Walter Graham, *Excavations at Olynthus, Part VIII: The Hellenic House: A Study of the Houses Found at Olynthus With a Detailed Account of Those Excavated in 1931 and 1934*, p. 99, figure 4. Copyright © 1938 by The Johns Hopkins Press. Reprinted with permission of The Johns Hopkins University Press.

3.2: Adapted from drawings by Edwin Thatcher, "The Open Rooms of the Terme del Foro at Ostia," *Memoirs of the American Academy in Rome* 24 (1956): 167–264.

3.7: Norman Neubauer

4.2 and 27.9: Dr. C. Julian Chen

4.7: British Library

5.4 and 5.5: The William Andrews Clark Memorial Library, University of California, Los Angeles

6.2: Science Source

6.3 and 6.4: Smithsonian Institution Archives

8.2, 8.3, and 8.4: Onondaga Historical Association, 321 Montgomery Street, Syracuse, NY, 13202

9.2: Bibliothèque nationale de France, Paris

10.1, 10.2, 10.3, 10.5, 10.6, 10.7, 10.8, 10.9, 10.10, 10.11, 10.12, and 10.13: Frank R. Shuman

11.7, 11.8, 11.9, and 11.10: Lawrence B. Romaine Trade Catalog Collection, Mss 107, Department of Special Collections, Davidson Library, University of California, Santa Barbara.

11.13: Los Angeles Public Library Photo Collection

11.17: Los Angeles County Museum of Natural History

12.2 and 12.8: William J. Bailey, Jr.

12.12: Reprinted with permission from *Popular Mechanics*, copyright © 1935 by the Hearst Corporation

14.1, 14.11, and 14.12: Professor Ichimatsu Tanashita

14.2: Gonen Yissar

14.5: John Yellot

14.6: The Solahart Corporation

14.8: M. D. Lewis Rome

14.13: Azuma-Koki Company, Tokyo

15.1, 15.2, 15.3, 15.4, 15.5, and 25.9: Maria Telkes Papers, Architecture and Environmental Design Library Special Collections, Arizona State University

15.6: Getty Images

16.2: Marc Emery, "Faust et Le Corbusier," *Novelle Revue Neuchateloise* 3 (Autumne 1984): 23.

16.3: Niedersachsisches Landesarchiv Stattsarchiv, Bueckeberg, F1 A xxx 28 E Nv.15, page 10

16.4: Landeshauptstadt Munchen, Stadarchive, NL – Vorherr -02-01

16.5: Margerete Sturm

16.6: Landeshauptstadt Munchen, Stadarchive, NL – Vorherr -33-01

16.7: Landeshauptstadt Munchen, Stadarchive, NL – Vorherr -15-01

16.10: Landeshauptstadt Munchen, Stadarchive, NL – Vorherr -37-01

16.12: Marc Emery, "Faust et Le Corbusier," *Novelle Revue Neuchateloise* 3 (Autumne 1984): 36.

16.13: Marc Emery, "Faust et Le Corbusier," *Novelle Revue Neuchateloise* 3 (Autumne 1984): 27.

16.14: Marc Emery, "Faust et Le Corbusier," *Novelle Revue Neuchateloise* 3 (Autumne 1984): 8–9.

16.15: Marc Emery, "Faust et Le Corbusier," *Novelle Revue Neuchateloise* 3 (Autumne 1984): 35.

17.3: Gehag Archive, Berlin

17.5: Stiftung Bauhaus-Dessau Archive

18.2: Dick Kent

18.3: Braun Research Library Collection, Autry National Center, Los Angeles, p. 17432

19.1, 19.5, and 19.7: Chicago History Museum

19.2: Royal Institute of British Architects

19.3: Wisconsin Historical Society, M99-085.box 8.fldr 26.001

19.4: Wisconsin Historical Society, M99-085.box 14.fldr16.002

19.8, 26.5, and 26.6: Jim Laukes

19.10: Wisconsin Historical Society, M75-026.box 4.huckins.001

20.1, 20.6, 20.7, and 20.14: Courtesy of MIT Historical Collections

20.2: Phillips Library Collection of the Peabody Essex Museum, Negative #9051

20.8: George Lof

20.9: Reprinted with permission from *Popular Science* © 1947 by Popular Science Publishing Company

20.15: Eleanor Raymond

20.17: Ray Manley

20.18: Michael Vaccaro

21.1 and 21.2: The IET Archives

21.3: King's College, London, Archives

21.4: Reprinted by permission of the CINDEX.Add.7657, R51063. Cambridge University Library, archive and manuscript collection.

21.6: Anthony Skelton

21.8: Reprinted with permission from *Popular Mechanics*, copyright © 1931 by the Hearst Corporation

21.9, 21.10, 21.11, 21.13, 21.14, 21.15, and 22.1: Property of AT&T Archives, Warren, New Jersey. Reprinted with permission of AT&T.

21.12: Audrey Chapin Svensson

22.5: Frederica Solar Mendl

22.6: Reprinted with permission of *Popular Science* magazine, copyright © 1956 Times-Mirror Magazines, Inc.

23.1, 23.2, and 23.3: Elliot Berman, Solar Power Corporation/Solar Power Limited

23.4: Solarex/BP

23.5: Good-All Manufacturing, Incorporated, a Corrpro Company

23.6, 23.7, and 23.8: John Oades

23.9 and 23.10: Telstra Corporation, Australia

25.1: Dr. Simon Sadler

25.3 and 25.4: Freeman Ford, founder, Fafco, Inc.

25.5, 26.1, and 26.9: Jon Hammond, partner, Indigo Architects

25.6 and 25.7: National Parks Service

25.8: David Wright

25.11, 25.12, 28.16, and 29.7: Pilkington Solar International, GMBH, Koln, Germany

25.13 and 25.14: Dr. J. Douglas Balcomb

25.15: Michael McCulley

26.4: Frederick Augustus Brooks, *An Introduction to Physical Microclimatology*, Davis, University of California, copyright © 1959

26.8: Robert Black

26.14: Robert Lechner

26.15: Lisa Frantzis

27.1: Dr. Ioannes Michaelides

27.3: Solar Austria

27.5 and 27.6: Professor Jan-Olof Dalenbach

27.7 and 27.8: Jess Heinemann

28.1: Clive Capps

28.2: Patrick Jourde

28.3 and 28.4: Richard Acker

28.5: Soluz, Incorporated

28.6 and 28.7: Solar Electric Light Company

28.9: Reprinted with permission from Wolfgang Palz, *Solar Electricity: An Economic Approach*, copyright © 1978 UNESCO

28.10: Markus Real, founder, Alpha Real

28.15: Ronald Reagan to Richard V. Allen, October 27, 1982, Danny Boggs File, Energy, Atomic Energy, Ronald Reagan Foundation and Library, Simi Valley, California

28.17: Craig Morris

29.1: Courtesy of Kiss+Cathcart, Architects, Sun Hung rendering; Fox & Fowle, Basebuilding Architects; the Durst Corporation, Developer

29.2: Courtesy of Exxon-Mobil Corporation

29.3: Courtesy of Dr. Anotonio Luque

29.4: From the collection of Donald E. Osborn. Original photograph by William A. Rhodes.

29.5: David Carlson

29.6: Reprinted with permission of Martin Andrew Green. Not to be reproduced without permission of copyright holder.

30.1: Independent Power Systems, +1 303 443 0115

30.2, 30.3. and 30.4: BrightSource Energy

30.6: Carmanah Industries

30.7: Steven Strong

30.8: Lyle Rawlins

30.9 and 30.10: Mali Aqua Viva, "Sahel Aqua Viva," c/o Foundation de France

Index

Page numbers given in *italics* indicate illustrations or material contained in their captions.

About the Author

An international expert on solar energy and forestry, John Perlin has lectured extensively on these topics in North America, Europe, Asia, and Australia. He mentors those involved in realizing photovoltaic, solar-hot-water, and energy-efficiency technologies at the University of California, Santa Barbara (UCSB), and coordinates the California Space Grant Consortium as a member of UCSB's Department of Physics. Perlin is the author of *A Forest Journey: The Story of Wood and Civilization* as well as *From Space to Earth: The Story of Solar Electricity*. *A Forest Journey* was chosen by Harvard University Press as a Classic in Science and World History and as one of the press's "100 Great Books." *Time* magazine chose the book as "a must read" for Earth Day. Perlin and Nobel Laureate Walter Kohn created the documentary film *The Power of the Sun*, narrated by John Cleese. He has also done multiple exhibits for the National Renewable Energy Laboratory that have traveled throughout the world. Perlin lives in Santa Barbara, California. His website is www.john-perlin.com.

About the Author

NEW WORLD LIBRARY is dedicated to publishing books and other media that inspire and challenge us to improve the quality of our lives and the world.

We are a socially and environmentally aware company, and we strive to embody the ideals presented in our publications. We recognize that we have an ethical responsibility to our customers, our staff members, and our planet.

We serve our customers by creating the finest publications possible on personal growth, creativity, spirituality, wellness, and other areas of emerging importance. We serve New World Library employees with generous benefits, significant profit sharing, and constant encouragement to pursue their most expansive dreams.

As a member of the Green Press Initiative, we print an increasing number of books with soy-based ink on 100 percent postconsumer-waste recycled paper. Also, we power our offices with solar energy and contribute to nonprofit organizations working to make the world a better place for us all.

Our products are available
in bookstores everywhere.
For our catalog, please contact:

New World Library
14 Pamaron Way
Novato, California 94949

Phone: 415-884-2100 or 800-972-6657
Catalog requests: Ext. 50
Orders: Ext. 52
Fax: 415-884-2199
Email: escort@newworldlibrary.com

To subscribe to our electronic newsletter, visit:
www.newworldlibrary.com

Environmental impact estimates were made using the Environmental Defense Fund Paper Calculator @ www.papercalculator.org.